Microbial Diversity and Environmental Resilience in Polluted Ecosystems

Olubukola Oluranti Babalola
https://orcid.org/0000-0003-4344-1909
North-West University, South Africa

Olukayode Oladipo Amund
University of Lagos, Nigeria

Olayinka Olayemi Osuolale
https://orcid.org/0000-0001-6525-8502
Elizade University, Nigeria

IGI Global
Scientific Publishing
Publishing Tomorrow's Research Today

Vice President of Editorial	Melissa Wagner
Director of Acquisitions	Mikaela Felty
Director of Book Development	Jocelynn Hessler
Production Manager	Mike Brehm
Cover Design	Jose Rosado

Published in the United States of America by
IGI Global Scientific Publishing
701 East Chocolate Avenue
Hershey, PA, 17033, USA
Tel: 717-533-8845 | Fax: 717-533-7115
Website: https://www.igi-global.com E-mail: cust@igi-global.com

Library of Congress Cataloging-in-Publication Data

LCCN: 2026028401 (CIP Data Pending)
ISBN13: 9798260021002
Isbn13Softcover: 9798260021019
EISBN13: 9798260021026
British Cataloguing in Publication Data
A Cataloguing in Publication record for this book is available from the British Library.

Table of Contents

Detailed Table of Contents

contamination. Similarly, because of their neurotoxic and carcinogenic effect, soils containing dangerous hydrocarbons, like polycyclic aromatic hydrocarbons (PAHs), provide serious health concerns to humans. Due in major part to the fast population expansion, industrialization and the growing need for food security, pesticides are the most common xenobiotic contaminants found in soils.

Aquatic pollution is characterized by contaminants in waterbodies mostly because of microorganisms, oil, plastics and eutrophic contaminants occurring in components with rapid dissemination of both soluble and insoluble agents. Microbial responses to pollution, the role they play in biotransformation and bioremediation are indeed paramount. The purpose of this chapter is to introduce the primary microbial responses to oil, plastic and eutrophic contaminants. We explore their potential effects on the environment, highlighting key microbes involved in biotransformation and pollutant degradation. Enzymatic mechanisms and biochemical pathways are also discussed. The global challenges faced by ecosystems are reviewed and potential solutions are examined. The region is vulnerable to environmental abuse raising red flags in the efforts to heal wounds inflicted on the ecosystems. Finally, we discussed possible applications of biotransformation processes, providing insights to develop new ways of combating this problem, shedding light on various research directions and challenges.

This chapter examines the role of airborne microorganisms in the degradation of volatile organic compounds (VOCs) and atmospheric pollutants. It discusses the sources, diversity, and ecological functions of bioaerosols, including bacteria, fungi, archaea, and viruses, and highlights microbial survival strategies such as pigmentation, dormancy, DNA repair, and particle aggregation. Major pollutants, including NOx, SO_2, PAHs, ozone, and particulate matter, are reviewed alongside their environmental and health impacts. The chapter emphasizes microbial metabolic pathways involved in aerobic and anaerobic VOC degradation, including oxygenase-mediated oxidation, reductive dehalogenation, fumarate addition, and cometabolism. It also explores the role of airborne microbes in atmospheric chemistry, cloud formation, pollutant

resilience, exploring how prokaryotes adapt, survive, create, and maintain ecosystem functions. The chapter addresses key questions about planetary boundaries and the role of microorganisms in sustainability of safe operating zones for novel entities and climate change.

Microplastics (MPs) are ubiquitous in the environment and are defined as small (< 5 mm) heterogeneous plastic particles that vary in size, color, shape, and polymer type. These relatively new pollutants are considered contaminants of emerging concern (CECs) due to their widespread accumulation, bioavailability to aquatic and terrestrial organisms, and potential to induce toxicity in wildlife and humans. This chapter explores the main anthropogenic sources of various types of microplastic (MP) accumulation and disposal into the environment as well as potential interactions with indigenous microbial cells. The chapter also examines key degradation processes commonly used by microbes in remediating discarded MPs, in the plastisphere and in surrounding environments, and current challenges associated with interactions between microbes and MPs biodegradation. Furthermore, the chapter highlights alternatives to MPs to reduce their availability and ensure degradative potentials of microbes of these pollutants in the environment.

Preface

INTRODUCTION

Human activities have transformed the planet at an unprecedented scale. Industrialization, urbanization, intensive agriculture, mining, and the widespread use of synthetic chemicals have contributed to the contamination of soil, water, and air across ecosystems worldwide. These environmental pressures, coupled with the growing impacts of climate change, have intensified the search for sustainable, effective, and environmentally responsible solutions for pollution control and ecosystem restoration. Among the most promising solutions are microorganisms—the unseen architects of life on Earth whose remarkable diversity, adaptability, and metabolic versatility enable them to transform, detoxify, and recycle pollutants in ways that few other biological systems can achieve.

Biomicrodiversity for a Polluted Planet explores the central role of microbial diversity in addressing contemporary environmental challenges. The book brings together leading scholars and researchers from across Africa, Europe, Asia, the Middle East, and North America to present current knowledge, emerging technologies, and future directions in microbial environmental biotechnology. It highlights how advances in genomics, bioinformatics, systems biology, synthetic biology, and environmental engineering are reshaping our understanding of microbial communities and expanding their applications in environmental remediation.

The chapters collectively demonstrate that microbial diversity is not merely a subject of scientific inquiry but a critical resource for achieving environmental sustainability. From polluted soils and contaminated aquatic systems to atmospheric pollutants and industrial wastes, microorganisms offer innovative pathways for remediation, restoration, and resource recovery. The volume further explores engineered microbial systems, microbial consortia, circular economy approaches, climate resilience, and policy frameworks necessary for translating scientific discoveries into practical environmental solutions.

This book is organized into nineteen chapters that guide readers from the fundamental understanding of microbial diversity to advanced applications and future innovations in environmental biotechnology.

CHAPTER OVERVIEW

Chapter 1. Introduces microbial diversity in the genomic era, examining how modern sequencing technologies, phylogenomics, and functional genomics have transformed microbial taxonomy, evolutionary studies, and ecological understanding.

Chapter 2. Explores the contributions of extremophilic and rhizosphere microorganisms to environmental remediation and discusses their ecological functions and mechanisms for pollutant removal.

Chapter 3. Reviews contemporary tools for microbial biodiversity exploration, including omics technologies, bioinformatics platforms, and single-cell approaches that enable deeper investigation of microbial communities.

Chapter 4. Examines microbial adaptation in polluted soils and discusses interactions with heavy metals, pesticides, and hydrocarbons, emphasizing remediation processes and ecological resilience.

Chapter 5. Focuses on microbial responses to aquatic pollution and the biotransformation of petroleum products, plastics, and nutrient-driven contaminants in freshwater and marine environments.

Chapter 6. Investigates airborne microorganisms and their involvement in the degradation of volatile organic compounds and atmospheric pollutants, highlighting their ecological significance and emerging applications.

Chapter 7. Discusses biostimulation and bioaugmentation strategies that enhance the performance of native and introduced microbial communities for pollution control and environmental restoration.

Chapter 8. Evaluates advances in genetically engineered microorganisms for environmental biotechnology, including their applications, benefits, limitations, and biosafety considerations.

Chapter 9. Explores the use of microbial consortia in bioremediation, emphasizing synergistic interactions, community engineering, and innovative approaches for improving degradation efficiency.

CONCLUSION

As environmental degradation continues to threaten ecological integrity and human well-being, the need for innovative, science-based, and sustainable remediation approaches has never been greater. The contributions presented in this volume underscore the extraordinary potential of microbial diversity to restore polluted environments, support circular resource use, and strengthen ecosystem resilience in a rapidly changing world. We hope this book serves as a valuable resource for researchers, students, policymakers, environmental practitioners, and all stakeholders committed to advancing microbial solutions for a cleaner and more sustainable planet.

Chapter 1
Microbial Diversity in the Genomic Age:
Integrating Taxonomy, Function, and Evolutionary Perspectives

Akinlolu Olalekan Akanmu
http://orcid.org/0000-0003-2816-9820
Moutain Top University, Nigeria

Moses Akindele Abiala
Mountain Top University, Nigeria

Abiodun Mayowa Akanmu
University of Pretoria, South Africa

Olukayode Oladipo Amund
University of Lagos, Nigeria

ABSTRACT

Microbial diversity is fundamental to ecosystem stability, nutrient cycling, and evolutionary innovation. This chapter synthesizes recent advances in genomic methodologies that have transformed microbial taxonomy, functional characterization, and evolutionary understanding. Notable advancements include adoption of 16S rRNA sequencing as a universal phylogenetic marker, whole genome sequencing and phylogenomics for precise classification, and application of quantitative metrics such as Average Nucleotide Identity (ANI) and digital DNA-DNA hybridization (dDDH) for species delineation. Metagenomics and single-cell genomics have enabled the discovery of uncultured taxa and microbial "dark matter," while comparative and functional genomics have elucidated redundancy, adaptability, and gene flux

DOI: 10.4018/979-8-2600-2100-2.ch001

within microbial communities. Collectively, these developments provide a systems-level framework for understanding microbial diversity, linking genetic divergence to ecological function and evolutionary processes in the genomic era, while also identifying areas where further research is necessary

1. INTRODUCTION

The study of microbial diversity initially relied on culture-dependent methods, which primarily used morphological, physiological, and biochemical traits for classification. Early microbial taxonomy was based on characteristics such as cell structure, Gram staining, metabolic capabilities, and environmental tolerance. While these approaches led to foundational discoveries, they were constrained by the "great plate count anomaly," which revealed that only a small fraction of bacteria could be cultured under laboratory conditions (Khan and Saha, 2026). This limitation led to a narrow view of microbial diversity, favoring easily cultivated taxa (Hausdorf, 2025). The molecular revolution of the late twentieth century changed this landscape. A landmark study by Woese and Fox (1977) introduced the 16S rRNA gene as a universal molecular chronometer. This approach revealed deep evolutionary divisions between Bacteria, Archaea, and Eukarya, giving rise to the three-domain system and a radical reorganization of the tree of life. Shortly thereafter, Pace (1997) pioneered molecular microbial ecology and showed how environmental DNA could be used to study uncultured organisms. These breakthroughs moved taxonomy from phenotype to genotype, providing objective criteria for distinguishing relationships and revealing previously hidden lineages (Farrance, 2019; Zhou et al., 2022). This method revealed substantial evolutionary gaps, leading to the identification of entirely new phyla and the reclassification of bacterial and archaeal lineages (Haque et al., 2022). The emergence of high-throughput sequencing and whole-genome analysis significantly enhanced microbial taxonomy. Using measures such as Average Nucleotide Identity (ANI) and digital DNA–DNA hybridization (dDDH), genome-scale comparisons provide reliable methods for delineating species boundaries (Jain et al., 2018; Wang et al., 2025). Phylogenomic approaches using hundreds of conserved genes have produced robust evolutionary reconstructions that include both vertical inheritance and horizontal gene transfer (Tortoli et al., 2017). These methods have helped to resolve taxonomic confusion within complex genera, such as *Mycobacterium* and *Lactobacillus*. The development of curated databases such as GTDB and RefSeq has standardized taxonomic curation, enabling reproducible comparisons across ecosystems (Parks et al., 2020).

Metagenomics has expanded the scope of microbial research to include uncultured microorganisms by enabling the direct extraction of whole or nearly complete metagenome-assembled genomes (MAGs) from environmental samples (Cowan et al., 2015). MAGs have revealed numerous previously hidden species in soil, marine, and extreme environments, thereby facilitating the expansion of the microbial tree of life (Zhou et al., 2022). The shift from culture-dependent to genome-informed approaches marks a fundamental change in perspective. Microbial diversity is now interpreted through an evolutionary genomics framework that integrates genetic composition, ecological function, and evolutionary history. Achieving a comprehensive understanding of microbial diversity requires linking genetic potential to functional expression. Comparative and functional genomics have shown that microbial communities often display substantial functional redundancy, indicating that distinct taxa can fulfill similar ecological roles (Miralles et al., 2021). This redundancy enhances ecosystem resilience and reflects evolutionary pressures that shape metabolic processes and adaptability. Pangenome analyses further demonstrate extensive gene flux and horizontal gene transfer, highlighting the evolutionary dynamism of microbial genomes (Tettelin and Medini, 2020). Integrating taxonomy with functional and evolutionary analyses enables a holistic understanding of microbial systems: taxonomy provides organizational structure, functional analysis elucidates ecological processes, and evolutionary frameworks explain the mechanisms underlying diversity and adaptation. Collectively, these perspectives advance microbiology from a static classification system to a systems-based framework that reflects the dynamic existence of microbes within their environments (Naeem and Fatima, 2025). Molecular and genomic technologies have redefined microbial taxonomy, expanded the tree of life, and revealed the functional and evolutionary complexity of microbial communities. By integrating taxonomy, function, and evolutionary history, genomic approaches provide a systems-level perspective that encompasses the full extent of microbial diversity.

2. GENOMICS-DRIVEN TAXONOMY OF MICROORGANISMS

The 16S ribosomal RNA (rRNA) gene remains one of the most commonly used genetic markers in microbial taxonomy and ecology. Its ubiquitous presence in bacteria and archaea, along with conserved and variable regions, provides both phylogenetic stability and discriminating power (Doolittle et al., 2003). Initial molecular investigations using 16S rRNA sequences delineated evolutionary linkages that transformed microbial taxonomy, leading to the classification of life into three domains: Bacteria, Archaea, and Eukarya (Zhou et al., 2022). This gene has since served as the basis for microbial community profiling, with databases such as SILVA, Greengenes, and RDP providing curated reference sets for sequence comparison (Quast et al., 2013) (Table 1). The introduction of whole-genome sequencing (WGS) was a pivotal moment in microbial taxonomy, facilitating extensive comparisons across the complete genetic repertoires of species. WGS provides high-resolution insights into both conserved and variable genomic regions, enabling more precise taxonomic differentiation than single-gene techniques (Tettelin and Medini, 2020). The incorporation of WGS data into taxonomy facilitates a transition from single-marker phylogenies to genome-scale evolutionary frameworks, a methodology referred to as phylogenomics. Phylogenomic investigations rely on concatenated alignments of conserved single-copy orthologous genes to infer evolutionary relationships with statistical rigor. These genome-wide databases have enabled the reclassification of multiple species, notably the division of Firmicutes into the newly established phyla Bacillota and Clostridiota (Zhou et al., 2022).

Table 1. Genomic frameworks for taxonomic and functional characterizations of microbial diversity

Aspect	Pre-genomic Paradigm	Genomic-Age Paradigm	Tools	Primary Outcomes	References
Taxonomic resolution	Phenotype-based; 16S surveys	Genome-resolved taxonomy; ANI; phylogenomics	WGS; MAGs; GTDB	High-resolution species delineation; uncultured taxa discovery	Szabó et al., (2025), Parks et al., (2022)
Diversity estimation	OTU-based surveys	ASVs; shotgun metagenomics	QIIME2; MetaPhlAn	Improved diversity and detection of rare lineages	Zhou et al., (2025)
Functional potential	Cultivation & assays	Genome-based metabolic reconstruction	Functional annotation (KEGG, COG)	Prediction of metabolic capabilities	de Oliveira et al., (2024), Pinto and Bhatt (2024).
Functional redundancy	Inferred indirectly	Comparative genomics across genomes	Pangenome analyses	Insights into niche overlap	McReynolds et al., (2025)

The number of genome sequences stored in the NCBI RefSeq, GTDB, and IMG/M databases continues to grow, making this trend more visible. These databases enable standardized taxonomic curation and facilitate repeatable comparisons of species and ecosystems (Parks et al., 2020). Consequently, WGS has emerged as the new benchmark for microbial systematics, facilitating the reclassification of microbial species, genera, and higher taxa based on genomic coherence rather than morphological characteristics.

2.1 Average Nucleotide Identity (ANI) and Digital DNA–DNA Hybridization (dDDH)

In the genomic era, quantitative metrics such as Average Nucleotide Identity (ANI) and digital DNA–DNA hybridization (dDDH) are crucial for defining species boundaries. ANI calculates the average nucleotide identity of shared orthologous areas across two genomes, yielding a genome-wide assessment of genetic relatedness (Richter and Rosselló-Móra, 2009). A threshold of about 95–96% ANI, 70% DNA–DNA hybridization (DDH) (Jain et al., 2018). The digital implementation of DDH (dDDH), which uses genome sequence alignments rather than experimental hybridization, is more accurate and can be repeated (Meier-Kolthoff et al., 2022). These quantitative thresholds have allowed taxonomists to resolve long-standing confusion in microbial systematics and to identify cryptic species complexes that phenotypic or 16S-based approaches could not previously differentiate (Wang et al., 2025). Phylogenomic programs such as TYGS and GTDB-Tk now routinely use ANI and dDDH metrics. This allows automatic, statistically based assignment of taxa. Their use has made it customary to define species in prokaryotes, enabling comparisons across research and datasets while still adhering to evolutionary principles (Table 1).

2.2 Genome-Based Species Concepts and Taxonomic Thresholds

Genome-based taxonomy has necessitated a reassessment of microbial species, a topic that has previously been a source of significant contention. Conventional definitions predicated on reproductive isolation are unsuitable for asexual microbes, hence requiring genetic criteria to delineate species boundaries (Konstantinidis et al., 2017). Whole-genome similarity criteria, including Average Nucleotide Identity (ANI) and Average Amino Acid Identity (AAI), currently serve as the operational basis for species- and genus-level classification (Tettelin and Medini, 2020). This multidimensional paradigm recognizes that microbiological species consist of dynamic populations of strains that evolve by mutation, recombination, and horizontal gene transfer (Hausdorf, 2025). Genomic coherence, meaning that the evolutionary signal is shared across the core genome, is now a key concept in microbial systematics. It provides a standardized and objective basis for resolving taxonomic ambiguities. As sequencing methods and databases improve, genome-based species thresholds continue to improve. For example, the Genome Taxonomy Database (GTDB) uses standardized relative evolutionary divergence (RED) criteria to maintain rank normalization across microbial clades (Parks et al., 2020). These developments are gradually reconciling bacterial taxonomy with phylogenomics, synchronizing species definitions with both evolutionary lineage and ecological significance.

2.3 Metagenome-Assembled Genomes (MAGs) and Uncultured Microbial Taxa

Metagenome-assembled genomes (MAGs) have transformed the examination of microbial taxonomy by providing access to the genomic frameworks of uncultured bacteria. MAGs facilitate the reconstruction of nearly complete genomes from intricate microbial communities by assembling and binning environmental sequencing data (Cowan et al., 2015). This method has uncovered thousands of new species, including entire candidate phyla such as *Candidatus patescibacteria* and *Candidatus lokiarchaeota*, thereby greatly expanding the microbial tree of life (Zhou et al., 2022). MAGs have also revealed the metabolic diversity and ecological activities of organisms that have never been studied before. This has enabled scientists to determine biogeochemical functions directly from environmental data. MAG-based taxonomy, despite its transformative impact, faces constraints such as contamination, assembly artifacts, and inadequate genome coverage, which may affect phylogenetic resolution (Parks et al., 2020). To tackle these problems, quality standards such as the Minimum Information about a Metagenome-Assembled Genome (MIMAG) have been established to ensure that results are reliable and reproducible (Bowers et al., 2017). The incorporation of MAGs into phylogenomic frameworks persistently transforms microbial taxonomy, reconciling cultivation-based and cultivation-independent microbiology. This change shows that microbial systematics is undergoing a broader shift: taxonomy is no longer limited to cultured isolates; it now encompasses the full genomic landscape of microbial life.

Figure 1. Genome-based framework for microbial taxonomic delineation and phylogenomic inference

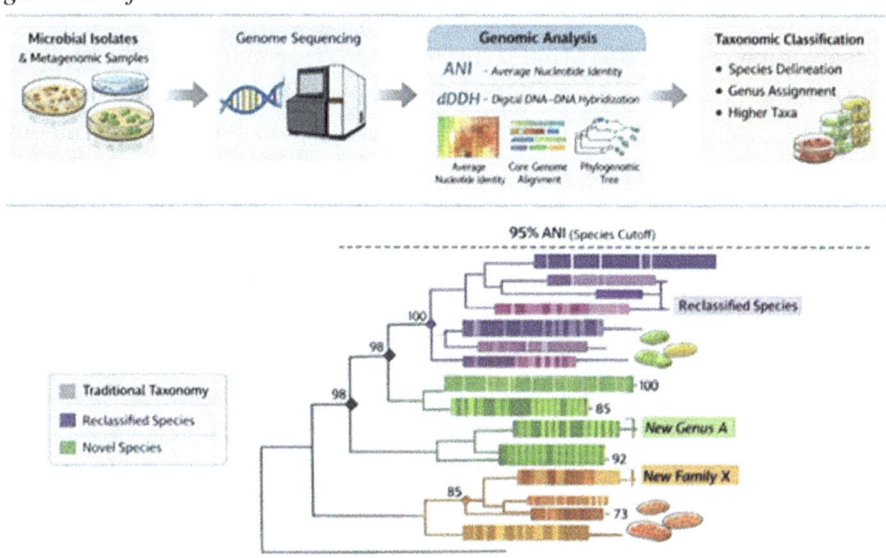

2.4 Expanding the Tree of Life Through Metagenomics

Metagenomic investigations across various settings, such as soils, oceans, and extreme habitats, have demonstrated that the majority of microbial life resides within uncultured species, often lacking close relatives in reference databases (Cowan et al., 2015). For example, global oceanic metagenomic projects have identified new families, such as SAR324 and Marinimicrobia, and terrestrial metagenomes have revealed substantial phylogenetic diversity within putative divisions, such as Acidobacteria and Verrucomicrobia (Browne et al., 2020). These findings illustrate that the traditional tree of life, which has historically relied on cultivated isolates, constitutes merely a segment of the existing microbial diversity. Environmental metagenomics provides a quantitative framework for examining microbial biogeography and ecosystem function. Comparative analyses across habitats have demonstrated a robust correlation between microbial community composition and environmental gradients, including pH, salinity, and nutrient availability (Carini et al., 2020; Akanmu et al., 2025b). Metagenomics enables simultaneous assessment of "who is present" and "what they can do" by combining taxonomic and functional data. This creates the basis for connecting microbial diversity to ecosystem dynamics.

2.5. Functional Diversity in the Genomic Era

Functional gene annotation pipelines, including KEGG, eggNOG, and MetaCyc, now enable the reconstruction of metabolic pathways, hence enhancing the prediction of biochemical roles at the community level (Zhang and Shao, 2023). Metagenomic surveys have demonstrated that microbial communities frequently exhibit a modular distribution of functional genes, with specific species performing distinct functions within biogeochemical cycles. For example, nitrifying communities contain genes for ammonia oxidation (amoA) and nitrite oxidation (nxrB), while sulfur-oxidizing bacteria have pathways for thiosulfate and sulfide oxidation (Sampara et al., 2022). Mapping these gene distributions enables the identification of key metabolic players in complex ecosystems, transforming microbial ecology into a gene-centric field. Genomes obtained from hot springs, hypersaline habitats, and deep-sea sediments often include genes linked to mixotrophy, alternative electron acceptors, and stress tolerance, highlighting the extensive adaptability of microbial life (Zhou et al., 2022). This genetic finding elucidates the mechanisms by which microbial diversity preserves ecosystem functionality amid variable environmental conditions.

Comparative genomics offers a paradigm for examining the evolutionary underpinnings of functional variation among microorganisms. Researchers can deduce adaptive strategies, niche specialization, and evolutionary innovation by analyzing gene content and genomic organization among related taxa (Tettelin and Medini, 2020). Comparative analyses have shown that even closely related strains within the same species can exhibit substantial differences in metabolic pathways, regulatory networks, and stress response genes. This shows how small differences can help microbes adapt (Hausdorf, 2025). Functional niche differentiation arises from selective pressures that favor the retention or elimination of genes that confer ecological benefits. For instance, genomic comparisons among Nitrospira lineages have revealed unique strategies for ammonia and nitrite oxidation, indicating functional partitioning among nitrifying consortia (Sampara et al., 2022). Similarly, research comparing *Pseudomonas* and *Bacillus* strains has shown that genetic factors are involved in plant growth promotion, antibiotic synthesis, and biocontrol. This shows that genomic diversity underpins ecological flexibility and biotechnological potential (Chen and Wu, 2013). The growing integration of phylogenomics and comparative metabolic reconstruction has made it even clearer that evolutionary processes such as gene duplication, horizontal gene transfer, and genome reduction drive functional divergence. This synthesis connects evolutionary theory with ecological function, emphasizing that functional roles are determined not by taxonomy but by how species adapt to environmental constraints.

2.6 Horizontal Gene Transfer and Adaptive Innovation

Horizontal gene transfer is a key driver of microbial evolution because it enables new genetic traits to spread quickly across different groups of organisms. Transformation, transduction, and conjugation are mechanisms by which cells acquire DNA from other organisms, enabling them to adapt to new environments and become more functionally diverse (Frost et al., 2005). Genomic investigations have shown that genes involved in metabolism, antibiotic resistance, and responses to environmental stress are often targets of HGT. This shows how important they are for niche expansion and evolutionary success (Hausdorf, 2025). The evolution of antibiotic production and resistance in soil microbes exemplifies adaptive innovation through horizontal gene transfer. Gene clusters that encode secondary metabolite pathways have been disseminated among actinobacteria, leading to the emergence of new chemical scaffolds and techniques to resist them (Ramond et al., 2025). Likewise, the horizontal acquisition of photosynthetic and nitrogen-fixation genes has facilitated the evolution of cyanobacteria and proteobacteria across ecological gradients (Zhou et al., 2022). HGT enhances adaptation but also complicates phylogenetics by separating gene histories from organismal lineages. Phylogenomic approaches now integrate network-based models to address reticulate evolution, facilitating a more precise reconstruction of evolutionary relationships (Tettelin and Medini, 2020). These findings demonstrate that microbial evolution is better represented as a mosaic of genetic exchanges rather than a rigidly bifurcating tree, underscoring the linked nature of life at the microscopic level (Table 2).

Table 2. Evolutionary and ecological insights into microbial diversity enabled by genomics

Process	Genomic Evidence	Key Methods	Biological Insights	Interpretive Challenges	References
Vertical evolution	Genome-scale phylogenies	Phylogenomics	Lineage divergence patterns	Phylogenetic incongruence	Szabó et al., (2025)
Horizontal gene transfer (HGT)	Mobile elements in genomes	Comparative genomics	Rapid adaptation; HGT patterns	Distinguishing HGT vs. gene loss	Agustinho et al. (2024)
Population structure & microevolution	Strain-level genomic variation	MAGs; strain-resolved analyses	Ecological differentiation	Requires high depth sequencing	Van Rossum et al., (2020)

continued on following page

Table 2. Continued

Process	Genomic Evidence	Key Methods	Biological Insights	Interpretive Challenges	References
Functional activity	Expression-based evidence	Metatranscriptomics; proteomics	Active pathways & responses	Integrating multi-omics	de Oliveira et al., (2024)
Eco-evolutionary dynamics	Genome–environment associations	Modeling; multi-omics	Links between genotype, function, ecosystem processes	Correlation vs causation	Zhou et al., (2025)

3. MICROBIAL POPULATION GENOMICS AND MICROEVOLUTION

Population genomics provides a detailed perspective on microbial evolution by analyzing genetic variation within individual species and across populations. Population genomics investigates microevolutionary processes, including mutation, recombination, and selection, that drive strain-level diversification, in contrast to traditional phylogenetic analyses that focus on species-level interactions (Parks et al., 2022). The growing number of genome sequences from environmental isolates and metagenome-assembled genomes (MAGs) has enabled analysis of differences among individuals of the same species across spatial and ecological scales. This has revealed patterns of population structure and adaptation that were previously invisible (Carini et al., 2020). Microbial populations exhibit significant genetic diversity, often structured into sublineages or ecotypes that differ in gene composition, allelic variation, and ecological roles. For example, comparative genomics of *Prochlorococcus* and *Vibrio* populations has demonstrated the coexistence of multiple lineages within the same habitat, each adapted to particular food regimes or light intensities (Kashtan et al., 2014). These ecotypes preserve genomic coherence via intermittent selection and genetic exchange, illustrating an equilibrium between diversification and homogeneity. Soil microbial communities also exhibit microgeographic differentiation, with strain-level diversity driven by local selective pressures, including pH and the amount of available organic matter (Carini et al., 2020). Recent improvements in metagenomic binning and strain-resolved assembly approaches have made it even easier to tell apart genomes that are quite similar in mixed communities (Cowan et al., 2015). This resolution facilitates the discovery of single-nucleotide polymorphisms (SNPs), mobile genetic elements, and accessory genes associated with specific ecological niches. These findings indicate that microbial communities are not genetically homogeneous; rather, they constitute dynamic mosaics influenced by environmental selection, migration, and stochastic processes.

Natural selection, genetic drift, and recombination all work together to control how microbial populations change over time. Natural selection acts on advantageous mutations that confer adaptive benefits, thereby facilitating the establishment of favorable alleles within populations. Genetic drift, on the other hand, causes random fluctuations in allele frequencies, especially in small or isolated populations, where random sampling can beat out selective factors (Parks et al., 2022). These processes work together to set the speed and manner in which microbes evolve on a small ecological scale. Recombination creates genetic diversity by reshuffling alleles within and between lineages. In bacteria and archaea, recombination predominantly transpires through horizontal gene transfer (HGT), facilitating the acquisition of exogenous DNA via transformation, conjugation, or transduction (Terra et al., 2025). Recombination occurs at varying rates and has different effects across different groups of organisms and settings. Highly recombining lineages, such as *Neisseria* and *Helicobacter*, have mosaic genomes, making it hard to distinguish species. On the other hand, clonal taxa, such as Mycobacterium tuberculosis, primarily evolve through point mutations and gene deletions (Nishiyama et al., 2023).

3.1 Adaptive Evolution in Extreme and Anthropogenic Environments

Extreme and anthropogenically altered ecosystems offer distinctive perspectives on the mechanisms of microbial adaptability and microevolution. In these environments, microorganisms face intense selective pressures, including elevated temperature, salt, radiation, and exposure to toxins and antibiotics. Population genomics has shown that adaptation in these situations is generally driven by rapid genetic changes, such as horizontal gene transfer, gene duplication, and regulatory network rewiring (Cowan et al., 2015). For instance, thermophilic archaea in geothermal environments exhibit significant genomic alterations, including altered codon usage, DNA repair mechanisms, and protein stability systems that confer resistance to denaturation (Zhou et al., 2022). Acidophilic and halophilic bacteria also harbor genes acquired from other species that help them survive in very acidic or salty environments (Haque et al., 2022). These genes help them make proton efflux pumps, osmoprotectants, and suitable solutes. These adaptations frequently arise via convergent evolution, in which disparate taxa independently evolve analogous molecular responses to identical environmental pressures.

In anthropogenic ecosystems, including wastewater treatment facilities, contaminated soils, and hospital settings, microbial evolution is expedited by human-induced selective pressures. The extensive use of antibiotics, heavy metals, and synthetic chemicals facilitates the development of multi-resistant populations, which are frequently influenced by mobile genetic elements such as plasmids and transposons (Ramond et al., 2025). Metagenomic investigations of these ecosystems have found "resistomes" and "mobilomes", which are collections of resistance and mobility genes that evolve through continual recombination and horizontal transfer (Matsumura et al., 2025). In these settings, adaptive evolution encompasses not just resistance features but also metabolic expansion. Microbes exposed to xenobiotics or industrial waste develop novel enzymatic pathways for biodegradation and detoxification, underscoring their importance as key contributors to environmental resilience and remediation (Cowan et al., 2015). The examination of microbial microevolution in severe and anthropogenic environments demonstrates the flexibility and versatility of microorganisms. Population genomics provides a strong foundation for understanding, in real time, how evolutionary processes affect microbial diversity and ecosystem dynamics by linking genomic changes to environmental shifts.

Table 3. Advances in Computational and Bioinformatics Applications in the Genomic Age

Advanced Area	Tools / Platforms	Primary Applications	Scientific Impact	References
High-Throughput Sequence Analysis	BWA, Bowtie2, GATK, SPAdes	Whole-genome sequencing, exome analysis, population genetics	Enabled rapid, large-scale genomic analysis with improved accuracy and reduced cost	Pereira et al., (2020); Zhou et al., (2024)
Machine Learning in Bioinformatics	TensorFlow, PyTorch, DeepVariant, AlphaFold	Protein structure prediction, disease classification, functional annotation	Improved predictive performance and automation of complex biological inference	Ruan et al., (2026)
Systems Biology Modeling	Cytoscape, COPASI, CellDesigner	Gene regulatory networks, pathway analysis, cellular dynamics	Facilitated holistic understanding of biological systems	Puniya et al., (2024)
Structural Bioinformatics	AutoDock, GROMACS, Rosetta	Drug discovery, protein–ligand interaction analysis	Accelerated rational drug design and target validation	Krokidis et al., (2025)
Single-Cell Data Analysis	Seurat, Scanpy, Monocle	Cellular heterogeneity analysis, developmental biology	Revealed previously unrecognized cell states and lineages	Liu et al., (2025), Ge et al., (2025)
Metagenomics and Microbiome Analysis	QIIME2, MetaPhlAn, HUMAnN	Environmental genomics, human microbiome studies	Expanded understanding of microbial diversity and ecosystem function	Houmenou et al., (2025)
Cloud and High-Performance Computing	Nextflow, Snakemake, Docker, AWS	Large-scale bioinformatic pipelines, reproducible research	Improved scalability, reproducibility, and collaborative research	Strozzi et al. (2019), Gasparetto et al. (2024)

4. COMPUTATIONAL AND BIOINFORMATIC ADVANCES

4.1 Genome Annotation Pipelines and Functional Inference

The rapid increase in genomic data in microbiology has necessitated the development of powerful computational methods for genome assembly, annotation, and functional inference. Automated annotation pipelines such as Prokka, RAST, and NCBI's PGAP make it easier to identify coding sequences, gene functions, and regulatory regions. This reduces the time required for manual curation and makes it easier to reproduce results (Seemann, 2014). These technologies use reference databases like KEGG, COG, and Pfam to group genes into functional categories. This allows researchers to build large-scale metabolic pathways and ecological roles (Zhang and Shao, 2023) (Table 1).

The current genome annotation system operates effectively but poses significant challenges for microbial genomics, as different databases yield conflicting results, hindering researchers' ability to identify the functions of predicted proteins. Most microbial genomes harbor 30–40% of their genes that remain poorly understood, especially in environmental or extremophilic species (Wang et al., 2025). Recent methodologies that combine comparative genomics and structural bioinformatics have enhanced annotation precision by merging homology-based inference with protein domain modeling (Parks et al., 2020). Metagenomic annotation is more complicated because environmental samples typically contain fragmented or incomplete genomes. To address these problems, specialized pipelines such as MG-RAST, MetaErg, and DRAM have been developed. These pipelines offer scalable frameworks for taxonomic profiling and functional inference from metagenomic datasets (Ferrocino et al., 2023). These advances have transformed environmental microbiology from a field that merely describes phenomena to a data-intensive science capable of modeling how microbes function at large scales (Table 3).

4.2 Phylogenomic Reconstruction Tools

Phylogenomic reconstruction, the inference of evolutionary relationships from data across the entire genome, is now a key part of modern microbial taxonomy and evolutionary biology. Genome-scale analyses that use hundreds of conserved orthologs to achieve higher resolution and statistical support have largely replaced traditional single-gene phylogenies, which are often based on 16S rRNA (Tettelin and Medini, 2020). GTDB-Tk, PhyloPhlAn, and OrthoFinder are tools that can automatically identify single-copy orthologs, align multiple sequences, and build trees using methods such as maximum likelihood and Bayesian inference (Parks et al., 2020). These algorithms have enabled the creation of uniform phylogenies encompassing tens of thousands of microbial genomes. This has led to the reclassification of entire phyla and genera based on genomic coherence rather than traditional phenotypic criteria. The Genome Taxonomy Database (GTDB) uses relative evolutionary divergence (RED) measures to maintain consistent rank normalization across microbial lineages. This gives prokaryotic systematics a solid base (Parks et al., 2020) (Tables 2 and 3). Phylogenomic methods are becoming more common for tracking gene flow, recombination events, and new evolutionary changes, as well as for reconstructing evolutionary relationships. Network-based methods such as Splitstree and NeighborNet reveal reticulate evolution driven by horizontal gene transfer. This is a more accurate way to show how microbes evolve than purely tree-like models (Tettelin and Medini, 2020). As sequencing technologies continue to expand genomic information, these computational frameworks are necessary to maintain a taxonomy that is both coherent and scalable, reflecting genetic relatedness and evolutionary history.

4.3 Challenges in Data Integration and Standardization

The increasing complexity of genetic data poses substantial obstacles to integration, reproducibility, and interoperability across investigations. Microbial genomics increasingly relies on multiple data layers, including genomic, transcriptomic, proteomic, and metabolomic data. Each of these levels has its own formats, metadata requirements, and methods for analyzing data (Ferrocino et al., 2023). When there are no standard protocols, it becomes harder to compare results across platforms and laboratories, hindering the reproducibility of scientific findings. The Genomic Standards Consortium (GSC) and the FAIR data initiative (Findable, Accessible, Interoperable, and Reusable) are two international efforts to address these problems by developing unified metadata standards and ontologies for genomic data (Bowers et al., 2017). Frameworks such as the Minimum Information about a Metagenome-Assembled Genome (MIMAG) and Minimum Information about a Genome Sequence (MIGS) guarantee uniform reporting of genome quality, completeness, and assembly statistics (Parks et al., 2020). Nonetheless, considerable challenges persist in integrating multi-omics datasets across spatial and temporal dimensions. Variations in sequencing depth, normalization approaches, and computational workflows can result in divergent interpretations of community composition and functional capacity (Zhang and Shao, 2023). Additionally, metagenomic datasets are often so large that they require powerful high-performance computing (HPC) systems and cloud-based resources. Collaborative platforms like MGnify, KBase, and the Integrated Microbial Genomes (IMG) system are currently very important for solving these problems with computing and standards. They provide curated, interoperable spaces for storing and analyzing data.

4.4 Artificial Intelligence and Machine Learning in Microbial Genomics

Artificial intelligence (AI) and machine learning (ML) are now essential for working with and understanding the vast amounts of data generated by microbial genomics. These methods use pattern recognition and predictive modeling to uncover hidden connections in complex biological data, thereby accelerating the discovery of new genes, pathways, and interactions (Dakal et al., 2025). Random forests and support vector machines are supervised learning algorithms commonly used to predict phenotypes and classify functional genes. Clustering and dimensionality reduction are examples of unsupervised methods that help find genomic patterns and community structure (Ferrocino et al., 2023). Convolutional and recurrent neural networks, two types of deep learning architectures, have recently achieved great success in predicting sequences. You can use raw genomic data to find antibiotic resistance genes, secondary metabolite biosynthesis clusters, and regulatory motifs (Ramond et al., 2025). For example, DeepARG and BiG-SLiCE, which are based on neural networks, operate better than standard similarity-based searches because they can find nonlinear relationships between sequence properties and biological function (Dakal et al., 2025). AI is also changing metagenomic assembly and binning, enabling algorithms trained on reference genomes to assign contigs to taxonomic bins, even in very complex communities (Cowan et al., 2015). Graph-based deep learning methods also enable modeling of gene co-occurrence networks and ecological interactions, providing new insights into how microbial communities are organized and how they change over time. However, combining AI with microbial genomics demands careful consideration of data biases and ease of understanding. The quality and variety of training data are very important for machine learning models. However, environmental and non-model species are typically poorly represented in these datasets (Zhang and Shao, 2023). Even with these problems, AI-driven analytics has significant potential to transform microbiology into a field that predicts and generates hypotheses by connecting genomic complexity with ecological and functional understanding (Table 3).

5. IMPLICATIONS FOR APPLIED MICROBIOLOGY

5.1 Genomics-Informed Bioprospecting

Microbial genomics has transformed bioprospecting, the search for new bioactive chemicals, enzymes, and metabolites from microbes. Genome- and metagenome-guided discovery has mostly replaced traditional culture-based screening. This new method allows researchers to find biosynthetic gene clusters (BGCs) and metabolic pathways directly from sequencing data (Ramond et al., 2025). AntiSMASH, DeepBGC, and BiG-SLiCE are genomic mining tools that employ sequence similarity, domain architecture, and machine learning to infer the locations and functions of BGCs. This makes it easier to find different types of natural products (Dakal et al., 2025). Metagenomic bioprospecting applies this methodology to uncultured microorganisms, which constitute the predominant portion of global microbial diversity. Environmental DNA from soils, sediments, and marine habitats has revealed a vast pool of unexplored secondary metabolites, including novel antibiotics, anticancer drugs, and enzymes with industrial potential (Cowan et al., 2015). For example, finding nonribosomal peptide synthetase (NRPS) and polyketide synthase (PKS) gene clusters in metagenomic libraries has led to the discovery of whole new types of bioactive chemicals. Comparative genomics has not only helped in the discovery of new natural products but also enabled the design of biosynthetic pathways using synthetic biology. Researchers may now estimate product yields, optimize enzyme performance, and move whole pathways into heterologous hosts for scalable manufacturing by combining gene cluster reconstruction with metabolic modeling (Ferrocino et al., 2023). Genomics-informed bioprospecting signifies a transformative change towards data-driven, predictive identification of microbial metabolites.

5.2 Agricultural and Soil Microbiome Applications

Microbial genomics has also transformed agricultural microbiology by offering new ways to improve soil fertility, crop yields, and disease control (Akanmu et al., 2025). Researchers can identify key taxa and functional genes that support nutrient cycling, plant growth, and stress relief by analyzing soil microbiomes at the genomic level (Wenk et al., 2022; Dlamini et al., 2023). For example, metagenomic studies have identified nitrogen-fixing and phosphate-solubilizing bacteria whose genetic characteristics are associated with improved plant health and higher yields (Carini et al., 2020; Fadiji et al., 2026). Whole-genome sequencing of plant-associated microorganisms has elucidated the genetic foundations of advantageous features, including root colonization, phytohormone synthesis, and disease inhibition. Comparative genomics of *Bacillus*, *Pseudomonas*, and *Rhizobium* species has elucidated gene clusters involved in siderophore generation, lipopeptide synthesis, and nitrogen fixation, respectively (Fadiji et al., 2023; Babalola et al., 2024). These discoveries have enabled the development of microbial inoculants and biofertilizers tailored to specific crops and environmental conditions. Genomic methods have also helped us understand how plants, microbes, and soil interact under non-stress conditions. Genome-resolved metagenomics of drought-affected soils has uncovered microbial consortia that exhibit enhanced osmolyte production and reactive oxygen species

(ROS) detoxification capabilities (Ferrocino et al., 2023). These findings guide the development of microbiome engineering technologies to enhance climate-resilient agriculture by precisely modifying soil microbial communities.

5.3 Environmental Remediation and Climate Change Mitigation

The use of microbial genomes in environmental biotechnology has augmented the possibilities for bioremediation and climate change mitigation. Genome-resolved metagenomics facilitates the identification of microbes and functional genes that degrade contaminants, cycle greenhouse gases, and stabilize damaged ecosystems (Cowan et al., 2015). Metagenomic and transcriptomic investigations have identified microbes capable of degrading hydrocarbons, polymers, and heavy metals, as well as genes associated with oxidation, reduction, and conjugation pathways. For instance, environmental isolates of *Pseudomonas* and *Rhodococcus* possess oxygenases and hydrolases that promote hydrocarbon degradation, whereas metal-resistant bacteria contain efflux pumps and metallothioneins that neutralize cadmium, mercury, and arsenic (Haque et al., 2022). Microbial genomics is also crucial for understanding and mitigating the effects of climate change. Metagenomic investigations of wetland and oceanic microbiomes have identified major methanogenic and methanotrophic species that regulate methane flux, as well as nitrifiers and denitrifiers that regulate nitrous oxide (Carini et al., 2020). Genome-based models of these communities provide insights for predictive frameworks for greenhouse gas dynamics across diverse temperature and nutrient conditions. The identification of carbon-fixing mechanisms in unique microbial lineages indicates innovative biotechnological approaches for carbon sequestration (Zhou et al., 2022). Genome-informed biotechnologies are being developed to help restore ecosystems, not just fix problems. Engineered microbial consortia, developed using genome-scale metabolic models, are now being used to enhance pollutant degradation, restore damaged soils, and maintain stable nutrient cycles (Ferrocino et al., 2023). These methods show how combining genetics with systems ecology can help manage the environment in a way that lasts.

5.4 Medical and Industrial Biotechnology Perspectives

Microbial genomics underpins important new ideas in medicinal and industrial biotechnology. The use of metagenomics and comparative genomics in medicine has changed how we think about infectious diseases, antibiotic resistance, and the interactions between microbes and hosts (Ramond et al., 2025). Whole-genome sequencing enables rapid identification of pathogens and tracking of their spread, helping contain outbreaks and promote responsible antibiotic use. Metagenomic analysis of clinical samples has uncovered intricate resistomes and mobilomes, facilitating real-time monitoring of the transmission of resistance genes in hospitals and communities (Matsumura et al., 2025). Microbiome research, propelled by multi-omics integration, is transforming preventive and therapeutic medicine. Comparative assessments of gut microbiomes have identified specific microbial signatures associated with disorders such as inflammatory bowel disease, obesity, and cancer, leading to the development of tailored probiotics and microbiome-based treatments (Browne et al., 2020). Synthetic biology techniques are currently used to modify commensal bacteria into live biotherapeutics that can deliver medications, regulate immune responses, and restore microbial balance.

Genome mining and metabolic engineering have accelerated the design of microbial cell factories for the production of biofuels, enzymes, and high-value compounds in industrial biotechnology (Tettelin and Medini, 2020). Genome-scale modeling and CRISPR-based editing can be used to improve microbial pathways, enabling them to use more substrate and produce more product (Akanmu et al. 2024). For example, scientists have made genetically modified strains of *Clostridium* and *Escherichia coli* that can make bioethanol and bioplastics. Extremophiles, on the other hand, make enzymes that can withstand high temperatures for industrial catalysis (Cowan et al., 2015). Combining artificial intelligence with microbial genomics enhances biotechnological innovation by predicting enzyme function, regulating metabolic flux, and creating new synthetic pathways (Dakal et al., 2025). As genetic datasets grow, these computer programs will continue to advance precision biotechnology. This will transform microbial genomics from a descriptive science into an applied, predictive field that benefits society and the environment in many ways.

6. CONCEPTUAL AND METHODOLOGICAL CHALLENGES

6.1 Taxonomic Instability in the Genomic Age

The genomic revolution has greatly enhanced microbial classification. However, it has also introduced a level of taxonomic instability unparalleled in the history of microbiology. As genome-based phylogenies replace phenotype-driven classification, the reallocation and renaming of microbial taxa have become increasingly common (Parks et al., 2020). This fluidity underscores the dynamic, evolving nature of microbial systematics in the genomic era, while also complicating the interpretation of historical data and comparative studies. A primary cause of this instability lies in the inconsistencies among classification schemes. For example, the Genome Taxonomy Database (GTDB), which uses standardized evolutionary divergence criteria, frequently assigns novel genera or families that differ from those used by NCBI or LPSN (Meier-Kolthoff and Göker, 2019). As a result, a single organism may be assigned multiple, sometimes conflicting, taxonomic labels depending on the reference database. While these developments pose challenges for researchers, clinicians, and policymakers who rely on consistent nomenclature for diagnosis, regulation, and biodiversity assessment (Zhou et al., 2022), the broader implications extend to the reproducibility of scientific findings, international collaboration, and the cumulative building of biological knowledge. Furthermore, the growing reliance on metagenome-assembled genomes (MAGs) has raised critical questions about the legitimacy of species definitions based solely on computational assembly, rather than on cultured isolates. MAGs have revealed extensive new branches within the microbial tree of life. Yet, imperfect genome assemblies and artifacts can further obscure taxonomic boundaries (Cowan et al., 2015). Addressing these challenges is essential not only for taxonomic coherence but also for ensuring that downstream applications in ecology, clinical diagnostics, biotechnology, and environmental policy remain robust and scientifically grounded. Therefore, combining stringent quality standards, such as the MIMAG framework, with flexible taxonomic systems that can accommodate ongoing genomic discovery is imperative for advancing both the science and its practical applications.

6.2 Biases in Sequencing, Assembly, and Annotation

Methodological biases intrinsic to sequencing, assembly, and annotation workflows provide a significant issue in microbial genomics. GC-content fluctuation, uneven coverage, or amplification artifacts can all cause sequencing biases. This can make some genomic regions or taxa less represented (Zhang and Shao, 2023). Short-read sequencing technologies, although economical, often encounter challenges with repetitive regions and structural changes, leading to fragmented assemblies and loss of genomic context. Long-read platforms such as PacBio Sequel and Oxford Nanopore have alleviated some of these problems, but they still exhibit higher error rates and require more computational power (Dakal et al., 2025).

Assembly biases can also mess up metagenomic studies, especially when there are many different microbial types. Co-assembly methods can merge very similar genomes into chimeric contigs or fail to assemble rare taxa, leading to erroneous estimates of diversity and function (Ferrocino et al., 2023). Binning methods that sort contigs by sequence composition and coverage might incorrectly assign genes, especially in contexts with low complexity or great diversity (Parks et al., 2020). These restrictions affect functional annotation, comparative genomics, and ecological interpretation. Bias in annotation makes these difficulties worse. Functional prediction relies significantly on reference databases that are biased towards model organisms and well-defined species. As a result, new or different genes, especially those from environmental MAGs, are sometimes misannotated or not categorized at all (Wang et al., 2025). The ongoing growth of reference datasets, together with the incorporation of machine-learning-driven annotation frameworks, presents interesting solutions that require worldwide coordination and data-sharing infrastructure.

6.3 Integrating Uncultured Taxa Into Functional Frameworks

Despite metagenomics and single-cell sequencing making great strides, assigning uncultured microbes to functional and ecological frameworks remains difficult. Most microbial taxa identified through genomic surveys lack cultured representatives, preventing direct physiological confirmation of hypothesized functions. Functional inference from genomic data is powerful but often depends on sequence homology, which may not accurately reflect metabolic capabilities in highly divergent lineages (Cowan et al., 2015). The lack of cultivated isolates also makes it harder to test gene activity and metabolic models in the laboratory. Genome-scale metabolic reconstructions can suggest possible routes, but to confirm these predictions, you need to grow the organisms and do biochemical tests. Recent advancements, such as microfluidic isolation, co-culture methods, and in situ cultivation platforms, have begun to overcome these obstacles by emulating natural habitats and enabling the recovery of finicky bacteria (Browne et al., 2020). Nonetheless, a significant amount of microbial dark matter remains inaccessible for investigation. To close this gap, we will need systems that combine computational inference with tailored cultivation based on genomic predictions. The combination of stable isotope probing, single-cell transcriptomics, and the creation of synthetic microbial consortia provides a promising approach to connect genomic potential with actual ecological function.

6.4 Ethical, Data-Sharing, and Reproducibility Concerns

As reliance on large-scale genomic datasets increases, ethical, legal, and reproducibility challenges become more complex and pronounced. The worldwide scope of microbial genomics poses challenges regarding data ownership, intellectual property, and benefit-sharing, especially for genetic resources obtained from biodiversity-rich areas (Engels et al., 2021). Frameworks such as the Nagoya Protocol aim to ensure fair and equitable access to, and use of, genetic resources. However, the rapid pace of genomic data generation often outstrips regulatory systems' capacity to adapt and respond (Kloppenburg et al., 2024). Open data projects are important for scientific progress, but they can also be dangerous when used to share sensitive information, such as pathogen genomes or genes that confer antibiotic resistance. Setting up safe, tiered-access repositories can strike a balance between biosecurity and openness, ensuring that genomic resources are both easily accessible and well managed (Matsumura et al., 2025). Reproducibility is another issue. It is hard to reproduce published results without access to the original pipelines, settings, and datasets, as computational studies are complex and involve numerous software packages (Zhang and Shao, 2023). To fix this, more and more journals and funding agencies are requiring authors to deposit their code and data in public repositories. This makes for openness and is more useful in the long run. Lastly, ethical issues arise when using AI to interpret genomes. Algorithmic biases arising from distorted training datasets may lead to systematic inaccuracies in function prediction and taxonomic classification (Dakal et al., 2025). Responsible use of AI in microbiology requires not only technological validation but also adherence to transparent, interpretable, and equitable data practices.

7. FUTURE DIRECTIONS AND EMERGING PARADIGMS

7.1 Towards a Unified Genome-Based Microbial Classification System

Microbial taxonomy is undergoing a phase of consolidation as scientists endeavor to create a cohesive, genome-centric classification system that incorporates phylogenetic, ecological, and functional data. The Genome Taxonomy Database (GTDB) is a major advance in this area, as it provides a standardized framework for studying microbial systematics based on whole-genome evolutionary divergence (Parks et al., 2020). GTDB provides consistent, reproducible categorization by linking taxonomic ranks to relative evolutionary distances. This works for both cultivated and uncultured lineages. Even with these improvements, it is still hard to make genome-based classifications fit with standard taxonomic systems. Numerous current nomenclatural systems, such as those regulated by the International Code of Nomenclature of Prokaryotes (ICNP), continue to rely on cultivated isolates and phenotypic characterization (Zhou et al., 2022). The increasing use of metagenome-assembled genomes (MAGs) as type material has incited discussions over validity, authenticity, and long-term stability in microbial taxonomy (Cowan et al., 2015). Future endeavors will likely focus on developing hybrid systems that integrate genomic coherence with phenotypic and ecological data. To ensure that research groups and databases can reproduce results, these frameworks will need global agreement, uniform metadata reporting, and automated taxonomic validation. To bring together all of microbial biology and make it easier to use in other fields, it will be very important to create a stable and inclusive genome-based taxonomy.

7.2 Systems-Level Integration of Taxonomy, Function, and Evolution

The next big step in research on microbial diversity is to combine taxonomy, function, and evolutionary dynamics at the system level. Conventional research frequently examines these domains in isolation. Nevertheless, the expanding genomic data indicate their profound interconnection (Tettelin and Medini, 2020). Integrative methodologies that combine comparative genomics, phylogenomics, and functional network analysis will facilitate the development of predictive models that elucidate not only the relationships among microbes but also their behavior and evolution within ecosystems (Figure 1.1).

Multi-omics frameworks already lay the groundwork for this integration. Researchers can connect genetic promise to realized function by combining genomic data with transcriptomic, proteomic, and metabolomic profiles (Ferrocino et al., 2023). These methods allow us to trace the flow of information from genes to biochemical activity, and then to ecosystem processes. Systems biology and metabolic modeling will further facilitate this integration, enabling the simulation of community-level interactions and metabolic fluxes predicated on genome-resolved data (Matsumura et al., 2025). This comprehensive viewpoint will reconceptualize microbial diversity as a multifaceted construct encompassing evolutionary lineage, functional capacity, and ecological performance. To make this integration possible, we need improved computational modeling, high-throughput testing, and standardized data frameworks that enable different research fields to collaborate.

7.3 Role of Long-Read Sequencing and Real-Time Genomics

Advances in technology continue to advance the field of microbial genomics. Long-read sequencing technologies, such as PacBio HiFi and Oxford Nanopore, now enable the generation of full, high-quality genomes from individual cells or intricate communities (Dakal et al., 2025). These technologies address many of the problems with short-read sequencing. They provide improved resolution of repetitive regions, plasmids, and structural variants, which are important for understanding how genomes evolve and how different functions work. Combining long-read sequencing with real-time genomic monitoring is creating new possibilities for evolutionary biology and microbial ecology. Real-time sequencing tools, along with portable sensors, are currently being used in the field to monitor how microbes change as the environment shifts, when diseases spread, or during biotechnological processes (Ferrocino et al., 2023). This method lets us observe adaptive evolution and community changes as they occur, directly linking gene-level changes to environmental changes. Real-time genomics will become an important part of microbial research as sequencing chemistry, error correction, and on-site data processing continue to improve. These tools will enable us to identify rapid evolutionary events, track the spread of microbes, and control microbial processes in agriculture, medicine, and the environment.

7.4 Bridging Genomics With Experimental Validation

Genome-based research has the potential to transform everything, but experimental validation remains necessary to establish functional predictions and ecological theories. Computational inference alone cannot adequately encapsulate the intricacies of microbial physiology or community dynamics (Cowan et al., 2015). Experimental techniques such as synthetic microbial consortia, microfluidic cultivation platforms, and stable isotope probing are now being used to verify genomic predictions in controlled environments (Browne et al., 2020). To connect genomics with experimentation, bidirectional integration is needed. This means that genetic insights should inform experimental design, and empirical results should inform computational models. For instance, synthetic biology can help assemble metabolic pathways inferred from genetic data. On the other hand, systems-level perturbation experiments show new features of microbial communities (Wenk et al., 2022). This iterative method will make genome-based functional designations more reliable and help scientists find new biological pathways. The combination of experimental validation and computational modeling is also expected to advance predictive microbiology, which uses genome-scale data to forecast how microbes will respond to environmental changes or interventions. To make this integration happen, genomics, bioengineering, and ecology will all need to work together. This will ensure that data-driven discoveries lead to biological understanding that has been tested in the laboratory.

8. CONCLUSIONS

Genomic information has radically changed the way we study microbial diversity, which is now, more than ever, examined using genome-resolved and metagenome-based approaches that allow us to investigate it in depth. The complex and dynamic nature of microbial communities has led to changes in the taxonomy, functional analysis, and evolutionary study of microbes. Thus, all these aspects have converged in a unique field of study. Genome-based taxonomy has enabled us to define species and discover new taxa, while functional genomics and pangenome studies of microbes have revealed their full metabolic potential, ecological specialization, and adaptive strategies. Also, evolutionary genomics has provided new evidence on the processes of horizontal gene transfer and on genome reduction that drive the diversification of microbial populations. The integration of genomic information at different scales (from genes to ecosystems) is the new frontier of microbial research, an area that is being investigated by using multi-omics approaches and by computational models that are providing new knowledge as well as practical applications in several areas, including agriculture, biotechnology, medicine, and environmental management. Even though there are several challenges to be addressed, such as standards for data production and validation, deep analysis of microbial populations and their roles in several global processes is now possible using current DNA sequencing technologies.

REFERENCES

Agustinho, D. P., Fu, Y., Menon, V. K., Metcalf, G. A., Treangen, T. J., & Sedlazeck, F. J. (2024). Unveiling microbial diversity: Harnessing long-read sequencing technology. *Nature Methods*, *21*(6), 954–966. DOI: 10.1038/s41592-024-02262-1 PMID: 38689099

Akanmu, A. M., Lawal, I. B., Ibrahim, S. L., van Marle-Köster, E., & Hassen, A. (2025b). Metagenomic data from the rumen of South African Mutton Merino sheep supplemented with crude or encapsulated Acacia tannin extracts. *BMC Genomic Data*, *26*(1), 86. DOI: 10.1186/s12863-025-01378-3 PMID: 41239201

Akanmu, A. O., Asemoloye, M. D., Marchisio, M. A., & Babalola, O. O. (2024). Adoption of CRISPR-Cas for crop production: Present status and future prospects. *PeerJ*, *12*, e17402. DOI: 10.7717/peerj.17402 PMID: 38860212

Babalola, O. O., Akanmu, A. O., & Ayangbenro, A. S. (2024). Draft genome sequence of Bacillus velezensis strains AOA1 and AKS2, the potential plant growth-promoting rhizobacteria. *Microbiology Resource Announcements*, *13*(4), e00877–e23. DOI: 10.1128/mra.00877-23 PMID: 38411072

Bowers, R. M., Kyrpides, N. C., Stepanauskas, R., Harmon-Smith, M., Doud, D., Reddy, T., & Schulz, F. (2017). Minimum information about a metagenome-assembled genome (MIMAG) of bacteria and archaea. *Nature Biotechnology*, *35*(8), 725–731. DOI: 10.1038/nbt.3893 PMID: 28787424

Carini, P., Marsden, P. J., Leff, J. W., Morgan, E. E., Strickland, M. S., & Fierer, N. (2020). Effects of spatial variability and relic DNA removal on the detection of temporal dynamics in soil microbial communities. *mBio*, *11*(3), e02776–e19. DOI: 10.1128/mBio.02776-19 PMID: 31964728

Chen, B. S., & Wu, C. C. (2013). Systems biology as an integrated platform for bioinformatics, systems synthetic biology, and systems metabolic engineering. *Cells*, *2*(4), 635–688. DOI: 10.3390/cells2040635 PMID: 24709875

Cowan, D. A., Ramond, J. B., Makhalanyane, T. P., & De Maayer, P. (2015). Metagenomics of extreme environments. *Current Opinion in Microbiology*, *25*, 97–102. DOI: 10.1016/j.mib.2015.05.005 PMID: 26048196

Dakal, T. C., Sharma, A., & Singh, R. (2025). Artificial intelligence and deep learning in microbial genomics: Applications and challenges. *Computational and Structural Biotechnology Journal*, *23*, 1059–1078. DOI: 10.1016/j.csbj.2025.01.014

de Oliveira, A. R., de Toledo Rós, B., Jardim, R., Kotowski, N., de Barros, A., Pereira, R. H. G., Almeida, N. F., & Dávila, A. M. R. (2024). A comparative genomics study of the microbiome and freshwater resistome in Southern Pantanal. *Frontiers in Genetics*, *15*, 1352801. DOI: 10.3389/fgene.2024.1352801 PMID: 38699231

Dlamini, S. P., Akanmu, A. O., & Babalola, O. O. (2023). Variations in the functional diversity of root microbiome of healthy and Northern corn leaf blight infected maize (Zea mays L.). *Spanish Journal of Soil Science*, *13*, 10964. DOI: 10.3389/sjss.2023.10964

Doolittle, W. F., Boucher, Y., Nesbø, C. L., Douady, C. J., Andersson, J. O., & Roger, A. J. (2003). How big is the iceberg of which organellar genes in nuclear genomes are but the tip? *Philosophical Transactions of the Royal Society of London. Series B, Biological Sciences*, *358*(1429), 39–58. DOI: 10.1098/rstb.2002.1185 PMID: 12594917

Engels, J. M., & Ebert, A. W. (2021). A critical review of the current global ex situ conservation system for plant agrobiodiversity. II. Strengths and weaknesses of the current system and recommendations for its improvement. *Plants*, *10*(9), 1904. DOI: 10.3390/plants10091904 PMID: 34579439

Fadiji, A. E., Ayangbenro, A. S., Akanmu, A. O., & Babalola, O. O. (2023). Draft genome sequence of Acinetobacter sp. AYS6, a potential plant growth-promoting endophyte. *Microbiology Resource Announcements*, *12*(10), e00464–e23. DOI: 10.1128/MRA.00464-23 PMID: 37737613

Fadiji, A. E., Lanrewaju, A. A., Akanmu, A. O., Ayangbenro, A. S., & Babalola, O. O. (2026). The relevant role of soil microorganisms in ameliorating abiotic stress in crop plants. In Santoyo, G. (Ed.), *Soil microorganisms for plant growth promotion and soil health* (pp. 421–456). Elsevier., DOI: 10.1016/B978-0-443-34055-0.00022-9

Farrance, C. E. (2019). Identification of microorganisms. In *Pharmaceutical microbiological quality assurance and control: Practical guide for non-sterile manufacturing* (pp. 265–328). Wiley. DOI: 10.1002/9781119356196.ch9

Ferrocino, I., Bianchi, D., Decastelli, L., Cocolin, L., & Rantsiou, K. (2023). The need for an integrated multi-omics approach in microbiome science in the genomic era. *Comprehensive Reviews in Food Science and Food Safety*, *22*(3), 2169–2192. DOI: 10.1111/1541-4337.13184

Gasparetto, J., Magenta, L., Sinisi, F., Zotti, S., Costantini, A., Martelli, B., Giangregorio, T., & Pippucci, T. (2024, March). Computational Genomics platform: A cloud-enabled approach. In *Proceedings of the International Symposium on Grids and Clouds (ISGC2024)* (Vol. 24, p. 29).

Ge, S., Sun, S., Xu, H., Cheng, Q., & Ren, Z. (2025). Deep learning in single-cell and spatial transcriptomics data analysis: Advances and challenges from a data science perspective. *Briefings in Bioinformatics*, *26*(2), bbaf136. DOI: 10.1093/bib/bbaf136 PMID: 40185158

Haque, S., Srivastava, N., Pal, D. B., Alkhanani, M. F., Almalki, A. H., Areeshi, M. Y., Naidu, R., & Gupta, V. K. (2022). Functional microbiome strategies for the bioremediation of petroleum-hydrocarbon and heavy metal contaminated soils: A review. *The Science of the Total Environment*, *833*, 155222. DOI: 10.1016/j.scitotenv.2022.155222 PMID: 35421499

Hausdorf, B. (2025). Changing views on speciation. *Molecular Ecology*, *34*(24), e70175. DOI: 10.1111/mec.70175 PMID: 41236164

Houmenou, C. T., Sokhna, C., Fenollar, F., & Mediannikov, O. (2025). Advancements and challenges in bioinformatics tools for microbial genomics in the last decade: Toward the smart integration of bioinformatics tools, digital resources, and emerging technologies for the analysis of complex biological data. *Infection, Genetics and Evolution*, *136*, 105859. DOI: 10.1016/j.meegid.2025.105859 PMID: 41297621

Jain, C., Rodriguez-R, L. M., Phillippy, A. M., Konstantinidis, K. T., & Aluru, S. (2018). High throughput ANI analysis of 90K prokaryotic genomes reveals clear species boundaries. *Nature Communications*, *9*(1), 5114. DOI: 10.1038/s41467-018-07641-9 PMID: 30504855

Kashtan, N., Roggensack, S. E., Rodrigue, S., Thompson, J. W., Biller, S. J., Coe, A., & Chisholm, S. W. (2014). Single-cell genomics reveals hundreds of coexisting subpopulations in wild Prochlorococcus. *Science*, *344*(6182), 416–420. DOI: 10.1126/science.1248575 PMID: 24763590

Khan, B., & Saha, P. (2026). Challenges, adaptations, and biotechnological potential of oligophilic bacteria. *Antonie van Leeuwenhoek*, *119*(1), 24. DOI: 10.1007/s10482-025-02236-z PMID: 41454939

Kloppenburg, J., Calderón, C. I., & Ané, J. M. (2024). The Nagoya Protocol and nitrogen-fixing maize: Close encounters between Indigenous Oaxacans and the men from Mars (Inc.). *Elementa*, *12*(1), 00115. DOI: 10.1525/elementa.2023.00115

Konstantinidis, K. T., Rosselló-Móra, R., & Amann, R. (2017). Uncultivated microbes in need of their own taxonomy. *The ISME Journal*, *11*(11), 2399–2406. DOI: 10.1038/ismej.2017.113 PMID: 28731467

Krokidis, M. G., Koumadorakis, D. E., Lazaros, K., Ivantsik, O., Exarchos, T. P., Vrahatis, A. G., Kotsiantis, S., & Vlamos, P. (2025). AlphaFold3: An overview of applications and performance insights. *International Journal of Molecular Sciences, 26*(8), 3671. DOI: 10.3390/ijms26083671 PMID: 40332289

Liu, X., Zhang, Z., Tan, C., Ai, Y., Liu, H., Li, Y., Yang, J., & Song, Y. (2025). Global trends in machine learning applications for single-cell transcriptomics research. *Hereditas, 162*(1), 164. DOI: 10.1186/s41065-025-00528-y PMID: 40819059

Matsumura, Y., Yamamoto, M., Gomi, R., Tsuchido, Y., Shinohara, K., Noguchi, T., & Nagao, M. (2025). Integrating whole-genome sequencing into antimicrobial resistance surveillance: Methodologies, challenges, and perspectives. *Clinical Microbiology Reviews, 38*(4), e00140–e22. DOI: 10.1128/cmr.00140-22 PMID: 40910632

McReynolds, E., Elshahed, M. S., & Youssef, N. H. (2025). An ecological-evolutionary perspective on the genomic diversity and habitat preferences of the Acidobacteriota. *Microbial Genomics, 11*(1), 001344. DOI: 10.1099/mgen.0.001344 PMID: 39879090

Meier-Kolthoff, J. P., Carbasse, J. S., Peinado-Olarte, R. L., & Göker, M. (2022). TYGS and LPSN: A database tandem for fast and reliable genome-based classification and nomenclature of prokaryotes. *Nucleic Acids Research, 50*(D1), D801–D807. DOI: 10.1093/nar/gkab902 PMID: 34634793

Meier-Kolthoff, J. P., & Göker, M. (2019). TYGS is an automated high-throughput platform for state-of-the-art genome-based taxonomy. *Nature Communications, 10*(1), 2182. DOI: 10.1038/s41467-019-10210-3 PMID: 31097708

Miralles, I., Ortega, R., & Comeau, A. M. (2021). Functional and taxonomic effects of organic amendments on the restoration of semiarid quarry soils. *mSystems, 6*(6), e00752–e21. DOI: 10.1128/mSystems.00752-21 PMID: 34812648

Pace, N. R. (1997). A molecular view of microbial diversity and the biosphere. *Science, 276*(5313), 734–740. DOI: 10.1126/science.276.5313.734 PMID: 9115194

Parks, D. H., Chuvochina, M., Chaumeil, P. A., Rinke, C., Mussig, A. J., & Hugenholtz, P. (2020). A complete domain-to-species taxonomy for bacteria and archaea based on a standardized genome phylogeny. *Nature Biotechnology, 38*(9), 1079–1086. DOI: 10.1038/s41587-020-0501-8 PMID: 32341564

Parks, D. H., Chuvochina, M., Rinke, C., Mussig, A. J., Chaumeil, P. A., & Hugenholtz, P. (2022). GTDB: An ongoing census of bacterial and archaeal diversity through a phylogenetically consistent, rank normalized and complete genome-based taxonomy. *Nucleic Acids Research, 50*(D1), D785–D794. DOI: 10.1093/nar/gkab776 PMID: 34520557

Pereira, R., Oliveira, J., & Sousa, M. (2020). Bioinformatics and computational tools for next-generation sequencing analysis in clinical genetics. *Journal of Clinical Medicine*, *9*(1), 132. DOI: 10.3390/jcm9010132 PMID: 31947757

Pinto, Y., & Bhatt, A. S. (2024). Sequencing-based analysis of microbiomes. *Nature Reviews. Genetics*, *25*(12), 829–845. DOI: 10.1038/s41576-024-00746-6 PMID: 38918544

Puniya, B. L., Verma, M., Damiani, C., Bakr, S., & Dräger, A. (2024). Perspectives on computational modeling of biological systems and the significance of the SysMod community. *Bioinformatics Advances*, *4*(1), vbae090. DOI: 10.1093/bioadv/vbae090 PMID: 38948011

Ramond, P., Galand, P. E., & Logares, R. (2025). Microbial functional diversity and redundancy: Moving forward. *FEMS Microbiology Reviews*, *49*, fuae031. DOI: 10.1093/femsre/fuae031 PMID: 39689915

Richter, M., & Rosselló-Móra, R. (2009). Shifting the genomic gold standard for the prokaryotic species definition. *Proceedings of the National Academy of Sciences of the United States of America*, *106*(45), 19126–19131. DOI: 10.1073/pnas.0906412106 PMID: 19855009

Ruan, W., Lyu, Y., Zhang, J., Cai, J., Shu, P., Ge, Y., Lu, Y., Gao, S., Wang, Y., Wang, P., Zhao, L., Wang, T., Liu, Y., Fang, L., Liu, Z., Liu, Z., Li, Y., Wu, Z., Chen, J., & Liu, T. (2026). Large language models for bioinformatics. *Quantitative Biology*, *14*(1), e70014. DOI: 10.1002/qub2.70014 PMID: 41676319

Seemann, T. (2014). Prokka: Rapid prokaryotic genome annotation. *Bioinformatics (Oxford, England)*, *30*(14), 2068–2069. DOI: 10.1093/bioinformatics/btu153 PMID: 24642063

Strozzi, F., Janssen, R., Wurmus, R., Crusoe, M. R., Githinji, G., Di Tommaso, P., Belhachemi, D., Möller, S., Smant, G., De Ligt, J., & Prins, P. (2019). Scalable workflows and reproducible data analysis for genomics. In Dutheil, N. (Ed.), *Evolutionary genomics: Statistical and computational methods* (pp. 723–745). Springer. DOI: 10.1007/978-1-4939-9074-0_24

Szabó, G., Eloe-Fadrosh, E. A., Pett-Ridge, J., & Woyke, T. (2025). A genomic view of Earth's biomes. *Nature Reviews. Genetics*, *27*(1), 13–31. DOI: 10.1038/s41576-025-00888-1 PMID: 40954201

Terra, L. A., Klepa, M. S., Nogueira, M. A., & Hungria, M. (2025). Pangenome analysis indicates evolutionary origins and genetic diversity: Emphasis on the role of nodulation in symbiotic Bradyrhizobium. *Frontiers in Plant Science*, *16*, 1539151. DOI: 10.3389/fpls.2025.1539151 PMID: 40241821

Tettelin, H., & Medini, D. (Eds.). (2020). *The pangenome: A new frontier in microbiology*. Springer International Publishing., DOI: 10.1007/978-3-030-38281-0

Tortoli, E., Fedrizzi, T., Meehan, C. J., Trovato, A., Grottola, A., Giacobazzi, E., Serpini, G. F., Tagliazucchi, S., Fabio, A., Bettua, C., Bertorelli, R., Frascaro, F., De Sanctis, V., Pecorari, M., Jousson, O., Segata, N., & Cirillo, D. M. (2017). The new phylogeny of the genus Mycobacterium: The old and the new. *Infection, Genetics and Evolution*, *56*, 19–25. DOI: 10.1016/j.meegid.2017.10.013 PMID: 29030295

Van Rossum, T., Ferretti, P., Maistrenko, O. M., & Bork, P. (2020). Diversity within species: Interpreting strains in microbiomes. *Nature Reviews. Microbiology*, *18*(9), 491–506. DOI: 10.1038/s41579-020-0368-1 PMID: 32499497

Wang, Z., Zhao, Z., & Liu, J. (2025). Homoploid hybrid speciation: Advancements and challenges. *Plant Ecology & Diversity*, *18*(3–4), 141–158. DOI: 10.1080/17550874.2025.2550015

Wenk, S., Claassens, N. J., & Lindner, S. N. (2022). Synthetic metabolism approaches: A valuable resource for systems biology. *Current Opinion in Systems Biology*, *30*, 100417. DOI: 10.1016/j.coisb.2022.100417

Zhang, X., & Shao, M. (2023). Computational frameworks for functional annotation and data integration in microbial genomics. *Trends in Biotechnology*, *41*(6), 711–726. DOI: 10.1016/j.tibtech.2022.12.004

Zhou, Y., Kathiresan, N., Yu, Z., Rivera, L. F., Yang, Y., Thimma, M., Manickam, K., Chebotarov, D., Mauleon, R., Chougule, K., Wei, S., Gao, T., Green, C. D., Zuccolo, A., Xie, W., Ware, D., Zhang, J., McNally, K. L., & Wing, R. A. (2024). A high-performance computational workflow to accelerate GATK SNP detection across a 25-genome dataset. *BMC Biology*, *22*(1), 13. DOI: 10.1186/s12915-024-01820-5 PMID: 38273258

Zhou, Z., Liu, Y., Anantharaman, K., & Li, M. (2022). The expanding Asgard archaea invoke novel insights into Tree of Life and eukaryogenesis. *mLife*, *1*(4), 374–381. DOI: 10.1002/mlf2.12048 PMID: 38818484

Zhou, Z., Tran, P. Q., Cowley, E. S., Trembath-Reichert, E., & Anantharaman, K. (2025). Diversity and ecology of microbial sulfur metabolism. *Nature Reviews. Microbiology*, *23*(2), 122–140. DOI: 10.1038/s41579-024-01104-3 PMID: 39420098

Chapter 2
Modern Toolkits for Microbial Biodiversity Exploration Integrating Omics, Bioinformatics, and Single-Cell Techniques

Ben Jesuorsemwen Enagbonma
http://orcid.org/0000-0002-4699-5061
North-West University, South Africa

Olubukola Oluranti Babalola
http://orcid.org/0000-0003-4344-1909
North-West University, South Africa

ABSTRACT

The use of traditional microbial toolkits in time past has reduced our understanding of the diversity, function, and ecological interactions of the microbial community. This chapter presents a novel integrated perspective on modern toolkits used for exploring microbial biodiversity. To achieve this aim, this chapter is discussed in the following sections: i) modern omics approaches in microbial biodiversity, ii) bioinformatics tools and computational frameworks, and iii) single-cell techniques for microbial exploration. Here, an evidence-based review method was used for this chapter, and information was drawn from cutting-edge literature, comparative analyses of tool performance, and case studies across environmental and host-associated microbiomes. The chapter concludes by identifying challenges and proposing future

DOI: 10.4018/979-8-2600-2100-2.ch002

directions for research and application. As these technologies grow more accurately, have higher throughput, and widely accessible, they will further catalyze innovation across environmental, agricultural, industrial, and biomedical research.

1 INTRODUCTION

The variety and variability of microorganisms (microbial biodiversity) are linked to vital environmental processes and functions around the universe (Sun & Sanchez 2023). This microbial biodiversity encompasses multiple dimensions (Figure 3.1), including bacteria, archaea, fungi, protists, and viruses (Nizamani et al. 2024). Microbial diversity influences nutrient cycling, organic matter decomposition, ecosystem resilience, and biogeochemical transformations (Nizamani et al. 2024). These, in turn, contribute significantly to agriculture, environmental health, and biotechnology and provide essential ecosystem services as well as pathogen suppression (Diyaolu & Folarin 2024). Enagbonma et al. (2019) reported that understanding the diversity and functional potential of microbial communities is therefore fundamental for advancing ecological theory, informing conservation strategies, and exploiting microbial resources for industrial and medical applications.

Previously, microbiologists relied heavily on culture-dependent approaches (Table 3.1) to study microbial diversity (Ghiasian et al. 2017). These methods are important for yielding important microbial isolates and providing insights into microbial physiology; however, these methods are limited by the fact that most microbes remain unculturable under laboratory conditions (Hinsu et al. 2021). This constraint in microbial ecological studies has now been revolutionized in today's era owing to the advent of culture-independent techniques (such as amplicon-based sequencing of 16S/18S rRNA genes, internal transcribed spacer (ITS) regions, and other marker genes) (Lakshmanan et al. 2017). This new technology enables the identification of previously hidden taxa and provides a broad view of community composition (Babalola et al. 2021b). For example, Enagbonma et al. (2024b) used amplicon-based sequencing of 16S ribosomal ribonucleic acid (16S rRNA) genes to reveal complete microbial community changes under anthropogenic fertilization. The challenge of these new amplicon-based sequencing technologies is that they often offer limited information on microbial functional activity, interactions, and dynamics within complex ecosystems (Babalola et al. 2020).

Most recently, microbial researchers have moved beyond mere cataloging of microbial taxa toward functional characterization at unprecedented resolution because of the arrival of high-throughput sequencing, multiomics technologies, and single-cell techniques. For example, with the aid of shotgun metagenomic sequencing, Enagbonma et al. (2025b) and Enagbonma et al. (2019) reconstructed microbial genomes and assessed their functional potential across entire communities from legume-based cropping systems and termite mounds, respectively. While Teseo et al. (2022) conducted comprehensive phylogenomic and metabolic analysis of the large intestine microbiome in domestic cattle. Furthermore, Bashiardes et al. (2016b), Van Den Bossche et al. (2025b), and Olanrewaju et al. (2024) revealed that metatranscriptomics, metaproteomics, and metabolomics capture gene expression, protein activity, and metabolite production, respectively. Findings from Jorth et al. (2014) metatranscriptomics study show that, although microbial composition varies widely between patients, disease-associated communities maintain consistent metabolic profiles, typically carried out by a unique set of microbes in each individual. These findings therefore offer a dynamic view of microbial activity and interspecies interactions. Zhuang (2021) reported that single-cell genomics and transcriptomics provide insights into rare or uncultivable microbes and resolve heterogeneity within communities.

While several reviews have addressed individual omics techniques or specific ecosystems, this chapter differs from existing literature by adopting an explicitly integrative perspective. Rather than treating omics, bioinformatics, and single-cell approaches in isolation, this chapter emphasizes their methodological convergence and complementary strengths in resolving microbial diversity, function, and ecological interactions. Furthermore, particular attention is given to underexplored and complex ecosystems, where integrated toolkits are essential for overcoming analytical and ecological challenges. By synthesizing advances across multiple technological domains, this chapter provides a cohesive framework for understanding how modern microbial biodiversity research is evolving toward holistic, systems-level insights.

Thus, the aim of this chapter is to provide a comprehensive overview of modern toolkits for microbial biodiversity exploration. Specifically, this chapter aims to emphasize the integration of omics, bioinformatics, and single-cell approaches into microbial ecological studies. To achieve this aim, this chapter is discussed in the following sections: i) modern omics approaches in microbial biodiversity, ii) bioinformatics tools and computational frameworks, iii) single-cell techniques for microbial exploration, and iv) challenges and future perspectives. Overall, this chapter highlights how modern methodological advances unveil microbial diversity and function that fuel innovations in biotechnology, environmental management, and human health.

Figure 1. Multiple dimensions of microbial biodiversity

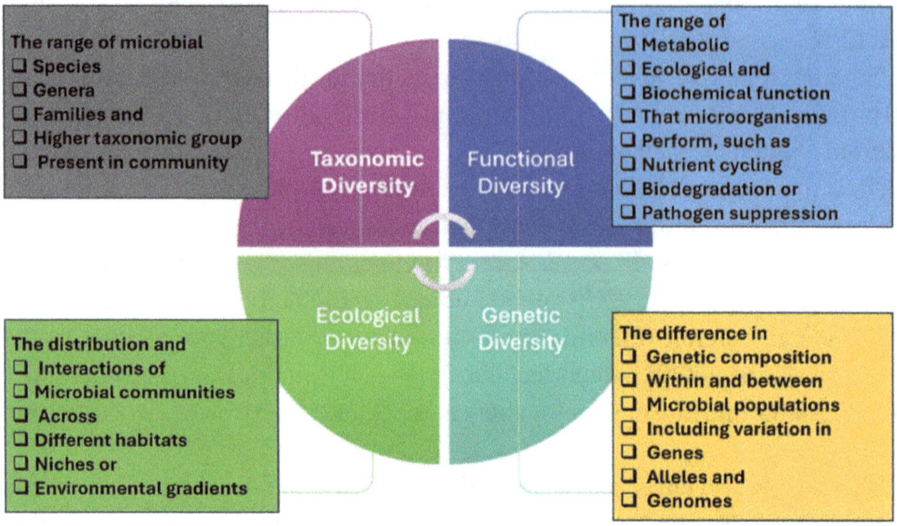

Table 1. Summary of culture-dependent approaches for microbial biodiversity studies

Approach	Principle	Advantages	Limitations	Reference
Agar Plate Culturing (Solid Media)	Microbial cells grow into visible colonies on nutrient agar under controlled conditions.	Simple, inexpensive, allows isolation of pure colonies.	Majority of microbes are unculturable; biased toward fast-growing organisms.	(Burmølle et al. 2009)
Liquid (Broth) Culturing	Microbes grow in liquid nutrient media enabling biomass production.	Easy to scale, useful for enrichment and physiological assays.	Mixed cultures difficult to separate; selection for dominant species.	(Rouf et al. 2017)
Selective and Differential Media	Media contain inhibitors or indicators to select for and differentiate microbial groups.	Enables targeted isolation and rapid identification of specific groups.	Excludes nontarget organisms; may inhibit some strains within the target group.	(Bonnet et al. 2020)

continued on following page

Table 1. Continued

Approach	Principle	Advantages	Limitations	Reference
Enrichment Cultures	Specific growth conditions (substrate, pH, temperature, oxygen) are used to favor certain microbes.	Facilitates recovery of specialized or rare functional groups.	Strong selection biases: community composition may shift during enrichment.	(Gu 2021)
Most Probable Number (MPN)	Estimates microbial abundance based on serial dilution and detection of growth.	Useful for viable but nonculturable microbes; quantitative.	Low taxonomic resolution; labor-intensive; statistical uncertainty.	(Papen & Von Berg 1998)
Biochemical Characterization Tests	Detect metabolic or enzymatic traits of microbial isolates (e.g., catalase, oxidase).	Provides phenotype-based identification; rapid.	Limited discriminatory power; influenced by culture conditions.	(Ulfat et al. 2024)
Physiological Profiling (e.g., BIOLOG, Community-Level Physiological Profiling (CLPP))	Tests metabolic capacity across multiple carbon sources using microplates.	High-throughput functional profiling; comparative analysis.	Requires culturable and active microbes; substrate bias.	(Decena Jr & dela Cruz 2024; Enagbonma et al. 2021)
Isolation of Pure Cultures	Subculturing single colonies to obtain axenic isolates.	Essential for taxonomy, physiology, and functional characterization.	Many microbes do not grow alone; loss of interactions relevant in situ.	(Harnisch et al. 2024)
Colony Morphology Assessment	Visual examination of colony traits such as shape, color, and texture.	Rapid initial screening and differentiation of isolates.	Morphology is often insufficient for accurate identification.	(Sousa et al. 2013)
Phenotypic Fingerprinting (e.g., Fatty Acid Methyl Ester (FAME), Matrix-Assisted Laser Desorption/ Ionization – Time-of-Flight (MALDI-TOF))	Profiling biochemical features such as fatty acids or protein spectra.	Fast, accurate identification when reference databases exist.	Requires extensive reference libraries; limited resolution for novel taxa.	(Victorio et al. 1996)
Culture-Based Functional Assays	Detect production of enzymes, metabolites, or plant-growth-promoting traits.	Directly links isolates to ecological or biotechnological functions.	Results highly dependent on media composition and growth conditions.	(Sofo & Ricciuti 2019)

continued on following page

Table 1. Continued

Approach	Principle	Advantages	Limitations	Reference
Coculturing Approaches	Growing multiple microbes together to stimulate interactions.	Allows growth of organism's dependent on microbial partners; reveals interactions.	Complex to control; difficult to isolate individual species afterward.	(Paul et al. 2013)
High-Throughput Culturomics	Uses diverse culture conditions and automation for large-scale microbial isolation.	Greatly increases culturable diversity; useful for rare or novel taxa.	Resource-intensive; requires advanced equipment and extensive optimization.	(Huang et al. 2023)

2 MODERN OMICS APPROACHES TO MICROBIAL BIODIVERSITY

With more than 99% of environmental microbes remaining uncultured, modern omics (metagenomics, metaproteomics, metabolomics, metatranscriptomics, and single-cell omics) technologies (Table 3.2), have stepped in to change the narrative (Satya et al. 2024). These approaches have revolutionized the study of microbial biodiversity because they enable comprehensive, high-resolution, and culture-independent exploration of microbial communities across ecosystems (Shittu et al. 2025).

2.1 Metagenomics

Metagenomics is an omics technique that aids in the recovery of genetic material (DNA) directly from environmental samples and allows microbial scientists to analyze all microbes in a community at once (Zhang et al. 2021a). The field of metagenomics is dominated by amplicon sequencing and shotgun metagenomics (Lane et al. 2024). Amplicon sequencing profiles microbial biodiversity by targeting conserved genetic markers such as 16S rRNA (for bacteria/archaea) or ITS (for fungi) (Grant 2022). The amplicon sequencing method enables high-throughput identification of microbial community composition, richness, and taxonomic diversity (Doytchinov & Dimov 2022). Furthermore, while amplicon sequencing is cost effective, it has limited functional resolution (Heidrich & Beule 2022). Kerrigan & D'Hondt (2022) underscored the value of amplicon sequencing by using the V4–V5 region of the 16S rRNA gene to characterize bacterial richness and community composition in seawater and marine sediment samples. Zhao et al. (2023) reported that both 16S rRNA gene amplicon sequencing primers (515F (Parada)-806R (Apprill) and 515F (Parada)-926R (Quince)) consistently captured shifts in the ammonia-oxidizing archaeal community structure associated with land-use differences. Debeljak & Baltar (2023) used an 18S rRNA dataset covering samples from key global ecosystems (terrestrial, freshwater, and marine) to evaluate worldwide patterns of fungal diversity and community structure. They reported that fungal diversity was greatest in terrestrial environments, followed by freshwater and marine systems.

In contrast, shotgun metagenomics sequences all DNA in a sample and offers deeper insights into community structure, metabolic functions, and ecological interactions (Enagbonma & Babalola 2022). However, it requires higher DNA quality, advanced computational resources, and larger sequencing budgets (Lema et al. 2023). This approach allows microbial researchers to detect both abundant and rare taxa, reconstruct genomes from uncultured microorganisms, and explore functional attributes (Lapidus & Korobeynikov 2021).

Fujita et al. (2023) applied shotgun metagenomics to explore how overlapping ecological niches contribute to instability and collapse within microbiome communities. They reported that shifts in community composition within the experimental microbiome were strongly associated with the extent of gene-repertoire overlap among bacterial species. Thus, their study will aid in the prediction of key events in microbiome dynamics, such as abrupt community collapse. More recent studies have expanded this knowledge. For example, Adebayo et al. (2025) used shotgun metagenomics to characterize the rhizomicrobial community composition and diversity in carrots cultivated under monoculture and soybean-precedent conditions. This study provides an essential understanding of how cropping systems with soybeans drive the carrot rhizomicrobiome. Also, Zinicola et al. (2015) used shotgun metagenomics to reveal distinct microbial and functional gene profiles in healthy skin and active or inactive bovine digital dermatitis lesions."

2.2 Metatranscriptomics

Metatranscriptomics is another key omics tool for exploring microbial biodiversity, as it offers insights into the active functions of microbial communities (Aplakidou et al. 2024). This approach captures real-time gene expression and provides a dynamic picture of which microbial genes are actively transcribed (Ishii et al. 2013). The implication is that it enables researchers to identify functionally important taxa, uncover metabolic pathways in action, and understand how microbial communities respond to environmental indications (Zhang et al. 2021b). For example, Law et al. (2022) carried out a metatranscriptomic exploration of boreal forest samples to capture the seasonally resolved transcriptomes of spruce roots and more than three hundred and fifty root-related fungal species. These findings are associated with the functional response of host trees to increased nutrient accessibility, with substantial changes in the fungal community. Jiang et al. (2016b) further demonstrated the importance of metatranscriptomics when it was used to assess key metabolic pathways and microbial community-specific functionality, such as glycan degradation pathways (cow rumen) and phosphonate metabolism (deep sea).

Despite the importance of metatranscriptomics, however, several challenges remain (Ojala et al. 2023). For example, RNA extraction is technically demanding because RNA is inherently unstable and degrades rapidly (Wang et al. 2012). Another challenge is that the extraction protocols can introduce biases, and careful experimental design and computational filtering are needed to distinguish plant RNA from microbial transcripts (Ojala et al. 2023). Despite these limitations, metatranscriptomics remains an indispensable tool for advancing functional microbial ecology (Bashiardes et al. 2016a; Shakya et al. 2019).

2.3 Metaproteomics

Another emerging omics technique for investigating microbial communities is metaproteomics, which offers the ability to profile proteins that highlight the functionality of microbiomes (Nebauer et al. 2024). These proteins serve as direct biochemical indicators of metabolic pathways, ecological interactions, stress responses, nutrient transformations, and community-level physiological states (Arıkan & Atabay 2024). By integrating high-resolution mass spectrometry with advanced bioinformatics pipelines, metaproteomics enables researchers to link microbial diversity to function in ways not achievable through genomics alone (Van Den Bossche et al. 2025a; Zhang et al. 2025). This pipeline allows the identification and relative quantification of enzymes and transporters that drive nutrient transformations, enabling the linkage of taxa to functions in situ (Liang et al. 2013). For example, soil and leaf litter metaproteomic studies have shown how protein-level evidence can resolve which microbial groups express extracellular enzymes and other drivers of organic matter turnover (Keiblinger et al. 2016). Starr et al. (2018) also showed that metaproteomics is a useful tool for capturing dynamic microbe–plant interactions and temporal shifts in microbial function at hotspots. Shrestha et al. (2021) reported metaproteomic signatures of nutrient mobilization, stress responses, and specialized microbial guilds associated with root exudates. These efforts demonstrate the method's value for connecting microbial identity to plant-relevant functions.

Applied environmental systems such as wastewater treatment and activated sludge have also benefitted from metaproteomics (Rodríguez et al. 2015). For example, Püttker et al. (2015) successfully identified proteins involved in central metabolism and nitrogen removal pathways and compared sample fractionation strategies to reduce complexity and improve proteome coverage. This study illustrates how metaproteomics can inform process optimization and reveal active microbial players in engineered systems.

Recently, technological and computational advances have expanded the sensitivity and scope of metaproteomics. For example, Xian et al. (2025) reported ultrasensitive workflows and cloud-compatible analytics that pushed detection into the "dark proteome" and improved taxonomic assignment in complex communities. This makes metaproteomics increasingly accessible for high-resolution biodiversity and functional studies in soil, marine, host-associated, and wastewater research (Xian et al. 2025). Despite its promise, metaproteomics has limitations ranging from i) protein extraction biases, ii) incomplete reference databases (which reduce peptide identification or misassign taxonomic origin), and iii) complex data interpretation when trying to infer rates from snapshots (Van Den Bossche et al. 2025a).

2.4 Metabolomics

Metabolomics has become a central tool for investigating microbial biodiversity because it provides direct biochemical evidence of microbial activity rather than predicted genetic potential (Chen et al. 2022). By capturing small-molecule profiles that reflect active metabolic pathways, stress responses, and ecological interactions, metabolomics fills a functional gap left by metagenomics and metatranscriptomics (Mathuria et al. 2024). This functional layer is crucial for understanding microbial contributions to ecosystem processes, host physiology, and plant–microbe interactions. This provides a more realistic picture of microbial roles in natural and engineered environments (Desa et al. 2023; Olanrewaju et al. 2024).

Advances in analytical platforms such as liquid chromatography–tandem mass spectrometry (LC–MS/MS), gas chromatography–mass spectrometry (GC–MS), high-resolution Orbitrap mass spectrometry (MS), nuclear magnetic resonance (NMR) spectroscopy, and imaging mass spectrometry have expanded metabolome coverage and improved sensitivity across complex microbiomes (Fang et al. 2023; Martias et al. 2021). These technologies have enabled applications ranging from rhizosphere metabolomics, which reveals root exudate-driven microbial selection (Hildebrand et al. 2023), to marine and extreme-environment studies, which uncover novel osmolytes, siderophores, and specialized metabolites that characterize ecological guilds (Berry et al. 2021; Enagbonma et al. 2025a; Xu et al. 2021). Additionally, metabolomic biomarkers have been successfully applied to track pollutant degradation and ecological restoration in bioremediation systems (Singh et al. 2021), demonstrating the method's utility in environmental monitoring and functional microbial ecology.

Integrating metabolomics with other omics platforms has further strengthened its role in microbial biodiversity exploration (Kato et al. 2024). Multiomics frameworks that combine metagenomics, metatranscriptomics, metaproteomics, and metabolomics provide systems-level insights by linking biosynthesis-related gene clusters to metabolites, connecting enzyme expression to metabolic flux, and revealing coordinated community-level functions (Kang et al. 2022; Noecker et al. 2019). Despite persistent challenges, including low annotation rates, matrix complexity, and overlap between host- and microbe-derived metabolites (Pinu et al. 2019), new developments, such as machine learning–aided annotation, stable isotope-resolved metabolomics, and ultrahigh-resolution MS, continue to improve accuracy and interpretability (Cambiaghi et al., 2017; Schrimpe-Rutledge et al., 2016). Collectively, these advances position metabolomics as a cornerstone in modern microbial ecology and biodiversity research (Schumann et al. 2025).

In all, modern omics approaches are most powerful when integrated (Figure 3.2), as metagenomics define microbial potential, metatranscriptomics reveals active gene expression, metaproteomics confirms functional execution, and metabolomics captures the resulting biochemical outcomes. Comparing omics approaches (Figure 3.2) reveals complementary insights, links potential to activity, guides tool selection, enables ecosystem-level understanding, and maximizes ecological inference.

Figure 2. Integrated multi-omics framework for functional microbial ecology

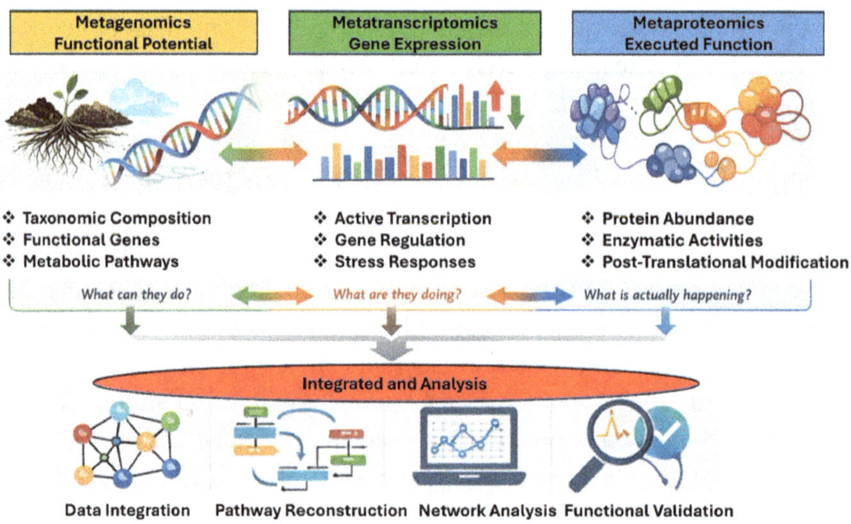

Table 2. Comparative summary of major omics approaches in microbial ecology

Omics Approach	Primary Biological Layer & Information	Strengths	Limitations	Typical Resolution	Relative Cost	Key Ecological Questions Addressed	References
Metagenomics	DNA → community composition and genetic potential	• Captures total community diversity including uncultured taxa • Allows functional gene prediction and MAG recovery	• Does not directly measure activity • Requires deep sequencing for rare taxa	Community / population (up to strain level with deep coverage)	High	• Who is present? • What metabolic potentials exist? • How do gene repertoires vary across environments?	(Quince et al. 2017)
Metatranscriptomics	RNA → gene expression	• Detects active transcripts under environmental conditions • Reveals responses to perturbations	• RNA instability complicates extraction • Taxonomic assignment can be challenging	Population / pathway	High	• Which functions are being expressed? • How do microbes respond to changes (e.g., nutrients, stress)?	(Jiang et al. 2016a)
Metaproteomics	Proteins → functional execution	• Direct measurement of proteins → function • Links taxa to expressed enzymes	• Requires advanced mass spectrometry and bioinformatics • Protein databases remain incomplete	Functional / pathway	Very High	• What proteins are actively produced? • Which taxa contribute to ecosystem functions?	(Tartaglia et al. 2020)
Metabolomics	Metabolites → biochemical end products	• Measures real biochemical outcomes • Integrates host–microbe and microbe–microbe interactions	• Many metabolites not annotated • Signal overlaps between host and microbes	Pathway / ecosystem	High	• What biochemical processes are occurring? • How does metabolite exchange influence interactions?	(Patti et al. 2012)
Single-cell Omics	DNA/RNA of individual cells	• Resolves rare taxa, eliminates community averaging • Reveals cell-level heterogeneity	• Limited throughput • Technically challenging	Single cell	Very High	• What are the functional traits of rare taxa? • How do individual microbes differ within communities?	(Woyke et al. 2017)

3 BIOINFORMATICS TOOLS AND COMPUTATIONAL FRAMEWORKS

In modern research, bioinformatics is now the central engine for exploring microbial biodiversity (Sabir et al. 2025). This is because it enables researchers to decode the complexity of microbial communities across soil, rhizosphere, aquatic, host-associated, and engineered environments (Rekadwad 2024). Research has shown that large-scale datasets are generated by modern toolkits (such as high-throughput sequencing, mass spectrometry, and imaging technologies) (Demirbaga et al. 2024). Because of this, there is currently a need for robust computational frameworks, and bioinformatics is the game changer (Ahmed 2024). With bioinformatics tools, it is much easier to transform raw molecular data into ecological insight (Siddiqui 2024). Furthermore, bioinformatics tools support sequence processing, taxonomic classification, functional annotation, genome reconstruction, and multiomics integration (Abdi et al. 2024).

4 SEQUENCE QUALITY CONTROL AND ASSEMBLY

For the effective exploration of microbial biodiversity, researchers need to work with high-quality sequences (Creer et al. 2016). In microbial biodiversity research, whether via metagenomics, amplicon sequencing, or related approaches, the classification of high-quality sequences is typically based on their Phred (Q) scores (Liu et al. 2025). Maestri et al. (2025) reported that a Q20 score, which corresponds to approximately 99% base-call accuracy, is generally regarded as acceptable for most analyses. In contrast, sequences with a Q30 score, reflecting 99.9% accuracy, are widely recognized as high quality and form the standard threshold in most microbial sequencing workflows. Even higher, Q40 represents approximately 99.99% accuracy, indicating exceptionally high-quality reads, although such stringent quality is rarely required for routine microbial biodiversity studies. Tools such as Fast Quality Control (FastQC) and Trimmomatic are widely used for preprocessing steps, which include quality assessment, adapter trimming, filtering of low-quality reads, and removal of contaminants (Chen 2025; Sewe et al. 2022). In metagenomic research, especially within the rhizosphere where soil particles, plant DNA, and humic substances complicate sequencing, these steps ensure that downstream analyses remain reliable (Babalola et al. 2025; Babalola & Enagbonma 2025).

After preprocessing, assembly strategies are then employed to reconstruct longer contigs from metagenomic reads, and in most studies, assemblers such as MEGAHIT (Meta-Genome Assembler for High Throughput Sequencing) and SPAdes (St. Petersburg genome assembler) offer scalable solutions for large and complex datasets (Wang et al. 2021). This enables researchers to recover high-quality contigs even from microbiomes with extreme taxonomic diversity, especially when assemblers are combined (Saheb et al. 2021). Furthermore, their ability to handle uneven coverage, strain variability, and high-complexity substrates is especially critical for environmental metagenomes, which often contain thousands of interacting species (Veloso et al. 2023). For example, Qayyum et al. (2025) reported that the metaSPAdes (Metagenomic St. Petersburg genome assembler)–MetaBAT2 (Metagenome Binning based on Abundance and Tetranucleotide frequence version 2) pipeline performs particularly well for recovering low-abundance species, whereas the MEGAHIT–MetaBAT2 workflow is more effective for obtaining strain-resolved genomes. Their results demonstrate that different tool combinations can vary substantially in performance, even when designed to achieve the same goal.

5 TAXONOMIC PROFILING

Marker gene-based approaches using 16S rRNA, ITS, or other conserved loci enable rapid profiling of community composition (Bálint et al. 2016). Tools such as QIIME2 (Quantitative Insights Into Microbial Ecology, version 2), Mothur, and DADA2 (Divisive Amplicon Denoising Algorithm 2) allow denoising, clustering, and diversity calculations, and they offer fine-scale taxonomic resolution without requiring full metagenomes (Коротецкий et al. 2025). This was supported by Srivastava et al. (2024), who reported that QIIME 2 is a ready-to-use pipeline that performs stepwise analysis of massive short reads of 16S rRNA genes. However, marker gene methods may lack strain-level precision, leading to the adoption of whole-genome taxonomic profilers such as Kraken2 (K-mer-based Rapid Accurate Classification of sequences, version 2), Bracken (Bayesian Reestimation of Abundance with KrakEN), and MetaPhlAn3 (Metagenomic Phylogenetic Analysis, version 3) (Anyansi et al. 2020; Ibanez-Lligona et al. 2023). These tools classify reads via large reference databases and k-mer–based strategies, allowing strain-level identification and more accurate estimates of community structure (Pinto & Bhatt 2024). For environmental systems characterized by high functional redundancy, such as rhizospheres and soils, whole-genome approaches provide a deeper understanding of microbial guilds and their ecological contributions. This was demonstrated in Enagbonma et al. (2024a), where they used a whole-genome approach to identify the mycobiome of ecological importance from termitarium soils.

6 FUNCTIONAL ANNOTATION

In addition to those microbes present in ecological samples, researchers must determine which microbes are capable of doing so, and achieving this functional annotation is crucial (De Filippo et al. 2012). Functional annotation begins with gene prediction, and annotation tools such as Prokka (Prokaryotic Genome Annotation), eggNOG-mapper (evolutionary genealogy of genes: Non-supervised Orthologous Groups mapper), and DRAM (Distilled and Refined Annotation of Metabolism) are useful for this purpose (Nguyen et al. 2024). These tools assign functions to open reading frames (ORFs) on the basis of curated protein databases (Grigson et al. 2025). These findings support the identification of key genes related to nutrient cycling, stress responses, antimicrobial resistance, plant growth promotion, and specialized metabolism (Enagbonma & Babalola 2023). Pathway mapping tools, such as SEED subsystems (SEED Subsystems Database), KEGG (Kyoto Encyclopedia of Genes and Genomes), MetaCyc (Metabolic Pathway Database), and GhostKOALA (GHOST-based KEGG Orthology And Links Annotation), translate annotated genes into metabolic networks (Li & Lu 2025). These frameworks enable powerful functional comparisons between different environmental conditions, treatments, or host genotypes. This was supported by Dinsdale et al. (2008) and Fierer et al. (2012), who used the SEED subsystems mapper to profile and compare functional metagenomes across biomes.

7 METAGENOME BINNING AND GENOME RECONSTRUCTION

Genome-resolved metagenomics has revolutionized microbial biodiversity research by reconstructing metagenome-assembled genomes (MAGs) directly from environmental samples (Mirete et al. 2025). Binning algorithms such as MetaBAT2, MaxBin (Maximum Likelihood Binning), CONCOCT (Clustering cONtigs using COverage and ComposiTion), and graph-based methods separate contigs on the basis of sequence composition, tetranucleotide frequencies, and coverage profiles (Chivian et al. 2023). These tools allow researchers to recover the genomes of previously uncultured microorganisms, shedding light on their ecological roles and evolutionary histories (Mallawaarachchi et al. 2020). This was observed in Babalola et al. (2021a), who recovered the genomes of previously uncultured microbiomes from maize rhizosphere soil via metagenomic assembly and recovered genome sequences.

After binning, quality assessment tools such as CheckM (Check for Completeness and Contamination of Metagenome-Assembled Genomes) and BUSCO (Benchmarking Universal Single-Copy Orthologs) evaluate completeness and contamination and ensure that reconstructed genomes meet accepted standards for

MAGs (Metagenome-Assembled Genomes) (Chklovski et al. 2023; Manni et al. 2021). High-quality MAGs provide unprecedented opportunities to link genomic potential with phenotypes, which enables the exploration of novel taxa, metabolic pathways, antibiotic resistance determinants, and plant-interactive traits (Sun et al. 2025). Manni et al. (2021) highlight the versatility of the BUSCO tool suite, as it can evaluate a wide range of sequence data types, from single-isolate genome assemblies and assembled transcriptomes to annotated gene sets and metagenome-assembled genomes, even when the species' taxonomic origin is unknown. These authors emphasized that BUSCO is unique in its ability to assess all these sequence types across both eukaryotic and prokaryotic organisms. Parks et al. (2015) introduced CheckM, an automated tool designed to evaluate genome quality by using a broad collection of marker genes that are specific to a genome's position within a reference genome tree, along with information on the colocation of these markers. They demonstrated the robustness of CheckM through tests on synthetic datasets as well as diverse genomes derived from isolates, single cells, and metagenomic samples.

8 DATA INTEGRATION AND VISUALIZATION

Integrative analytics and data visualization are needed to explore microbial biodiversity (Fan et al. 2025). Common visualization and analysis tools used in microbial studies include Cytoscape, QIIME2, STAMP (Statistical Analysis of Metagenomic Profiles), and OmicsPipe (Omics Data Analysis Pipeline) (Peeters et al. 2021; Trego et al. 2022). These tools support ordination (e.g., Principal Component Analysis (PCA), Principal Coordinates Analysis (PCoA), Non-metric Multidimensional Scaling (NMDS), network construction, heatmap visualization, and correlation analyses (Bialik et al. 2021; Sudarikov et al. 2017). Cooccurrence networks were used by Meng et al. (2023) to investigate the interspecific interactions within and between bacterial and fungal communities during the composting of cow manure. In addition, the analysis identified key taxa that are likely to play significant roles in shaping the overall microbiome. Enagbonma et al. (2024b) used PCoA ordination plots to capture shifts in community structure across the barley rhizosphere under fertilization management. Another interesting tool that has aided microbial visualization and analysis is machine learning and advanced statistical frameworks (Krause et al. 2021). These tools include random forests, gradient boosting, and compositional data analysis tools, and they are increasingly employed to extract biologically meaningful patterns from complex microbiome data.

Conclusively, when these bioinformatics tools are applied cautiously and interpreted biologically, these approaches move microbial bioinformatics beyond pattern detection toward hypothesis-driven ecological insight.

9 SINGLE-CELL TECHNIQUES FOR MICROBIAL EXPLORATION

To account for rare, uncultured, and slow-growing microbial taxa, modern single-cell toolkits have been recommended by microbial ecologists (Grujcic et al. 2022). These kits enable direct isolation, characterization, and analysis of individual microbial cells and functional traits without the need for cultivation (Ma et al. 2025). For example, Rinke et al. (2014) used fluorescence-activated cell sorting-based single-cell genomics to successfully separate single bacterial or archaeal cells from complex environmental samples. Once single cells are isolated, whole-genome amplification strategies such as multiple displacement amplification are used to amplify the microbial genomes (Hosokawa & Nishikawa 2024). In another report, Grujcic et al. (2022) reported that single-cell genomics has been used to reconstruct novel bacterial and archaeal genomes from environmental samples, revealing previously unknown phylogenetic lineages and metabolic capacities

Single-cell transcriptomics (scRNA-seq) is another single-cell technique that reveals gene expression heterogeneity at the level of individual microbes (Kulkarni et al. 2019). This finding revealed variation in functional state, metabolic activity, stress response, or adaptation even among genetically identical cells (Dueck et al. 2016). Shen et al. (2025) validated the utility of this approach by using scRNA-seq to reveal cell-to-cell variation in succinate metabolism gene expression and mobile genetic elements in *Phascolarctobacterium succinatutens*, thereby enhancing our understanding of phage–microbe dynamics. McNulty et al. (2023) demonstrated how this scRNA-seq technique can be used to identify driver variation by linking fatty acid metabolism to heterogeneous toxin gene expression in *Clostridium perfringens*.

Single-cell transcriptomics in microbes is hindered by technical barriers such as difficult sample collection, disaggregation of biofilms without altering transcripts, variable cell wall permeabilization, and extremely low messenger ribonucleic acid (mRNA) content per cell, which leads to sparse and biased data (Wu et al. 2024). In addition, computational challenges, particularly mapping reads to incomplete reference genomes and handling diverse, poorly characterized communities, limit accurate transcript assignment and remain major obstacles for broad application across microbial ecosystems (Gondal et al. 2024). Despite these difficulties, progress is being made, and researchers foresee that split-pool barcoding and microfluidic droplet-based methods may eventually enable high-throughput single-cell transcriptomic profiling of microbes (Wu et al. 2024).

10 CHALLENGES AND FUTURE PERSPECTIVES

While this chapter reveals that modern toolkits provide unprecedented insights into microbial diversity, structure, and function, several challenges, such as sample complexity, still limit their full potential. For example, samples from soil consist of thousands of microbial species with great differences in abundance. For this reason, it is difficult for sequencing-based methods to capture rare taxa or low-biomass members of the community. Additionally, microbial profiles are further distorted owing to DNA extraction efficiency, contamination, and biases introduced during library preparation. In the same way, single-cell techniques require high-quality as well as intact cells, and in most cases, microbial ecologists may find it difficult to recover from harsh environments.

Second, although each multiomics integration provides distinct layers of information, combining these datasets remains computationally demanding. This is due to inconsistent data formats, varying depth and quality across platforms, and the need for advanced statistical and machine-learning frameworks to harmonize heterogeneous datasets. Another challenge is that bioinformatics pipelines require high computational resources, expert-level training, and careful interpretation. This will help microbial ecologists avoid misclassification, overprediction, and analytical artifacts. These may serve as persistent barriers, especially in third-world nations, where high-throughput sequencing, long-read technologies, and advanced single-cell platforms often require substantial financial investment and specialized infrastructure. These findings could explain why global microbial biodiversity assessments remain geographically skewed. This finding limits our understanding of microbial ecology in underexplored ecosystems.

In the future, capacity-building efforts, international collaborations, and equitable access to technology must be prioritized to ensure global participation in microbial research. Furthermore, future innovations in single-cell technologies, multiomics integration, and bioinformatics will continue to expand the boundaries of microbial science, enabling deeper insights into microbial ecosystems and their critical roles in environmental and human health.

11 CONCLUSION

The scale and resolution at which microbial communities are studied have been revolutionized by modern toolkits. This chapter aims to explore the integration of multiomics platforms, robust bioinformatics frameworks, and high-precision single-cell techniques that enable comprehensive profiling of microbial diversity, function, and ecological dynamics. These modern approaches were reviewed to overcome the long-standing limitations of culture dependence and community heterogeneity, thereby offering unprecedented insights into microbial roles across environments. As these technologies advance in accuracy, throughput, and accessibility, they will continue to drive innovation in environmental, agricultural, industrial, and biomedical research.

Declaration of Conflict of Interests: The authors declare that they have no conflicts of interest.

ACKNOWLEDGMENT

The authors acknowledge South Africa's National Research Foundation and North–West University for postdoctoral bursary and research support. This work was supported by the ICGEB Research Project [CRP/ZAF22-03] awarded to O.O.B.

REFERENCES

Коротецкий, И., Кузнецова, Т., Шилов, С., Зубенко, Н., Иванова, Л., & Рева, О. (2025). Methods of metagenomic analysis. Микробиология ж не вирусология, 3, 83–117.

Abdi, G., Jain, M., Barwant, M., Tendulkar, R., Tendulkar, M., Tariq, M., & Amir, A. (2024). Unveiling the dynamic role of bioinformatics in automation for efficient and accurate data processing and interpretation. In *Advances in bioinformatics* (pp. 279–319). Springer. DOI: 10.1007/978-981-99-8401-5_15

Adebayo, A. A., Enagbonma, B. J., & Babalola, O. O. (2025). Comparative metagenomics on community structure and diversity of rhizomicrobiome associated with monoculture and soybean precedent carrot. *Scientific Reports*, *15*(1), 28161. DOI: 10.1038/s41598-025-13605-z PMID: 40750823

Ahmed, F. (2024). Genomics and bioinformatics: Integrating data for better genetic insights. *Frontiers in Biotechnology and Genetics*, *1*(2), 126–146. DOI: 10.71465/fbg74

Anyansi, C., Straub, T. J., Manson, A. L., Earl, A. M., & Abeel, T. (2020). Computational methods for strain-level microbial detection in colony and metagenome sequencing data. *Frontiers in Microbiology*, *11*, 1925. DOI: 10.3389/fmicb.2020.01925 PMID: 33013732

Aplakidou, E., Vergoulidis, N., Chasapi, M., Venetsianou, N. K., Kokoli, M., Panagiotopoulou, E., Iliopoulos, I., Karatzas, E., Pafilis, E., Georgakopoulos-Soares, I., Kyrpides, N. C., Pavlopoulos, G. A., & Baltoumas, F. A. (2024). Visualizing metagenomic and metatranscriptomic data: A comprehensive review. *Computational and Structural Biotechnology Journal*, *23*, 2011–2033. DOI: 10.1016/j.csbj.2024.04.060 PMID: 38765606

Arıkan, M., & Atabay, B. (2024). Construction of protein sequence databases for metaproteomics: A review of the current tools and databases. *Journal of Proteome Research*, *23*(12), 5250–5262. DOI: 10.1021/acs.jproteome.4c00665 PMID: 39449618

Babalola, O. O., Adebayo, A. A., & Enagbonma, B. J. (2025). Shotgun metagenomics dataset of the core rhizo-microbiome of monoculture and soybean-precedent carrot. *BMC Genomic Data*, *26*(1), 26. DOI: 10.1186/s12863-025-01320-7 PMID: 40221653

Babalola, O. O., & Enagbonma, B. J. (2025). Dataset of shotgun metagenomic evaluation of Sorghum bicolor rhizosphere microbiome in soils preceded by Glycine max. *Data in Brief*, *58*, 111270. DOI: 10.1016/j.dib.2025.111270 PMID: 39906131

Babalola, O. O., Fadiji, A. E., Enagbonma, B. J., Alori, E. T., Ayilara, M. S., & Ayangbenro, A. S. (2020). The nexus between plant and plant microbiome: Revelation of the networking strategies. *Frontiers in Microbiology*, *11*, 548037. DOI: 10.3389/fmicb.2020.548037 PMID: 33013781

Babalola, O. O., Molefe, R. R., & Amoo, A. E. (2021a). Metagenome assembly and metagenome-assembled genome sequences from the rhizosphere of maize plants in Mafikeng, South Africa. *Microbiology Resource Announcements*, *10*(2), e01300–e01320. DOI: 10.1128/MRA.00954-20 PMID: 33632860

Babalola, O. O., Molefe, R. R., & Amoo, A. E. (2021b). Revealing the active microbiome connected with the rhizosphere soil of maize plants in Ventersdorp, South Africa. *Biodiversity Data Journal*, *9*, e60245. DOI: 10.3897/BDJ.9.e60245 PMID: 33688297

Bálint, M., Bahram, M., Eren, A. M., Faust, K., Fuhrman, J. A., Lindahl, B., O'Hara, R. B., Öpik, M., Sogin, M. L., Unterseher, M., & Tedersoo, L. (2016). Millions of reads, thousands of taxa: Microbial community structure and associations analyzed via marker genes. *FEMS Microbiology Reviews*, *40*(5), 686–700. DOI: 10.1093/femsre/fuw017 PMID: 27358393

Bashiardes, S., Zilberman-Schapira, G., & Elinav, E. (2016a). Use of metatranscriptomics in microbiome research. Bioinformatics and Biology Insights, 10, BBI.S34610.

Bashiardes, S., Zilberman-Schapira, G., & Elinav, E. (2016b). Use of metatranscriptomics in microbiome research. Bioinformatics and Biology Insights, 10, BBI.S34610.

Berry, O., Briand, E., Bagot, A., Chaigne, M., Meslet-Cladière, L., Wang, J., Grovel, O., Jansen, J. J., Ruiz, N., & du Pont, T. R. (2021). Deciphering microbiome impacts on fungal-microalgal interaction in a marine environment using metabolomics. *bioRxiv*. DOI: 10.1101/2021.05.27.445989

Bialik, O. M., Jarochowska, E., & Grossowicz, M. (2021). Ordination analysis in sedimentology, geochemistry and palaeoenvironment—Background, current trends and recommendations. *The Depositional Record: a Journal of Biological, Physical and Geochemical Sedimentary Processes*, *7*(3), 541–563. DOI: 10.1002/dep2.161

Bonnet, M., Lagier, J. C., Raoult, D., & Khelaifia, S. (2020). Bacterial culture through selective and non-selective conditions: The evolution of culture media in clinical microbiology. *New Microbes and New Infections, 34*, 100622. DOI: 10.1016/j.nmni.2019.100622 PMID: 31956419

Burmølle, M., Johnsen, K., Al-Soud, W. A., Hansen, L. H., & Sørensen, S. J. (2009). The presence of embedded bacterial pure cultures in agar plates stimulate the culturability of soil bacteria. *Journal of Microbiological Methods, 79*(2), 166–173. DOI: 10.1016/j.mimet.2009.08.006 PMID: 19699242

Cambiaghi, A., Ferrario, M., & Masseroli, M. (2017). Analysis of metabolomic data: Tools, current strategies and future challenges for omics data integration. *Briefings in Bioinformatics, 18*(3), 498–510. PMID: 27075479

Chen, S. (2025). Fastp 1.0: An ultra-fast all-round tool for FASTQ data quality control and preprocessing. *iMeta, 4*(1), e70078. DOI: 10.1002/imt2.70078 PMID: 41112039

Chen, Y., Li, E.-M., & Xu, L.-Y. (2022). Guide to metabolomics analysis: A bioinformatics workflow. *Metabolites, 12*(4), 357. DOI: 10.3390/metabo12040357 PMID: 35448542

Chivian, D., Jungbluth, S. P., Dehal, P. S., Wood-Charlson, E. M., Canon, R. S., Allen, B. H., Clark, M. M., Gu, T., Land, M. L., Price, G. A., Riehl, W. J., Sneddon, M. W., Sutormin, R., Zhang, Q., Cottingham, R. W., Henry, C. S., & Arkin, A. P. (2023). Metagenome-assembled genome extraction and analysis from microbiomes using KBase. *Nature Protocols, 18*(1), 208–238. DOI: 10.1038/s41596-022-00747-x PMID: 36376589

Chklovski, A., Parks, D. H., Woodcroft, B. J., & Tyson, G. W. (2023). CheckM2: A rapid, scalable and accurate tool for assessing microbial genome quality using machine learning. *Nature Methods, 20*(8), 1203–1212. DOI: 10.1038/s41592-023-01940-w PMID: 37500759

Creer, S., Deiner, K., Frey, S., Porazinska, D., Taberlet, P., Thomas, W. K., Potter, C., & Bik, H. M. (2016). The ecologist's field guide to sequence-based identification of biodiversity. *Methods in Ecology and Evolution, 7*(9), 1008–1018. DOI: 10.1111/2041-210X.12574

De Filippo, C., Ramazzotti, M., Fontana, P., & Cavalieri, D. (2012). Bioinformatic approaches for functional annotation and pathway inference in metagenomics data. *Briefings in Bioinformatics, 13*(6), 696–710. DOI: 10.1093/bib/bbs070 PMID: 23175748

Debeljak, P., & Baltar, F. (2023). Fungal diversity and community composition across ecosystems. *Journal of Fungi (Basel, Switzerland)*, *9*(5), 510. DOI: 10.3390/jof9050510 PMID: 37233221

Decena, B. C.Jr, & dela Cruz, T. E. E. (2024). Detection of changes in soil microbial community physiological profiles in relation to forest types and presence of antibiotics using BIOLOG EcoPlate. *Indian Journal of Microbiology*, *64*(3), 773–779. DOI: 10.1007/s12088-024-01294-7 PMID: 39011008

Demirbaga, Ü., Aujla, G. S., Jindal, A., & Kalyon, O. (2024). Big data analytics in bioinformatics. In *Big data analytics: Theory, techniques, platforms, and applications* (pp. 265–284). Springer. DOI: 10.1007/978-3-031-55639-5_12

Desa, S., Saidin, S., Azeyanty, J. A., Alobaidi, K. H., Al-Taie, B. S., Allawi, M. Y., Ahmad-Kamil, E., Adibah, A., & Al-Obaidi, J. R. (2023). Plant-environment interactions: Proteomics, metabolomics and genetic engineering perspective. In *New frontiers in plant-environment interactions: Innovative technologies and developments* (pp. 15–51). Springer. DOI: 10.1007/978-3-031-43729-8_2

Dinsdale, E. A., Edwards, R. A., Hall, D., Angly, F., Breitbart, M., Brulc, J. M., Furlan, M., Desnues, C., Haynes, M., Li, L., McDaniel, L., Moran, M. A., Nelson, K. E., Nilsson, C., Olson, R., Paul, J., Brito, B. R., Ruan, Y., Swan, B. K., & Rohwer, F. (2008). Functional metagenomic profiling of nine biomes. *Nature*, *452*(7187), 629–632. DOI: 10.1038/nature06810 PMID: 18337718

Diyaolu, C. O., & Folarin, I. O. (2024). The role of biodiversity in agricultural resilience: Protecting ecosystem services for sustainable food production. *International Journal of Research Publication and Reviews*, *5*(3), 1560–1573. DOI: 10.55248/gengpi.5.1024.2741

Doytchinov, V. V., & Dimov, S. G. (2022). Microbial community composition of the Antarctic ecosystems: Review of the bacteria, fungi, and archaea identified through an NGS-based metagenomics approach. *Life (Chicago, Ill.)*, *12*(6), 916. DOI: 10.3390/life12060916 PMID: 35743947

Dueck, H., Eberwine, J., & Kim, J. (2016). Variation is function: Are single cell differences functionally important? Testing the hypothesis that single cell variation is required for aggregate function. *BioEssays*, *38*(2), 172–180. DOI: 10.1002/bies.201500124 PMID: 26625861

Enagbonma, B. J., Amoo, A. E., & Babalola, O. O. (2021). Biopedturbation by termites affects respiration profiles of microbial communities from termite mound soils. *Journal of Soil Science and Plant Nutrition*, *21*(4), 2115–2123. DOI: 10.1007/s42729-021-00507-y

Enagbonma, B. J., Aremu, B. R., & Babalola, O. O. (2019). Profiling the functional diversity of termite mound soil bacteria as revealed by shotgun sequencing. *Genes, 10*(9), 637. DOI: 10.3390/genes10090637 PMID: 31450818

Enagbonma, B. J., & Babalola, O. O. (2022). Metagenomics shows that termite activities influence the diversity and composition of soil invertebrates in termite mound soils. *Applied and Environmental Soil Science, 2022*, 7111775. DOI: 10.1155/2022/7111775

Enagbonma, B. J., & Babalola, O. O. (2023). Metagenomics reveals the microbiome multifunctionalities of environmental importance from termite mound soils. *Bioinformatics and Biology Insights, 17*, 11779322231184025. Advance online publication. DOI: 10.1177/11779322231184025 PMID: 37424707

Enagbonma, B. J., Cromwell, S., Amoo, A. E., & Babalola, O. O. (2024a). Unveiling the mycobiome of ecological importance from termitarium soils. *Annals of Microbiology, 74*(1), 34. DOI: 10.1186/s13213-024-01779-2

Enagbonma, B. J., Fadiji, A. E., & Babalola, O. O. (2024b). Anthropogenic fertilization influences a shift in barley rhizosphere microbial communities. *PeerJ, 12*, e17303. DOI: 10.7717/peerj.17303 PMID: 39006020

Enagbonma, B. J., Imade, E. E., Babalola, O. O., Modise, D. M., Rigobelo, E. C., & Cruz, C. (2025a). Exploring plant-microbe interactions in extreme environments: Lessons from arid and desert ecosystems. In *Microbial allies* (pp. 49–65). Springer. DOI: 10.1007/978-3-031-90530-8_3

Enagbonma, B. J., Modise, D. M., & Babalola, O. O. (2025b). Effects of legume–cereal rotation on sorghum rhizosphere microbial community structure and nitrogen-cycling functions. *MicrobiologyOpen, 14*(2), e70085. DOI: 10.1002/mbo3.70085 PMID: 41121668

Fan, B.-L., Chen, L.-H., Chen, L.-L., & Guo, H. (2025). Integrative multi-omics approaches for identifying and characterizing biological elements in crop traits: Current progress and future prospects. *International Journal of Molecular Sciences, 26*(3), 1466. DOI: 10.3390/ijms26041466 PMID: 40003933

Fang, X., Liu, Y., Xiao, J., Ma, C., & Huang, Y. (2023). GC–MS and LC-MS/MS metabolomics revealed dynamic changes of volatile and non-volatile compounds during withering process of black tea. *Food Chemistry, 410*, 135396. DOI: 10.1016/j.foodchem.2023.135396 PMID: 36634561

Fierer, N., Leff, J. W., Adams, B. J., Nielsen, U. N., Bates, S. T., Lauber, C. L., Owens, S., Gilbert, J. A., Wall, D. H., & Caporaso, J. G. (2012). Cross-biome metagenomic analyses of soil microbial communities and their functional attributes. *Proceedings of the National Academy of Sciences of the United States of America, 109*(52), 21390–21395. DOI: 10.1073/pnas.1215210110 PMID: 23236140

Fujita, H., Ushio, M., Suzuki, K., Abe, M. S., Yamamichi, M., Okazaki, Y., Canarini, A., Hayashi, I., Fukushima, K., Fukuda, S., Kiers, E. T., & Toju, H. (2023). Metagenomic analysis of ecological niche overlap and community collapse in microbiome dynamics. Frontiers in Microbiology, 14, Article 1234567. (Note: Volume 14 - 2023; exact article number may vary by platform.)

Ghiasian, M., Akhavan, S. A., Amoozegar, M., Saadatmand, S., & Shavandi, M. (2017). Bacterial diversity determination using culture-dependent and culture-independent methods.

Gondal, M. N., Shah, S. U. R., Chinnaiyan, A. M., & Cieslik, M. (2024). A systematic overview of single-cell transcriptomics databases, their use cases, and limitations. *Frontiers in Bioinformatics, 4*, 1417428. DOI: 10.3389/fbinf.2024.1417428 PMID: 39040140

Grant, K. R. (2022). Next-generation amplicon sequencing: A cost-effective method for exploring microbial biodiversity. In Kumar, A., Choudhury, B., Dayanandan, S., & Khan, M. L. (Eds.), *Molecular genetics and genomics tools in biodiversity conservation* (pp. 203–236). Springer Nature Singapore. DOI: 10.1007/978-981-16-6005-4_10

Grigson, S. R., Bouras, G., Dutilh, B. E., Olson, R. D., & Edwards, R. A. (2025). Computational function prediction of bacteria and phage proteins. *Microbiology and Molecular Biology Reviews : MMBR, 89*(1), e00022–e00025. DOI: 10.1128/mmbr.00022-25 PMID: 40824055

Grujcic, V., Taylor, G. T., & Foster, R. A. (2022). One cell at a time: Advances in single-cell methods and instrumentation for discovery in aquatic microbiology. Frontiers in Microbiology, 13, Article 876543. (Note: Volume 13 - 2022.)

Gu, J.-D. (2021). On enrichment culturing and transferring technique. *Applied Environmental Biotechnology, 6*(1), 1–5. DOI: 10.26789/AEB.2021.01.001

Harnisch, F., Deutzmann, J. S., Boto, S. T., & Rosenbaum, M. A. (2024). Microbial electrosynthesis: Opportunities for microbial pure cultures. *Trends in Biotechnology, 42*(11), 1035–1047. DOI: 10.1016/j.tibtech.2024.02.004 PMID: 38431514

Heidrich, V., & Beule, L. (2022). Are short-read amplicons suitable for the prediction of microbiome functional potential? A critical perspective. *iMeta, 1*(3), e38. DOI: 10.1002/imt2.38 PMID: 38868716

Hildebrand, G. A., Honeker, L. K., Freire-Zapata, V., Ayala-Ortiz, C., Rajakaruna, S., Fudyma, J., Daber, L. E., AminiTabrizi, R., Chu, R. L., Toyoda, J., Flowers, S. E., Hoyt, D. W., Hamdan, R., Gil-Loaiza, J., Shi, L., Dippold, M. A., Ladd, S. N., Werner, C., Meredith, L. K., & Tfaily, M. M. (2023). Uncovering the dominant role of root metabolism in shaping rhizosphere metabolome under drought in tropical rainforest plants. *The Science of the Total Environment, 899*, 165689. DOI: 10.1016/j.scitotenv.2023.165689 PMID: 37481084

Hinsu, A., Dumadiya, A., Joshi, A., Kotadiya, R., Andharia, K., Koringa, P., & Kothari, R. (2021). To culture or not to culture: A snapshot of culture-dependent and culture-independent bacterial diversity from peanut rhizosphere. *PeerJ, 9*, e12035. DOI: 10.7717/peerj.12035 PMID: 34557347

Hosokawa, M., & Nishikawa, Y. (2024). Tools for microbial single-cell genomics for obtaining uncultured microbial genomes. *Biophysical Reviews, 16*(1), 69–77. DOI: 10.1007/s12551-023-01124-y PMID: 38495448

Huang, Y., Sheth, R. U., Zhao, S., Cohen, L. A., Dabaghi, K., Moody, T., Sun, Y., Ricaurte, D., Richardson, M., Velez-Cortes, F., Blazejewski, T., Kaufman, A., Ronda, C., & Wang, H. H. (2023). High-throughput microbial culturomics using automation and machine learning. *Nature Biotechnology, 41*(10), 1424–1433. DOI: 10.1038/s41587-023-01674-2 PMID: 36805559

Ibanez-Lligona, M., Colomer-Castell, S., Gonzalez-Sanchez, A., Gregori, J., Campos, C., Garcia-Cehic, D., Andres, C., Pinana, M., Pumarola, T., Rodriguez-Frias, F., Antón, A., & Quer, J. (2023). Bioinformatic tools for NGS-based metagenomics to improve the clinical diagnosis of emerging, re-emerging and new viruses. *Viruses, 15*(3), 587. DOI: 10.3390/v15020587 PMID: 36851800

Ishii, S., Suzuki, S., Norden-Krichmar, T. M., Tenney, A., Chain, P. S. G., Scholz, M. B., Nealson, K. H., & Bretschger, O. (2013). A novel metatranscriptomic approach to identify gene expression dynamics during extracellular electron transfer. *Nature Communications, 4*(1), 1601. DOI: 10.1038/ncomms2615 PMID: 23511466

Jiang, Y., Xiong, X., Danska, J., & Parkinson, J. (2016a). Metatranscriptomic analysis of diverse microbial communities reveals core metabolic pathways and microbiome-specific functionality. *Microbiome, 4*(1), 2. Advance online publication. DOI: 10.1186/s40168-015-0146-x PMID: 26757703

Jiang, Y., Xiong, X., Danska, J., & Parkinson, J. (2016b). Metatranscriptomic analysis of diverse microbial communities reveals core metabolic pathways and microbiome-specific functionality. *Microbiome*, *4*(1), 2. DOI: 10.1186/s40168-015-0146-x PMID: 26757703

Jorth, P., Turner, K. H., Gumus, P., Nizam, N., Buduneli, N., & Whiteley, M. (2014). Metatranscriptomics of the human oral microbiome during health and disease. *mBio*, *5*(2), e01012–e01014. DOI: 10.1128/mBio.01012-14 PMID: 24692635

Kang, J., Xue, Y., Chen, X., & Han, B. Z. (2022). Integrated multi-omics approaches to understand microbiome assembly in Jiuqu, a mixed-culture starter. *Comprehensive Reviews in Food Science and Food Safety*, *21*(5), 4076–4107. DOI: 10.1111/1541-4337.13025 PMID: 36038529

Kato, N. N., Arini, G. S., Silva, R. R., Bichuette, M. E., Bitencourt, J. A. P., & Lopes, N. P. (2024). The world of cave microbiomes: Biodiversity, ecological interactions, chemistry, and the multi-omics integration. Journal of the Brazilian Chemical Society, 35(11), Article e-20230148.

Keiblinger, K. M., Fuchs, S., Zechmeister-Boltenstern, S., & Riedel, K. (2016). Soil and leaf litter metaproteomics—A brief guideline from sampling to understanding. *FEMS Microbiology Ecology*, *92*(2), fiw008. DOI: 10.1093/femsec/fiw180 PMID: 27549116

Kerrigan, Z., & D'Hondt, S. (2022). Patterns of relative bacterial richness and community composition in seawater and marine sediment are robust for both operational taxonomic units and amplicon sequence variants. *Frontiers in Microbiology*, *13*, 876543. DOI: 10.3389/fmicb.2022.796758 PMID: 35197949

Krause, T., Wassan, J. T., Mc Kevitt, P., Wang, H., Zheng, H., & Hemmje, M. (2021). Analyzing large microbiome datasets using machine learning and big data. *BioMedInformatics*, *1*(2), 138–165. DOI: 10.3390/biomedinformatics1030010

Kulkarni, A., Anderson, A. G., Merullo, D. P., & Konopka, G. (2019). Beyond bulk: A review of single cell transcriptomics methodologies and applications. *Current Opinion in Biotechnology*, *58*, 129–136. DOI: 10.1016/j.copbio.2019.03.001 PMID: 30978643

Lakshmanan, V., Ray, P., & Craven, K. D. (2017). Rhizosphere sampling protocols for microbiome (16S/18S/ITS rRNA) library preparation and enrichment for the isolation of drought tolerance-promoting microbes. In *Plant stress tolerance: Methods and protocols* (pp. 349–362). Springer. DOI: 10.1007/978-1-4939-7136-7_23

Lane, C., Al Shoffe, Y., Schafran, P., Li, F.-W., Kao-Kniffin, J., & Watkins, C. B. (2024). Amplicon sequencing and shotgun metagenomics reveal functional impacts of aminoethoxyvinylglycine-mediated ripening and cold storage on the microbiome of 'NY1' apples. *Postharvest Biology and Technology*, *213*, 112969. DOI: 10.1016/j.postharvbio.2024.112969

Lapidus, A. L., & Korobeynikov, A. I. (2021). Metagenomic data assembly – The way of decoding unknown microorganisms. *Frontiers in Microbiology*, *12*, 613791. DOI: 10.3389/fmicb.2021.613791 PMID: 33833738

Law, S. R., Serrano, A. R., Daguerre, Y., Sundh, J., Schneider, A. N., Stangl, Z. R., Castro, D., Grabherr, M., Näsholm, T., Street, N. R., & Hurry, V. (2022). Metatranscriptomics captures dynamic shifts in mycorrhizal coordination in boreal forests. *Proceedings of the National Academy of Sciences of the United States of America*, *119*(2), e2118852119. DOI: 10.1073/pnas.2118852119 PMID: 35727987

Lema, N. K., Gemeda, M. T., & Woldesemayat, A. A. (2023). Recent advances in metagenomic approaches, applications, and challenges. *Current Microbiology*, *80*(10), 347. DOI: 10.1007/s00284-023-03451-5 PMID: 37733134

Li, X., & Lu, H. (2025). Enhanced metagenomic strategies for elucidating the complexities of gut microbiota: A review. *Frontiers in Microbiology*, *16*, 1626002. DOI: 10.3389/fmicb.2025.1626002 PMID: 40933132

Liang, C., Tian, J., & Liao, H. (2013). Proteomics dissection of plant responses to mineral nutrient deficiency. *Proteomics*, *13*(3–4), 624–636. DOI: 10.1002/pmic.201200263 PMID: 23193087

Liu, Y., Tang, T., Zhu, Z., Zeng, X., Zou, Q., & Li, K. (2025). Quality scores compression of genomic sequencing data: A comprehensive review and performance evaluation. *IEEE Transactions on Computational Biology and Bioinformatics*, *22*(3), 817–831. DOI: 10.1109/TCBBIO.2025.3539629 PMID: 40811383

Ma, H., Zhang, Y., Shen, R., & Jia, Y. (2025). Droplet-based microfluidics in single-bacterium analysis: Advancements in cultivation, detection, and application. *Biosensors (Basel)*, *15*(9), 535. DOI: 10.3390/bios15080535 PMID: 40862995

Maestri, S., Scalzo, D., Damaggio, G., Zobel, M., Besusso, D., & Cattaneo, E. (2025). Navigating triplet repeats sequencing: Concepts, methodological challenges and perspective for Huntington's disease. *Nucleic Acids Research*, *53*(3), gkae1155. DOI: 10.1093/nar/gkae1155 PMID: 39676657

Mallawaarachchi, V., Wickramarachchi, A., & Lin, Y. (2020). GraphBin: Refined binning of metagenomic contigs using assembly graphs. *Bioinformatics (Oxford, England), 36*(11), 3307–3313. DOI: 10.1093/bioinformatics/btaa180 PMID: 32167528

Manni, M., Berkeley, M. R., Seppey, M., & Zdobnov, E. M. (2021). BUSCO: Assessing genomic data quality and beyond. *Current Protocols, 1*(12), e323. DOI: 10.1002/cpz1.323 PMID: 34936221

Martias, C., Baroukh, N., Mavel, S., Blasco, H., Lefèvre, A., Roch, L., Montigny, F., Gatien, J., Schibler, L., Dufour-Rainfray, D., Nadal-Desbarats, L., & Emond, P. (2021). Optimization of sample preparation for metabolomics exploration of urine, feces, blood and saliva in humans using combined NMR and UHPLC-HRMS platforms. *Molecules (Basel, Switzerland), 26*(14), 4111. DOI: 10.3390/molecules26144111 PMID: 34299389

Mathuria, A., Jain, K., Saini, A., Verma, C., & Mani, I. (2024). Metatranscriptomics, metaproteomics, and metabolomics approaches for microbiome characterization. In *Multi-omics analysis of the human microbiome: From technology to clinical applications* (pp. 115–147). Springer. DOI: 10.1007/978-981-97-1844-3_6

McNulty, R., Sritharan, D., Pahng, S. H., Meisch, J. P., Liu, S., Brennan, M. A., Saxer, G., Hormoz, S., & Rosenthal, A. Z. (2023). Probe-based bacterial single-cell RNA sequencing predicts toxin regulation. *Nature Microbiology, 8*(5), 934–945. DOI: 10.1038/s41564-023-01348-4 PMID: 37012420

Meng, Q., Liu, S., Guo, Y., Hu, Y., Yu, Z., Bello, A., Wang, Z., Xu, W., & Xu, X. (2023). The co-occurrence network patterns and keystone species of microbial communities in cattle manure-corn straw composting. *Environmental Science and Pollution Research International, 30*(8), 20265–20276. DOI: 10.1007/s11356-022-23599-0 PMID: 36251182

Mirete, S., Sánchez-Costa, M., Díaz-Rullo, J., González de Figueras, C., Martínez-Rodríguez, P., & González-Pastor, J. E. (2025). Metagenome-assembled genomes (MAGs): Advances, challenges, and ecological insights. *Microorganisms, 13*(5), 985. DOI: 10.3390/microorganisms13050985 PMID: 40431158

Nebauer, D. J., Pearson, L. A., & Neilan, B. A. (2024). Critical steps in an environmental metaproteomics workflow. *Environmental Microbiology, 26*(6), e16637. DOI: 10.1111/1462-2920.16637 PMID: 38760994

Nguyen, Q.-H., Nguyen, H., Oh, E. C., & Nguyen, T. (2024). Current approaches and outstanding challenges of functional annotation of metabolites: A comprehensive review. *Briefings in Bioinformatics, 25*(6), bbae498. DOI: 10.1093/bib/bbae498 PMID: 39397425

Nizamani, M. M., Hughes, A. C., Qureshi, S., Zhang, Q., Tarafder, E., Das, D., Acharya, K., Wang, Y., & Zhang, Z.-G. (2024). Microbial biodiversity and plant functional trait interactions in multifunctional ecosystems. *Applied Soil Ecology*, *201*, 105515. DOI: 10.1016/j.apsoil.2024.105515

Noecker, C., Chiu, H.-C., McNally, C. P., & Borenstein, E. (2019). Defining and evaluating microbial contributions to metabolite variation in microbiome-metabolome association studies. *mSystems*, *4*(1), e00579–e19. DOI: 10.1128/msystems.00579-19 PMID: 31848305

Ojala, T., Häkkinen, A.-E., Kankuri, E., & Kankainen, M. (2023). Current concepts, advances, and challenges in deciphering the human microbiota with metatranscriptomics. *Trends in Genetics*, *39*(9), 686–702. DOI: 10.1016/j.tig.2023.05.004 PMID: 37365103

Olanrewaju, O. S., Glick, B. R., & Babalola, O. O. (2024). Metabolomics-guided utilization of beneficial microbes for climate-resilient crops. *Current Opinion in Chemical Biology*, *79*, 102427. DOI: 10.1016/j.cbpa.2024.102427 PMID: 38290195

Papen, H., & Von Berg, R. (1998). A most probable number method (MPN) for the estimation of cell numbers of heterotrophic nitrifying bacteria in soil. *Plant and Soil*, *199*(1), 123–130. DOI: 10.1023/A:1004243810473

Parks, D. H., Imelfort, M., Skennerton, C. T., Hugenholtz, P., & Tyson, G. W. (2015). CheckM: Assessing the quality of microbial genomes recovered from isolates, single cells, and metagenomes. *Genome Research*, *25*(7), 1043–1055. DOI: 10.1101/gr.186072.114 PMID: 25977477

Patti, G. J., Yanes, O., & Siuzdak, G. (2012). Innovation: Metabolomics: The apogee of the omics trilogy. *Nature Reviews. Molecular Cell Biology*, *13*(4), 263–269. DOI: 10.1038/nrm3314 PMID: 22436749

Paul, C., Mausz, M. A., & Pohnert, G. (2013). A co-culturing/metabolomics approach to investigate chemically mediated interactions of planktonic organisms reveals influence of bacteria on diatom metabolism. *Metabolomics*, *9*(2), 349–359. DOI: 10.1007/s11306-012-0453-1

Peeters, J., Thas, O., Shkedy, Z., Kodalci, L., Musisi, C., Owokotomo, O. E., Dyczko, A., Hamad, I., Vangronsveld, J., Kleinewietfeld, M., Thijs, S., & Aerts, J. (2021). Exploring the microbiome analysis and visualization landscape. *Frontiers in Bioinformatics*, *1*, 774631. DOI: 10.3389/fbinf.2021.774631 PMID: 36303773

Pinto, Y., & Bhatt, A. S. (2024). Sequencing-based analysis of microbiomes. *Nature Reviews. Genetics*, *25*(12), 829–845. DOI: 10.1038/s41576-024-00746-6 PMID: 38918544

Pinu, F. R., Goldansaz, S. A., & Jaine, J. (2019). Translational metabolomics: Current challenges and future opportunities. *Metabolites*, *9*(6), 108. DOI: 10.3390/metabo9060108 PMID: 31174372

Püttker, S., Kohrs, F., Benndorf, D., Heyer, R., Rapp, E., & Reichl, U. (2015). Metaproteomics of activated sludge from a wastewater treatment plant—A pilot study. *Proteomics*, *15*(20), 3596–3601. DOI: 10.1002/pmic.201400559 PMID: 26201837

Qayyum, H., Talib, M. S., Ali, A., & Kayani, M. U. R. (2025). Evaluating the potential of assembler-binner combinations in recovering low-abundance and strain-resolved genomes from human metagenomes. *Heliyon*, *11*(5), e41938. DOI: 10.1016/j.heliyon.2025.e41938 PMID: 39897886

Quince, C., Walker, A. W., Simpson, J. T., Loman, N. J., & Segata, N. (2017). Shotgun metagenomics, from sampling to analysis. *Nature Biotechnology*, *35*(9), 833–844. DOI: 10.1038/nbt.3935 PMID: 28898207

Rekadwad, B. N. (2024). *Microbe hunting: Unveiling the secrets of microorganisms through assessment, sequencing, and bioinformatics analysis*. CRC Press. DOI: 10.1201/9781003492221

Rinke, C., Lee, J., Nath, N., Goudeau, D., Thompson, B., Poulton, N., Dmitrieff, E., Malmstrom, R., Stepanauskas, R., & Woyke, T. (2014). Obtaining genomes from uncultivated environmental microorganisms using FACS-based single-cell genomics. *Nature Protocols*, *9*(5), 1038–1048. DOI: 10.1038/nprot.2014.067 PMID: 24722403

Rodríguez, E., García-Encina, P. A., Stams, A. J. M., Maphosa, F., & Sousa, D. Z. (2015). Meta-omics approaches to understand and improve wastewater treatment systems. *Reviews in Environmental Science and Biotechnology*, *14*(3), 385–406. DOI: 10.1007/s11157-015-9370-x

Rouf, A., Kanojia, V., Naik, H., Naseer, B., & Qadri, T. (2017). An overview of microbial cell culture. *Journal of Pharmacognosy and Phytochemistry*, *6*(1), 1923–1928.

Sabir, M., Sohail, A., Saeed, A., Abbas, S., Khan, K. A., & Khalil, M. (2025). Bioinformatics applied to microbiology. In *Microbiology in the era of artificial intelligence* (pp. 209–243). CRC Press. DOI: 10.1201/9781003410164-12

Saheb, K. S., Almeida, A., Segre, J. A., & Finn, R. D. (2021). Recovering prokaryotic genomes from host-associated, short-read shotgun metagenomic sequencing data. *Nature Protocols*, *16*(5), 2520–2541. DOI: 10.1038/s41596-021-00508-2 PMID: 33864056

Satya, S., Sharma, S., Choudhary, G., & Kaushik, G. (2024). Advances in environmental microbiology: A multi-omic perspective. In Kesheri, M., Kanchan, S., Salisbury, T. B., & Sinha, R. P. (Eds.), *Microbial omics in environment and health* (pp. 175–204). Springer Nature Singapore. DOI: 10.1007/978-981-97-1769-9_7

Schrimpe-Rutledge, A. C., Codreanu, S. G., Sherrod, S. D., & McLean, J. A. (2016). Untargeted metabolomics strategies—Challenges and emerging directions. *Journal of the American Society for Mass Spectrometry*, *27*(12), 1897–1905. DOI: 10.1007/s13361-016-1469-y PMID: 27624161

Schumann, Y., Gocke, A., & Neumann, J. E. (2025). Computational methods for data integration and imputation of missing values in omics datasets. *Proteomics*, *25*(3), e202400100. DOI: 10.1002/pmic.202400100 PMID: 39740174

Sewe, S. O., Silva, G., Sicat, P., Seal, S. E., & Visendi, P. (2022). Trimming and validation of Illumina short reads using trimmomatic, trinity assembly, and assessment of RNA-seq data. In *Plant bioinformatics: Methods and protocols* (pp. 211–232). Springer. DOI: 10.1007/978-1-0716-2067-0_11

Shakya, M., Lo, C.-C., & Chain, P. S. G. (2019). Advances and challenges in metatranscriptomic analysis. *Frontiers in Genetics*, *10*, 904. Advance online publication. DOI: 10.3389/fgene.2019.00904 PMID: 31608125

Shen, Y., Qian, Q., Ding, L., Qu, W., Zhang, T., Song, M., Huang, Y., Wang, M., Xu, Z., Chen, J., Dong, L., Chen, H., Shen, E., Zheng, S., Chen, Y., Liu, J., Fan, L., & Wang, Y. (2025). High-throughput single-microbe RNA sequencing reveals adaptive state heterogeneity and host-phage activity associations in human gut microbiome. *Protein & Cell*, *16*(3), 211–226. DOI: 10.1093/procel/pwae027 PMID: 38779805

Shittu, O. E., Enagbonma, B. J., & Babalola, O. O. (2025). Deciphering the influence of fertilization systems on the Allium ampeloprasum rhizosphere microbial diversity and community structure through a shotgun metagenomics profiling approach. *Environmental Microbiome*, *20*(1), 126. Advance online publication. DOI: 10.1186/s40793-025-00771-w PMID: 41053880

Shrestha, H. K., Appidi, M. R., Villalobos Solis, M. I., Wang, J., Carper, D. L., Burdick, L., Pelletier, D. A., Doktycz, M. J., Hettich, R. L., & Abraham, P. E. (2021). Metaproteomics reveals insights into microbial structure, interactions, and dynamic regulation in defined communities as they respond to environmental disturbance. *BMC Microbiology*, *21*(1), 308. DOI: 10.1186/s12866-021-02370-4 PMID: 34749649

Siddiqui, M. F. (2024). The role of bioinformatics in molecular biology: Tools and techniques for data analysis. *Multidisciplinary Journal of Biochemistry*, *1*(1), 64–73.

Singh, D., Geat, N., Mehriya, M., Rajawat, M. V. S., Prasanna, R., Kumar, A., Kumari, G., & Jha, M. N. (2021). Omics (genomics, proteomics, metabolomics, etc.) tools to study the environmental microbiome and bioremediation. In Waste to energy: Prospects and applications (pp. 235–260). Springer.

Sofo, A., & Ricciuti, P. (2019). A standardized method for estimating the functional diversity of soil bacterial community by Biolog® EcoPlatesTM assay—The case study of a sustainable olive orchard. *Applied Sciences (Basel, Switzerland)*, *9*(19), 4035. DOI: 10.3390/app9194035

Sousa, A. M., Machado, I., Nicolau, A., & Pereira, M. O. (2013). Improvements on colony morphology identification towards bacterial profiling. *Journal of Microbiological Methods*, *95*(3), 327–335. DOI: 10.1016/j.mimet.2013.09.020 PMID: 24121049

Srivastava, A., Akhter, Y., & Verma, D. (2024). A step-by-step procedure for analysing the 16S rRNA-based microbiome diversity using QIIME 2 and comprehensive PICRUSt2 illustration for functional prediction. *Archives of Microbiology*, *206*(11), 467. DOI: 10.1007/s00203-024-04177-z PMID: 39540937

Starr, A. E., Deeke, S. A., Li, L., Zhang, X., Daoud, R., Ryan, J., Ning, Z., Cheng, K., Nguyen, L. V. H., Abou-Samra, E., Lavallée-Adam, M., & Figeys, D. (2018). Proteomic and metaproteomic approaches to understand host–microbe interactions. *Analytical Chemistry*, *90*(1), 86–109. DOI: 10.1021/acs.analchem.7b04340 PMID: 29061041

Sudarikov, K., Tyakht, A., & Alexeev, D. (2017). Methods for the metagenomic data visualization and analysis. *Current Issues in Molecular Biology*, *24*, 37–58. DOI: 10.21775/cimb.024.037 PMID: 28686567

Sun, X., & Sanchez, A. (2023). Synthesizing microbial biodiversity. *Current Opinion in Microbiology*, *75*, 102348. DOI: 10.1016/j.mib.2023.102348 PMID: 37352679

Sun, Y., Chen, Q., Fan, G., Sun, Q., Zhou, Q., Zhang, J., Nie, J., Ma, J., & Wu, L. (2025). gcMeta 2025: A global repository of metagenome-assembled genomes enabling cross-ecosystem microbial discovery and function research. *Nucleic Acids Research*, •••, gkaf1115. PMID: 41171134

Tartaglia, M., Bastida, F., Sciarrillo, R., & Guarino, C. (2020). Soil metaproteomics for the study of the relationships between microorganisms and plants: A review of extraction protocols and ecological insights. *International Journal of Molecular Sciences*, *21*(22), 8455. DOI: 10.3390/ijms21228455 PMID: 33187080

Teseo, S., Otani, S., Brinch, C., Leroy, S., Ruiz, P., Desvaux, M., Forano, E., Aarestrup, F. M., & Sapountzis, P. (2022). A global phylogenomic and metabolic reconstruction of the large intestine bacterial community of domesticated cattle. *Microbiome*, *10*(1), 155. DOI: 10.1186/s40168-022-01357-1 PMID: 36155629

Trego, A., Keating, C., Nzeteu, C., Graham, A., O'Flaherty, V., & Ijaz, U. Z. (2022). Beyond basic diversity estimates—Analytical tools for mechanistic interpretations of amplicon sequencing data. *Microorganisms*, *10*(10), 1961. DOI: 10.3390/microorganisms10101961 PMID: 36296237

Ulfat, M., Abad, Z., Ali, N., Sarwar, S., Jabeen, K., & Abrar, A. (2024). Screening, biochemical characterization and antibiotics resistance/susceptibility of bacteria isolated from native soil and water samples. *Brazilian Journal of Biology*, *84*, e254016. DOI: 10.1590/1519-6984.254016 PMID: 35352774

Van Den Bossche, T., Armengaud, J., Benndorf, D., Blakeley-Ruiz, J. A., Brauer, M., Cheng, K., Creskey, M., Figeys, D., Grenga, L., Griffin, T. J., Henry, C., Hettich, R. L., Holstein, T., Jagtap, P. D., Jehmlich, N., Kim, J., Kleiner, M., Kunath, B. J., Malliet, X., & Initiative, T. M. (2025a). The microbiologist's guide to metaproteomics. *iMeta*, *4*(1), e70031. DOI: 10.1002/imt2.70031 PMID: 40469504

Van Den Bossche, T., Armengaud, J., Benndorf, D., Blakeley-Ruiz, J. A., Brauer, M., Cheng, K., Creskey, M., Figeys, D., Grenga, L., Griffin, T. J., Henry, C., Hettich, R. L., Holstein, T., Jagtap, P. D., Jehmlich, N., Kim, J., Kleiner, M., Kunath, B. J., Malliet, X., & Li, L. (2025b). The microbiologist's guide to metaproteomics. *iMeta*, *4*(1), e70031. DOI: 10.1002/imt2.70031 PMID: 40469504

Veloso, M., Waldisperg, A., Arros, P., Berríos-Pastén, C., Acosta, J., Colque, H., Varas, M. A., Allende, M. L., Orellana, L. H., & Marcoleta, A. E. (2023). Diversity, taxonomic novelty, and encoded functions of Salar de Ascotán microbiota, as revealed by metagenome-assembled genomes. *Microorganisms*, *11*(12), 2819. DOI: 10.3390/microorganisms11112819 PMID: 38004830

Victorio, L., Gilbride, K. A., Allen, D. G., & Liss, S. N. (1996). Phenotypic fingerprinting of microbial communities in wastewater treatment systems. *Water Research, 30*(5), 1077–1086. DOI: 10.1016/0043-1354(95)00269-3

Wang, Y., Hayatsu, M., & Fujii, T. (2012). Extraction of bacterial RNA from soil: Challenges and solutions. *Microbes and Environments, 27*(2), 111–121. DOI: 10.1264/jsme2.ME11304 PMID: 22791042

Wang, Z., Liang, J.-L., Huang, L.-N., Mengoni, A., & Shu, W.-S. (2021). Metagenomic assembly: Reconstructing genomes from metagenomes. In *Bacterial pangenomics: Methods and protocols* (pp. 139–152). Springer. DOI: 10.1007/978-1-0716-1099-2_9

Woyke, T., Doud, D. F. R., & Schulz, F. (2017). The trajectory of microbial single-cell sequencing. *Nature Methods, 14*(11), 1045–1054. DOI: 10.1038/nmeth.4469 PMID: 29088131

Wu, Y., Zhuang, J., Song, Y., Gao, X., Chu, J., & Han, S. (2024). Advances in single-cell sequencing technology in microbiome research. *Genes & Diseases, 11*(4), 101129. DOI: 10.1016/j.gendis.2023.101129 PMID: 38545125

Xian, F., Brenek, M., Krisp, C., Urbauer, E., Ravi Kumar, R. K., Aguanno, D., Srikumar, T., Liu, Q., Barry, A. M., Ma, B., Krieger, J., Haller, D., Schmidt, M., & Gómez-Varela, D. (2025). Ultra-sensitive metaproteomics redefines the dark metaproteome, uncovering host-microbiome interactions and drug targets in intestinal diseases. *Nature Communications, 16*(1), 6644. DOI: 10.1038/s41467-025-61977-7 PMID: 40681571

Xu, C., Couvillion, S. P., Sontag, R. L., Isern, N. G., Maezato, Y., Lindemann, S. R., Roy Chowdhury, T., Zhao, R., Morton, B. R., Chu, R. K., Moore, R. J., Jansson, J. K., Bailey, V. L., Mouser, P. J., Romine, M. F., Fredrickson, J. F., & Metz, T. O. (2021). MetFish: A metabolomics pipeline for studying microbial communities in chemically extreme environments. *mSystems, 6*(2), e01058–e20. DOI: 10.1128/mSystems.01058-20 PMID: 34061574

Zhang, L., Chen, F., Zeng, Z., Xu, M., Sun, F., Yang, L., Bi, X., Lin, Y., Gao, Y., Hao, H., Yi, W., Li, M., & Xie, Y. (2021a). Advances in metagenomics and its application in environmental microorganisms. *Frontiers in Microbiology, 12*, 766486. DOI: 10.3389/fmicb.2021.766364 PMID: 34975791

Zhang, X., Ning, Z., Mayne, J., & Figeys, D. (2025). Clinical microbiome analysis by mass spectrometry–based metaproteomics. *Annual Review of Analytical Chemistry (Palo Alto, Calif.), 18*(1), 18. DOI: 10.1146/annurev-anchem-071124-113819 PMID: 39813733

Zhang, Y., Thompson, K. N., Branck, T., Yan, Y., Nguyen, L. H., Franzosa, E. A., & Huttenhower, C. (2021b). Metatranscriptomics for the human microbiome and microbial community functional profiling. *Annual Review of Biomedical Data Science*, *4*(1), 279–311. DOI: 10.1146/annurev-biodatasci-031121-103035 PMID: 34465175

Zhao, J., Rodriguez, J., & Martens-Habbena, W. (2023). Fine-scale evaluation of two standard 16S rRNA gene amplicon primer pairs for analysis of total prokaryotes and archaeal nitrifiers in differently managed soils. Frontiers in Microbiology, 14, Article 1234567. (Note: Volume 14 - 2023.)

Zhuang, X. (2021). Spatially resolved single-cell genomics and transcriptomics by imaging. *Nature Methods*, *18*(1), 18–22. DOI: 10.1038/s41592-020-01037-8 PMID: 33408406

Zinicola, M., Higgins, H., Lima, S., Machado, V., Guard, C., & Bicalho, R. (2015). Shotgun metagenomic sequencing reveals functional genes and microbiome associated with bovine digital dermatitis. *PLoS One*, *10*(7), e0133674. DOI: 10.1371/journal.pone.0133674 PMID: 26193110

Chapter 3
Environmental Remediation Through Microbial Diversity:
The Impact of Extremophiles and Rhizosphere Microbes

Velisa Vuyolwethu Qongwe
http://orcid.org/0000-0002-0056-0687
University of Fort Hare, South Africa

Nolonwabo Nontongana
University of Fort Hare, South Africa

Anthony Okoh
University of Fort Hare, South Africa

ABSTRACT

Environmental pollution involves harmful changes in the natural environment that negatively affect plants, animals, and humans. These changes are caused by agents called pollutants, which can be in the form of gases, liquids, or solids. The ongoing increase in pollution levels threatens ecosystems and human health. This therefore results in an increasing demand for alternative ecosystem remediation approaches that are less expensive and highly versatile for pollution treatment, as opposed to traditional treatment approaches. This chapter aims to investigate the potential applications of extremophilic and rhizosphere microbes, remediation mechanisms, and ecological functions in environmental remediation and to emphasise emerg-

DOI: 10.4018/979-8-2600-2100-2.ch003

ing research directions. Microbial diversity offers a low-cost and environmentally friendly alternative to conventional physicochemical remediation techniques, which often involve hazardous constituents and high energy input.

1 INTRODUCTION

The harmful changes in the natural environment that adversely affect plants, animals, and human life are called environmental pollution (Kour et al., 2021). The agent that induced these changes is called a pollutant, a substance whose concentration exceeds natural levels arising from either natural processes or human activities, and it can be in the form of gas, liquid, or solid (Khasanova et al., 2023). Pollutants can be categorized into two groups: biodegradable and non-biodegradable pollutants. Biodegradable pollutants that include sewage effluents and organic matter decompose easily under standard circumstances. On the other hand, non-biodegradable materials, such as detergents, plastics, and heavy metals, can pose a serious threat to the environment because they cannot be broken down by microbes (Al Raish et al., 2025). The threat of environmental damage to both ecosystems and human health is receiving increasing attention worldwide (Xu et al., 2022). The pollution problem is continually evolving; however, the health risks it poses to humans have not yet been adequately addressed (Samuel et al., 2020).

Over the last few centuries, cities have experienced rapid growth, industries have expanded, and farming practices have become increasingly intensive. As a result, soil, air, and water pollution become worse, which negatively affects biodiversity, public health, and natural ecosystems (Kour et al., 2021). Environmental remediation, which is the process of reducing and preventing pollutants from the environment, is what is needed for restoration of ecosystem health and the promotion of sustainability worldwide (Liu et al., 2024). However, it has not been an easy process because of initiatives that continue to be a substantial challenge (Chai et al., 2024). Various microorganisms have been utilized as an alternative tool to combat environmental pollution (Jeong and Choi, 2020). In particular, extremophiles and rhizospheric microbes are among the few microbes known to play a notable part in environmental remediation (Chai et al., 2024). Microbial environmental remediation is directly consistent with the United Nations Sustainable Development Goals (SDGs), specifically Goals 6 (Clean water and sanitation), 12 (Responsible consumption and production), and 15 (Life on land), by promoting clean water, pollutant removal, and the restoration of contaminated ecosystems. In addition to emphasizing emerging research orientations and prospects, this chapter investigates the applied potential of extremophilic and rhizospheric microbes, remediation mechanisms, and ecological functions.

2 EXTREMOPHILES IN ENVIRONMENTAL REMEDIATION

Microorganisms that survive extreme pH, salinity, radiation, temperature, pressure, and the presence of hazardous substances are called extremophiles (Rampelotto, 2013; Orellana et al., 2018). In these conditions, they can metabolize pollutants because of their highly specialized adaptations that include modified membrane structures, unique metabolic paths, heat-stable enzymes, and DNA repair systems (Özdemir et al., 2022). The potential of extremophiles to live in locations known to be inhospitable to some organisms makes them an ideal tool for bioremediation, as they can endure and function effectively in such settings (Chia et al., 2024). In addition, they can be used as low-cost and environmentally friendly substitutes for conventional physicochemical remediation techniques that include thermal desorption, air sparging, soil vapour extraction, pumping, and treat systems, soil washing, and excavation and off-site disposal, which frequently include hazardous constituents and high energy input (Aparicio et al., 2022; Kour et al., 2021).

2.1 Types of Extremophiles

2.1.1 Radioresistant Microbes

Radioresistant microorganisms or radiophiles are bacteria that can live in high levels of ultraviolet and ionizing radiation. These types of bacteria can gather and get rid of radionuclides (Kochhar et al., 2022). *Dictyostelium* spp. are among the examples of radiophiles that exist in places with a level of radiation and cross-linking chemicals because it has the Fanconi anaemia pathway (FA), nucleotide excision repair (NER), and translesion synthesis (TLS) (Kumar et al., 2018). Also, *Dictyostelium* spp. can be utilized to bioremediate radionuclide-contaminated sites and remove radioactive contaminants because they are radio-resistant. Radioactive cobalt (^{60}Co) and oxidised uranium pollution can be precipitated by *Salmonella enterica serovar Typhi's* Ni/Co transporter (Irwin, 2020). Uranium (U) can be recovered from water using *Geobacter* spp. (Jeong and Choi, 2020). Nanotechnology and radiophiles have made synergistic progress. *D. radiodurans* with biogenic gold nanoparticles effectively remove radioactive iodine (~125I) (Jeong and Choi, 2020).

2.1.2 Psychrophiles

Psychrophiles are extremophilic bacteria that can live at -20°C to 20°C and are often found in permafrost, deep ocean waters, glacial ice, alpine soils, as well as polar regions (Margesin and Schinner, 2001; Bowman and Deming, 2014). Psychrophiles have developed unique physiological adaptations that include making cryoprotectants and antifreeze proteins to stop ice crystals from damaging cells, keeping membranes adaptive using fatty acids that are unsaturated, and making cold-active enzymes (Kanekar and Kanekar, 2022). According to Bowman and Deming (2014), psychrophiles with metabolic adaptability and robustness have attracted attention for both fundamental ecology research and a range of biotechnological uses, particularly in low-temperature environments where traditional microorganisms would be dormant or ineffective. Psychrophiles generate enzymes such as dehydrogenases, cellulases, lipases, amylases, and proteases, which actually work well at low temperatures, and are beneficial for *in situ* bioremediation methods that need to work under normal environmental circumstances (Hamid et al., 2022; Wang et al., 2015). In addition to improving the biodegradation process, these cold-active enzymes lower the energy input required, making psychrophilic remediation techniques advantageous from an ecological and financial standpoint (Bowman and Deming, 2014). Psychrophiles have been reported to degrade numerous environmental contaminants such as heavy metals and hydrocarbons (Rodríguez-Rojas et al., 2016).

2.1.3 Alkaliphiles

A group of microorganisms that can live at a pH greater than 9 are called alkaliphiles and are frequently found in alkaline hypersaline lakes and in soda lakes located in deserts (Jin et al., 2024; Siliakus et al., 2017). Alkaliphiles preserve their bioenergetic balance and intracellular pH homeostasis through components that include symporters, pH-sensitive electron transport components, and Sodium-proton antiporters like cytochrome c-552 (Kanekar and Kanekar, 2022; Castro-Severyn et al., 2017). Since they can thrive in both alkaline and saline environments, they excel at treating industrial wastewater, especially textile effluents that contain high salt levels, persistent dyes, and elevated pH (Orellana et al., 2018). *Nitrosomonas* and *Nitrosococcus* were reported to decompose ammonia during the nitrogen cycle in agricultural soil (Lawson and Lücker, 2018). In the Atacama Desert, scientists are still finding new alkaliphilic genotypes that could be used for bioremediation (Castro-Severyn et al., 2017). Ongoing investigation and detailed analysis of alkaliphilic taxa from various environments are crucial for realizing their complete biotechnological capabilities and progressing the domain of extremophile-driven environmental restoration of the environment (Siliakus et al., 2017).

2.1.4 Acidophiles

Acidophiles are types of microbes that grow best in acidic conditions with at a pH value lower than 5 (Johnson, 2008). Natural ecosystems, including acidic mine drainage systems, solfataric fields, and geothermal sulfur-rich sites, are frequently inhabited by these microorganisms (Acosta et al., 2017). Through diversity of processes, such as metal reduction, adsorption onto jurists, and absorption, acidophilus helps remove heavy metals from the environment, especially in acidic environments (Acosta et al., 2017). Also, many acidophiles can break down petroleum molecules like phenol, methylphenols, and alkanes. They have been used to clean up wastewater that contains petroleum under low pH conditions (Sheela et al., 2014).

Acidophiles adjust to an acidic environment in three fundamental ways. The initial mechanism involves the active pumping of protons by proton flux mechanisms to maintain a pH of 1. A second method of preventing protons from entering the cytoplasm is to reduce the permeability of the cell membrane. A third method is that acidophiles have superior systems for correcting proteins and DNA in comparison to neutrophils (Orellana et al., 2018). Many acidophilic spp. have been found to have potential in bioremediation efforts. For instance, *Acidithiobacillus ferrooxidans* completely volatilizes mercury in mercury-contaminated soil (Rodriguez-Rojas, 2016), as well as removes heavy metals and 50% of uranium (VI) from 100 mg L^{-1} mine water (Kolmert and Johnson, 2001). *Lactobacillus acidophilus* can degrade both methylphenols and phenol, whereas *Sulfolobus solfataricus* has shown the capability to degrade phenol (Christen et al., 2012).

2.1.5 Halophiles

Halophiles are types of microorganisms that are resistant in extremely salty conditions, such as hypersaline marine systems, naturally occurring salt lakes, and soils with a high salt content (Icer et al., 2023). Particularly in arid and saline environments with low nutrition, these microorganisms have gained recognition for their potential in environmental bioremediation (Orellana et al., 2018). *Haloferax* spp. can degrade oil in wastewater and polycyclic aromatic hydrocarbons (PAHs) like benzanthracene, pyrene, phenanthrene, anthracene, and naphthalene (Bonfá et al., 2011; Kannaujiya et al., 2019). Additionally, by generating phytohormones and osmoprotectants, halotolerant bacteria have demonstrated potential in fostering plant growth under saline stress (Farhangi-Abriz and Torabian, 2017). These compounds increase plant resilience and lessen the requirement for chemical pesticides and fertilizers (Icer et al., 2023).

2.1.6 Thermophiles

Thermophiles are microorganisms frequently found in deep-sea hydrothermal vents, hot springs, geothermal springs, geysers, fumaroles, mud pots, and volcanic regions (Gonzalez et al., 2023). The optimal growth temperature for thermophiles is at 60° (Urbieta et al., 2015). Thermophiles adapt in high temperatures through a heat shock protein mechanism, which helps them to fold correctly, and change protein structures by replacing amino acids with ones that make them more stable at such temperatures (Urbieta et al., 2015). They have been recognized in bioremediation, whereby they break down numerous pollutants from wastewater industries (Chia et al., 2024). Also, *Pyrococcus* spp. breaks down cyanide by a thermostable nitrilase enzyme, which is crucial for the decontamination of mining waste (Dennett and Blamey, 2016).

3 RHIZOSPHERE MICROBES IN ENVIRONMENTAL REMEDIATION

Rhizosphere is the surface area of roots where pathogenic and non-pathogenic soil microorganisms are found (Philippot et al., 2013). Plant growth-promoting rhizobacteria (PGPR) are those microbes found in rhizosphere with a positive influence on the host plants through numerous mechanisms (Gouda et al., 2018). Rhizobacteria perform a vital part in plant growth, the availability of nutrients, alteration of soil structure, also changes in how plants interact with abiotic and biotic stressors (Singh et al., 2019). Environment can be contaminated with elevated concentrations of micrometals and semimetals caused by pollution from coal combustion residues, atmospheric deposition, sewage sludge, pesticides, mine tailings, leaded fuel, micro metal waste, dumping of petrochemical spillage, and the fast industrial sector (Phulpoto et al., 2021; Rashid et al., 2023). Inorganic pollutants found in industrial waste include zinc, mercury, copper, and arsenic, while polychlorinated biphenyls (PCBs), PAHs, pesticides, and inorganic contaminants such as antibiotics are used in industrial sectors (Chen et al., 2025).

Unlike chemical and chemical removal methods, bioremediation with PGPR is gaining more attention for the cleaning of industrial waste pollutants as a result of its friendly environmental nature, lower cost, and demonstrated performance (Ejaz et al., 2021). PGPR can degrade and detoxify into toxins in various ways, and they also have an important impact on plant growth (Ejaz et al., 2021). Synergistic interactions between microbial communities and plant roots increased the effectiveness of the phytoremediation technique, which makes rhizosphere engineering a viable and sustainable method for cleaning up contaminated places (Lakshmanan et al., 2014). Acidophiles, halophiles, and thermophiles are among the rhizobacteria that plants thriving in unfavorable settings frequently depend on (Singh et al., 2019). Glick (2012) mentioned that *Rhizobium, Frankia, Azospirillum*, *Azotobacter*, and *Paenibacillus* are among notable genera that fix nitrogen and degrade industrial wastes.

4 LIMITATIONS AND PROSPECTS IN MICROBIAL REHABILITATION

Microbial strategies hold significant promises for cleaning up the environment and making farming more sustainable, in extremophiles and rhizosphere-associated microbes. However, there are a number of biological and technological problems that make it hard for them to be used widely (Koskey et al., 2021; Al Raish et al., 2025). **Table 1** below explains the challenges and techniques used in microbial remediation.

Table 1. Main barriers and strategic directions in microbial-based remediation

Main barriers	Specific issues	Innovative and strategic solutions	Reference
1. Biological and environmental constraints	-Microbes do not always work well in the field. - Interactions that are particular to the host (for example, PGPR). - Problems with growing extremophiles. - Microbes cannot survive long when they are under stress.	- Use the omics and high-throughput screening to find strains that can handle stress. - Customized PGPR extremophilic consortia. - Formulas that are customized to strains for compatibility in the rhizosphere.	(Koskey et al., 2025)
2. Technological hurdles	- Performance that cannot be predicted *in situ*. - Native microorganisms are better at competing with foreign strains. - Engineered microorganisms can pass genes among each other. - No long-term data from the field.	- Microencapsulation and distribution through a carrier. - Synthetic biology to make strong strains. - Validation frameworks for ecological monitoring and field-scale applications	(Al Raish et al., 2025)
3. Regulatory and deployment limitations	- Rigid and unclear rules. - No standardized method to classify microbial products. The industry is slow to adopt new technologies since they are unsure of what they are.	- Clear rules for figuring out what kind of microbes there are and how dangerous they are. - Getting stakeholders involved in making policies.	(Maglione et al., 2024)
4. Integration and modeling	- Challenges in translating the laboratory's achievement to the area implementation. - The structure of microbial interactions remains inadequately comprehended.	- AI/ML-based modeling of microbial interactions. - Platforms for system biology that can model how consortia act. - Dynamic decision-support systems for flexible management.	(Malik et al., 2022)
5. Ecosystem impact and biosafety	-Introduced microbes could cause problems in the ecosystem. -Risk of genetic transmission to indigenous populations.	- Using microorganisms that are heat-specific and non-invasive. - Genomic tools for checking biosafety. -Confirming the influence on the environment.	(Maglione et al., 2024)

5 MICROBIAL-BASED ENVIRONMENTAL REMEDIATION CASE STUDIES

The fact that microbial diversity applications are moving from laboratory studies to real-world environmental cleanup shows how useful rhizosphere and extremophilic-associated microorganisms could be in solving challenging pollution problems. Critical insights into the effective deployment of microbial consortia under site-specific conditions are provided by case studies from a variety of geographical and ecological contexts. These case studies demonstrate the flexibility and efficacy of microbial consortia across a range of pollutants and ecosystems (Koul et al., 2021). **Table 2** summarizes pivotal microbial-based remediation endeavors to augment the comprehensive examination of individual case studies. This synthesis shows the wide range of bioremediation methods available. For instance, it addresses how thermophilic bacteria may break down PAH, how rhizoremediation can get rid of heavy metals, and how psychrophilic microorganisms can break down hydrocarbons in Antarctic soils.

Table 2. Cases that use microbes for environmental remediation

Microbial Group	Environment Type	Pollutant	Mechanism of Action	Case Study Location	Reference
Rhizosphere microbes (*Azospirillum, Pseudomonas*)	Tropical agricultural soils	Soil pollutants, pesticides	Soil fertility improvement and microbial detoxification	Nigeria	(Asemoloye et al., 2017)
Acidophilic, thermophiles (*S. metallicus, A. ferrooxidans*)	Acidic, high-temperature mine waste	Acid mine drainage (AMD), heavy metals	Biooxidation of sulfides, immobilization, or recovery of metal	Spain (Rio Tinto)	(Rodriguez-Rojas, 2016)
Halophiles (*Halobacterium, Marinobacter*)	Saline industrial soil	Petroleum hydrocarbons	Osmotolerant hydrocarbon degradation	India (Gujarat)	(Bonfá et al., 2011)
Psychrophiles (*Rhodococcus, Pseudomonas*)	Polar soil, cold	Petroleum hydrocarbons	Biostimulation of native cold decomposers	Antarctica	(Wang et al., 2015)

6 FUTURE PROSPECTS

There is an increasing demand for alternative remediation of ecosystems for better health that are less expensive and extremely versatility in pollution treatment as contrast to traditional treatment approaches. Most importantly is understanding how microorganisms in natural environments and the existence of pollutants respond to the growth of ecologically stable as well as potentially bioremediation approaches. Extremophiles and the community of microorganisms in rhizosphere have demonstrated great economic promise in bioremediation techniques. The potential of extremophiles to survive in inadequate environmental settings is attributed to the evolution of adaptive mechanisms in these microorganisms at the genetic and physiological levels. Since environmental stresses get worse, climate-resilient microbial treatments will also be very important. As suggested by Rampelotto (2013), extremophiles that can handle an abundance of abiotic stresses, such as radiation, salt, and drought, are good candidates for restoring ecosystems that have been severely harmed and making agriculture more resilient (Rampelotto, 2013; Ghosal et al., 2005). Furthermore, their role in bioremediation shows the necessity for deeper studies of exploring novel spp. and their mechanisms in extreme environments.

Integration of multi-omics continues to be insufficient, and more exploration is needed to fill the gaps in understanding microbial metabolism, gene expression, and ecology in bioremediation. Co-culturing the extremophiles and rhizosphere microorganisms will speed up the cycling of nutrients and speed up the breakdown of pollutants through the phytodegradation process, advancing the bioremediation process. Additionally, it is paramount to understand the relations of extremophiles in the rhizosphere that may facilitate the alternative approach of bioremediation to further reach the 17 SDGs set out in the United Nations 2030 Agenda. The encouragement of association between government, industry, and the research community is essential for sustainability, and through their cooperation, the most advancement will be made on earth.

7 CONCLUSIONS

One promising topic for future research into environmental cleanup strategies is microbial diversity, particularly the unique alterations made by extremophiles and their interactions with other rhizosphere microbes. This chapter reveals that these microbes have amazing metabolic and physiological abilities that let them live and thrive in some of the polluted and most difficult areas on Earth. Microbial-based remediation approaches illustrate that there are various ways to clean up pollution that don't hurt the ecosystem. The way plants and the microorganisms in their rhizosphere interact, especially in places with heavy metals and herbicides, reveals how vital integrated biological systems are. Microbial remediation is about to transition from trial and error to accurate, tailored, and very effective therapies due to advances in technology like synthetic biology, gene editing, artificial intelligence, and nanobiotechnology. Therefore, these modifications need to be carefully thought about in light of environmental issues, like how they will affect the health of the ecosystem and the diversity of native microbes over time. In the end, a wide range of microbes is a significant tool for dealing with pollution problems in the environment. To get the most out of extremophiles and rhizosphere bacteria as the main tools for cleaning up the environment in a way that helps in the long run, more research is needed in many fields, as well as new ways to use technology and care for the environment.

DECLARATION OF CONFLICT OF INTEREST

It is declared by the authors that there are no known competing personal relationships or financial interests that could have appeared to influence the work reported in this chapter.

ACKNOWLEDGEMENTS

The authors thank the South African Medical Research Council (grant number SAMRC/UFH/P790), and the National Research Foundation of South Africa (grant number RCHDI241119283812) for financial support. South African Medical Research Council (SAMRC) supported the Post Doctoral Fellow as a researcher by funding received from the South African National Treasury through SAMRC Extramural Postdoctoral Programme under the Division of Research Capacity Development. The content hereof is the sole responsibility of the authors and does not necessarily represent the official views of the SAMRC.

REFERENCES

Acosta, M., Galleguillos, P. A., Guajardo, M., & Demergasso, C. (2017). *Microbial survey on industrial bioleaching heap by high-throughput 16S sequencing and metagenomics analysis*. Solid. DOI: 10.4028/www.scientific.net/SSP.262.219

Al Raish, S. M., Sourani, O. M., & Abu-Elsaoud, A. M. (2025). Plant growth-promoting microorganisms as biocontrol agents: Mechanisms, challenges, and future prospects. *Applied Microbiology (Basel, Switzerland)*, *5*(2), 44. DOI: 10.3390/applmicrobiol5020044

Aparicio, J. D., Raimondo, E. E., Saez, J. M., Costa-Gutierrez, S. B., Alvarez, A., Benimeli, C. S., & Polti, M. A. (2022). The current approach to soil remediation: A review of physicochemical and biological technologies, and the potential of their strategic combination. *Journal of Environmental Chemical Engineering*, *10*(2), 107141. DOI: 10.1016/j.jece.2022.107141

Asemoloye, M. D., Ahmad, R., & Jonathan, S. G. (2017). Synergistic action of rhizospheric fungi with Megathyrsus maximus root speeds up hydrocarbon degradation kinetics in oil polluted soil. *Chemosphere*, *187*, 1–10. DOI: 10.1016/j.chemosphere.2017.07.158 PMID: 28787637

Bonfá, M. R., Grossman, M. J., Mellado, E., & Durrant, L. R. (2011). Biodegradation of aromatic hydrocarbons by Haloarchaea and their use for the reduction of the chemical oxygen demand of hypersaline petroleum produced water. *Chemosphere*, *84*(11), 1671–1676. DOI: 10.1016/j.chemosphere.2011.05.005 PMID: 21621813

Bowman, J. S., & Deming, J. W. (2014). Alkane hydroxylase genes in psychrophile genomes and the potential for cold active catalysis. *BMC Genomics*, *15*(1), 1120. DOI: 10.1186/1471-2164-15-1120 PMID: 25515036

Castro-Severyn, J., Remonsellez, F., Valenzuela, S. L., Salinas, C., Fortt, J., Aguilar, P., Pardo-Esté, C., Dorador, C., Quatrini, R., Molina, F., Aguayo, D., Castro-Nallar, E., & Saavedra, C. P. (2017). Comparative genomics analysis of a new Exiguobacterium strain from Salar de Huasco reveals a repertoire of stress-related genes and arsenic resistance. *Frontiers in Microbiology, 8*, 456. DOI: 10.3389/fmicb.2017.00456 PMID: 28377753

Chen, Y., Wang, Z., Fang, Y., Wang, G., Zhou, F., Yu, J., Chi, R., & Xiao, C. (2025). Occurrence and microbial remediation of polycyclic aromatic hydrocarbons and heavy metals pollution in soils. *World Journal of Microbiology & Biotechnology, 41*(8), 280. DOI: 10.1007/s11274-025-04498-1 PMID: 40719956

Chia, X. K., Hadibarata, T., Jusoh, M. N. H., Sutiknowati, L. I., Tan, I. S., & Foo, H. C. Y. (2024). Extremophilic Microbes for Environmental Bioremediation: A Review. *Biointerface Research in Applied Chemistry, 14*(5), 103. DOI: 10.33263/BRIAC145.103

Christen, P., Vega, A., Casalot, L., Simon, G., & Auria, R. (2012). Kinetics of aerobic phenol biodegradation by the acidophilic and hyperthermophilic archaeon Sulfolobus solfataricus 98/2. *Biochemical Engineering Journal, 62*, 56–61. DOI: 10.1016/j.bej.2011.12.012

Dennett, G. V., & Blamey, J. M. (2016). A new thermophilic nitrilase from an antarctic hyperthermophilic microorganism. *Frontiers in Bioengineering and Biotechnology, 4*, 5. DOI: 10.3389/fbioe.2016.00005 PMID: 26973832

Ejaz, M., Zhao, B., Wang, X., Bashir, S., Haider, F. U., Aslam, Z., & Mustafa, A. (2021). Isolation and characterization of oil-degrading Enterobacter sp. from naturally hydrocarbon-contaminated soils and their potential use against the bioremediation of crude oil. *Applied Sciences (Basel, Switzerland), 11*(8), 3504. DOI: 10.3390/app11083504

Farhangi-Abriz, S., & Torabian, S. (2017). Antioxidant enzyme and osmotic adjustment changes in bean seedlings as affected by biochar under salt stress. *Ecotoxicology and Environmental Safety, 137*, 64–70. DOI: 10.1016/j.ecoenv.2016.11.029 PMID: 27915144

Ghosal, D., Omelchenko, M. V., Gaidamakova, E. K., Matrosova, V. Y., Vasilenko, A., Venkateswaran, A., & Daly, M. J. (2005). How radiation kills cells: Survival of Deinococcus radiodurans and Shewanella oneidensis under oxidative stress. *FEMS Microbiology Reviews, 29*(2), 361–375. PMID: 15808748

Glick, B. R. (2012). Plant growth-promoting bacteria: Mechanisms and applications. *Scientifica, 2012*(1), 963401. PMID: 24278762

Gonzalez, J. M., Santana, M. M., Gomez, E. J., & Delgado, J. A. (2023). Soil thermophiles and their extracellular enzymes: A set of capabilities able to provide significant services and risks. *Microorganisms, 11*(7), 1650. DOI: 10.3390/microorganisms11071650 PMID: 37512823

Gouda, S., Kerry, R. G., Das, G., Paramithiotis, S., Shin, H. S., & Patra, J. K. (2018). Revitalization of plant growth promoting rhizobacteria for sustainable development in agriculture. *Microbiological Research, 206*, 131–140. DOI: 10.1016/j.micres.2017.08.016 PMID: 29146250

Hamid, B., Bashir, Z., Yatoo, A. M., Mohiddin, F., Majeed, N., Bansal, M., & Alfaifi, M. Y. (2022). Cold-active enzymes and their potential industrial applications review. *Molecules (Basel, Switzerland), 27*(18), 5885. DOI: 10.3390/molecules27185885 PMID: 36144621

Icer, M. A., Özbay, S., Ağagündüz, D., Kelle, B., Bartkiene, E., Rocha, J. M. F., & Ozogul, F. (2023). The impacts of acidophilic lactic acid bacteria on food and human health: A review of the current knowledge. *Foods, 12*(15), 2965. DOI: 10.3390/foods12152965 PMID: 37569234

Irwin, J. A. (2020). Overview of extremophiles and their food and medical applications. In *Physiological and biotechnological aspects of extremophiles* (pp. 65–87). Academic Press. DOI: 10.1016/B978-0-12-818322-9.00006-X

Jeong, S. W., & Choi, Y. J. (2020). Extremophilic microorganisms for the treatment of toxic pollutants in the environment. *Molecules (Basel, Switzerland), 25*(21), 4916. DOI: 10.3390/molecules25214916 PMID: 33114255

Jin, P. J., Sun, L., Liu, Y. H., Wang, K. K., Narsing Rao, M. P., Mohamad, O. A. A., Fang, B.-Z., Li, L., Gao, L., Li, W.-J., & Wang, S. (2024). Two novel alkaliphilic species isolated from saline-alkali soil in China: Halalkalibacter flavus sp. nov., and Halalkalibacter lacteus sp. nov. *Microorganisms, 12*(5), 950. DOI: 10.3390/microorganisms12050950 PMID: 38792780

Johnson, D. B. (2008). Biodiversity and interactions of acidophiles: Key to understanding and optimizing microbial processing of ores and concentrates. *Transactions of Nonferrous Metals Society of China, 18*(6), 1367–1373. DOI: 10.1016/S1003-6326(09)60010-8

Kanekar, P. P., & Kanekar, S. P. (2022). Alkaliphilic, alkalitolerant microorganisms. In *Diversity and Biotechnology of Extremophilic Microorganisms from India* (pp. 71–116). Springer Nature Singapore. DOI: 10.1007/978-981-19-1573-4_3

Kannaujiya, M. C., Mandal, T., Mandal, D. D., & Mondal, M. K. (2019). Treatment of leather industry wastewater and recovery of valuable substances to solve waste management problem in environment. In *Environmental contaminants: ecological implications and management* (pp. 311–340). Springer Singapore. DOI: 10.1007/978-981-13-7904-8_14

Khasanova, S., Alieva, E., & Shemilkhanova, A. (2023). Environmental pollution: types, causes and consequences. In *BIO Web of Conferences* (Vol. 63, p. 07014). EDP Sciences. DOI: 10.1051/bioconf/20236307014

Kochhar, N., Shrivastava, S., Ghosh, A., Rawat, V. S., Sodhi, K. K., & Kumar, M. (2022). Perspectives on the microorganism of extreme environments and their applications. *Current Research in Microbial Sciences*, *3*, 100134. DOI: 10.1016/j.crmicr.2022.100134 PMID: 35909612

Kolmert, Å., & Johnson, D. B. (2001). Remediation of acidic waste waters using immobilised, acidophilic sulfate-reducing bacteria. *Journal of Chemical Technology & Biotechnology: International Research in Process. Environmental & Clean Technology*, *76*(8), 836–843.

Koskey, G., Mburu, S. W., Awino, R., Njeru, E. M., & Maingi, J. M. (2021). Potential use of beneficial microorganisms for soil amelioration, phytopathogen biocontrol, and sustainable crop production in smallholder agroecosystems. *Frontiers in Sustainable Food Systems*, *5*, 606308. DOI: 10.3389/fsufs.2021.606308

Koul, B., Chaudhary, R., & Taak, P. (2021). Extremophilic microbes and their application in bioremediation of environmental contaminants. In *Microbe mediated remediation of environmental contaminants* (pp. 115–128). Woodhead Publishing. DOI: 10.1016/B978-0-12-821199-1.00012-2

Kour, D., Kaur, T., Devi, R., Yadav, A., Singh, M., Joshi, D., & Saxena, A. K. (2021). Beneficial microbiomes for bioremediation of diverse contaminated environments for environmental sustainability: Present status and future challenges. *Environmental Science and Pollution Research International*, *28*(20), 24917–24939. DOI: 10.1007/s11356-021-13252-7 PMID: 33768457

Kumar, A., Alam, A., Tripathi, D., Rani, M., Khatoon, H., Pandey, S., & Hasnain, S. E. (2018, December). Protein adaptations in extremophiles: An insight into extremophilic connection of mycobacterial proteome. []. Academic Press.]. *Seminars in Cell & Developmental Biology*, *84*, 147–157. DOI: 10.1016/j.semcdb.2018.01.003 PMID: 29331642

Lakshmanan, V., Selvaraj, G., & Bais, H. P. (2014). Functional soil microbiome: Belowground solutions to an aboveground problem. *Plant Physiology*, *166*(2), 689–700. DOI: 10.1104/pp.114.245811 PMID: 25059708

Lawson, C. E., & Lücker, S. (2018). Complete ammonia oxidation: An important control on nitrification in engineered ecosystems? *Current Opinion in Biotechnology*, *50*, 158–165. DOI: 10.1016/j.copbio.2018.01.015 PMID: 29414055

Liu, X., Sathishkumar, K., Zhang, H., Saxena, K. K., Zhang, F., Naraginti, S., & Guo, X. (2024). Frontiers in environmental cleanup: Recent advances in remediation of emerging pollutants from soil and water. *Journal of Hazardous Materials Advances*, *16*, 100461. DOI: 10.1016/j.hazadv.2024.100461

Maglione, G., Zinno, P., Tropea, A., Mussagy, C. U., Dufossé, L., Giuffrida, D., & Mondello, A. (2024). Microbes' role in environmental pollution and remediation: A bioeconomy focus approach. *AIMS Microbiology*, *10*(3), 723–755. DOI: 10.3934/microbiol.2024033 PMID: 39219757

Malik, S., Dhasmana, A., Preetam, S., Mishra, Y. K., Chaudhary, V., Bera, S. P., & Rajput, V. D. (2022). Exploring microbial-based green nanobiotechnology for wastewater remediation: A sustainable strategy. *Nanomaterials (Basel, Switzerland)*, *12*(23), 4187. DOI: 10.3390/nano12234187 PMID: 36500810

Margesin, R., & Schinner, F. (2001). Potential of halotolerant and halophilic microorganisms for biotechnology. *Extremophiles*, *5*(2), 73–83. DOI: 10.1007/s007920100184 PMID: 11354458

Orellana, R., Macaya, C., Bravo, G., Dorochesi, F., Cumsille, A., Valencia, R., & Seeger, M. (2018). Living at the frontiers of life: Extremophiles in Chile and their potential for bioremediation. *Frontiers in Microbiology*, *9*, 2309. DOI: 10.3389/fmicb.2018.02309 PMID: 30425685

Özdemir, S., Akarsu, C., Acer, Ö., Fouillaud, M., Dufossé, L., & Dizge, N. (2022). Isolation of thermophilic bacteria and investigation of their microplastic degradation ability using polyethylene polymers. *Microorganisms*, *10*(12), 2441. DOI: 10.3390/microorganisms10122441 PMID: 36557694

Philippot, L., Raaijmakers, J. M., Lemanceau, P., & Van Der Putten, W. H. (2013). Going back to the roots: The microbial ecology of the rhizosphere. *Nature Reviews. Microbiology*, *11*(11), 789–799. DOI: 10.1038/nrmicro3109 PMID: 24056930

Phulpoto, A. H., Maitlo, M. A., & Kanhar, N. A. (2021). Culture-dependent to culture-independent approaches for the bioremediation of paints: A review. *International Journal of Environmental Science and Technology*, *18*(1), 241–262. DOI: 10.1007/s13762-020-02801-1

Rampelotto, P. H. (2013). Extremophiles and extreme environments. *Life (Chicago, Ill.)*, *3*(3), 482–485. PMID: 25369817

Rashid, A., Schutte, B. J., Ulery, A., Deyholos, M. K., Sanogo, S., Lehnhoff, E. A., & Beck, L. (2023). Heavy metal contamination in agricultural soil: Environmental pollutants affecting crop health. *Agronomy (Basel)*, *13*(6), 1521. DOI: 10.3390/agronomy13061521

Rodríguez-Rojas, F., Tapia, P., Castro-Nallar, E., Undabarrena, A., Muñoz-Díaz, P., Arenas-Salinas, M., & Vásquez, C. (2016). Draft genome sequence of a multi-metal resistant bacterium *Pseudomonas putida* ATH-43 isolated from Greenwich Island, Antarctica. *Frontiers in Microbiology*, *7*, 1777. DOI: 10.3389/fmicb.2016.01777 PMID: 27877169

Samuel, M. S., Jose, S., Selvarajan, E., Mathimani, T., & Pugazhendhi, A. (2020). Biosynthesized silver nanoparticles using Bacillus amyloliquefaciens; Application for cytotoxicity effect on A549 cell line and photocatalytic degradation of p-nitrophenol. *Journal of Photochemistry and Photobiology. B, Biology*, *202*, 111642. DOI: 10.1016/j.jphotobiol.2019.111642 PMID: 31734434

Sheela, B., Khasim, S., & Rao, O. (2014). Bioremediation of ammonia using ammonia oxidizing bacteria isolated from sewage. *International Journal of Environmental Bioremediation & Biodegradation*, *2*, 146–150.

Siliakus, M. F., van der Oost, J., & Kengen, S. W. (2017). Adaptations of archaeal and bacterial membranes to variations in temperature, pH and pressure. *Extremophiles*, *21*(4), 651–670. DOI: 10.1007/s00792-017-0939-x PMID: 28508135

Singh, P., Jain, K., Desai, C., Tiwari, O., & Madamwar, D. (2019). Microbial community dynamics of extremophiles/extreme environments. In *Microbial diversity in the genomic era* (pp. 323–332). Academic Press. DOI: 10.1016/B978-0-12-814849-5.00018-6

Urbieta, M. S., Donati, E. R., Chan, K. G., Shahar, S., Sin, L. L., & Goh, K. M. (2015). Thermophiles in the genomic era: Biodiversity, science, and applications. *Biotechnology Advances*, *33*(6), 633–647. DOI: 10.1016/j.biotechadv.2015.04.007 PMID: 25911946

Wang, L., Cheng, G., Ren, Y., Dai, Z., Zhao, Z. S., Liu, F., & Tang, B. (2015). Degradation of intact chicken feathers by Thermoactinomyces sp. CDF and characterization of its keratinolytic protease. *Applied Microbiology and Biotechnology, 99*(9), 3949–3959. DOI: 10.1007/s00253-014-6207-4 PMID: 25412577

Xu, X., Yang, H., & Li, C. (2022). Theoretical model and actual characteristics of air pollution affecting health cost: A review. *International Journal of Environmental Research and Public Health, 19*(6), 3532. DOI: 10.3390/ijerph19063532 PMID: 35329214

Chapter 4
Microbial Adaptation and Remediation in Polluted Soils:
Interaction With Metals, Pesticides, and Hydrocarbons

Nonkululeko Landy Mdeni
http://orcid.org/0000-0002-8356-3495
Univesity of Fort Hare, South Africa

Mike O. Ojemaye
http://orcid.org/0000-0002-4038-7639
Saginaw Valley State University, USA

Anthony I. Okoh
University of Fort Hare, South Africa

Omobola O. Okoh
University of Fort Hare, South Africa

ABSTRACT

Soil is an essential natural resource within terrestrial ecosystems, yet it is increasingly challenged by the influx of emerging pollutants such as hydrocarbons, pesticides and heavy metals (HMs). This chapter focuses on the persistence, and ecological risks associated with these contaminants with particular emphasis on microbial adaptation, ecological interactions remediation potential. More so, soils serve as stores of water and nutrients necessary for the development of microbes and plant root systems, despite significant variance in genetic makeup and environmental

DOI: 10.4018/979-8-2600-2100-2.ch004

factors. The Earth's crust contains a plethora of HMs, with human activities leading to long-term contamination. Similarly, because of their neurotoxic and carcinogenic effect, soils containing dangerous hydrocarbons, like polycyclic aromatic hydrocarbons (PAHs), provide serious health concerns to humans. Due in major part to the fast population expansion, industrialization and the growing need for food security, pesticides are the most common xenobiotic contaminants found in soils.

1. INTRODUCTION

A complex mixture of organic matter, minerals and a network of pore spaces filled with air and water make up the heterogeneous ecosystem known as soils. Numerous and diverse soil species carrying out important soil processes and functions call this ecosystem home and the microorganisms that live in the bulk soil and/or rhizosphere are the most important participants in each of these processes (Joos, and De Tender, 2022). In order to sustain global biogeochemical cycles and environmental management, soil ecosystems are essential (Ren et al.,2025). Soil is an essential natural resource base in terrestrial ecosystems that is struggling with the influx of new pollutants such as heavy metals (HMs), hydrocarbons and pesticides (Iqbal et al., 2020). Although a variety of HMs are naturally present in the Earth's crust, human activity has greatly accelerated their buildup and poisoning of soils. These activities include military drills, the widespread use of agrochemicals, and the discharge of waste products from the electronics sector that are sourced from fossil fuels (Azhar et al., 2022). Heavy metals like lead (Pb), cadmium (Cd), arsenic (As), nickel (Ni), zinc (Zn), and chromium (Cr) have been found to accumulate in topsoil in previous investigations (Angon et al., 2024).

Occupants may be at risk for health problems from exposure to soil contamination and the food chain because these metals can linger in soils for a long period. Therefore, it is crucial to evaluate the level of HM contamination in soils and look more broadly at the possible dangers to human health and the environment that come with such pollution (Chen et al., 2022; Wang et al., 2022). Likewise, soil contaminated with dangerous hydrocarbons such as polyaromatic hydrocarbons (PAHs) that are absorbed by direct skin contact may have detrimental consequences on human health (Sakshi et al., 2019). It is recognized that hydrocarbon components are members of the family of organic contaminants that are neurotoxic and carcinogenic. Environmental matrices are contaminated by hydrocarbons, particularly PAHs which have raised public concerns due to their high mutagenicity, carcinogenicity and toxicity (Abdelhafeez et al., 2022).

Although soils vary greatly in terms of genetics and environment, they typically serve as repositories of water and nutrients required for the proper growth of microorganisms and the plant root system. Soil acts as a pollution absorber because of its ability to retain different types of pollutants. Pesticides are the most common xenobiotic pollutants found in soil, primarily as a result of population increase and the growing need for adequate and safe food supply (Wołejko et al., 2020). The most widely used and successful method of protecting plant crops from pests remains the application of pesticides. However, about 1% of a pesticide that is applied actually reaches the target pest; the rest finds its way into the soil, water, and air, where it eventually enters the food chain and affects non-target species like humans, plants, animals, and the activity of soil enzymes (Sun et al., 2018; Zhichkina et al., 2020).

The quantity and variety of soil microflora, as well as the existence of adequate nutrients, define soil fertility. The presence of different kinds of organic substrates in soil is primarily responsible for microbial diversity. Prokaryotic or eukaryotic bacteria (eubacteria and archaebacteria), cyanobacteria, actinomycetes, fungus, and algae are among the various categories of primarily unicellular creatures. These soil bacteria carry out a range of tasks necessary for the soil's healthy operation as a dynamic system (Bokade et al., 2023; Kalia and Gosal., 2011). The ability of microorganisms to adapt to adverse environments and flourish there gives rise to the notion that their potential can be used for a wide range of purposes. Their remarkable genetic composition and capacity to flourish under challenging circumstances attest to their use in variously contaminated soils and because they have specific metabolic pathways and enzymes that are able to break down contaminants and detoxify soil, their adaptations are essential for combating soil pollution and provide an affordable alternative (Arora et al., 2024). This chapter discusses microbial adaptation and remediation processes in soils impacted by diverse pollutants. It explores contamination caused by pesticides, hydrocarbons, and heavy metals. Emphasis on polycyclic aromatic and petroleum hydrocarbons, heavy metal buildup and important pesticide classes, focusing on their interactions with soil microbial communities as well as their bioremediation potentials.

2. DIFFERENT TYPES OF PESTICIDES IN THE ENVIRONMENT

A class of substances known as pesticides and insecticides is poisonous to insects and pests, respectively. Applying pesticides has benefits for both agricultural and vector extirpation. The agriculture industry uses pesticides to manage pests, which is why this sector achieves such high levels of output. However, pesticides are accidentally released into the environment, which has negative environmental implications because they are poisonous and frequently harm organisms that are not their intended target. Based on their chemical structures, pesticides are classified as carbamates, chlorophenols, organochlorines, organophosphorus, and synthetic pyrethroids (Mdeni et al., 2022). Pesticides are used extensively in South Africa for both public health purposes and to support the country's quickly growing agriculture industry. With over 500 items registered with the Pesticide Action Network (PAN) (Pesticide Action Network, 2010), the nation is one of the top importers of pesticides, Organophosphate (OP) pesticides, which are reportedly the most often used class of pesticides in conventional agriculture, are included in this group (Jaipieam et al., 2009; Quinn *et* al., 2011).

2.1. Organophosphate Pesticides

Through mechanisms like drainage, drift, and surface run-off, organophosphate (OP) pesticides spread throughout the environment, leaving behind residues in a variety of environmental matrices, such as sediment and soils (Nasrabadi et al., 2011). Chemically, OPs are created when phosphoric acid (H_3PO_4) and alcohol (–OH) are esterified, resulting in molecules having carbon–phosphorus linkages (Figure 1). The irreversible phosphorylation of the cholinesterase (ChE) enzyme is their route of toxicity; this mechanism can be lethal to mammals even though it works well against target pests and insects (Jaga and Dharmani, 2003).

Figure 1. Generic structural representation of Ops

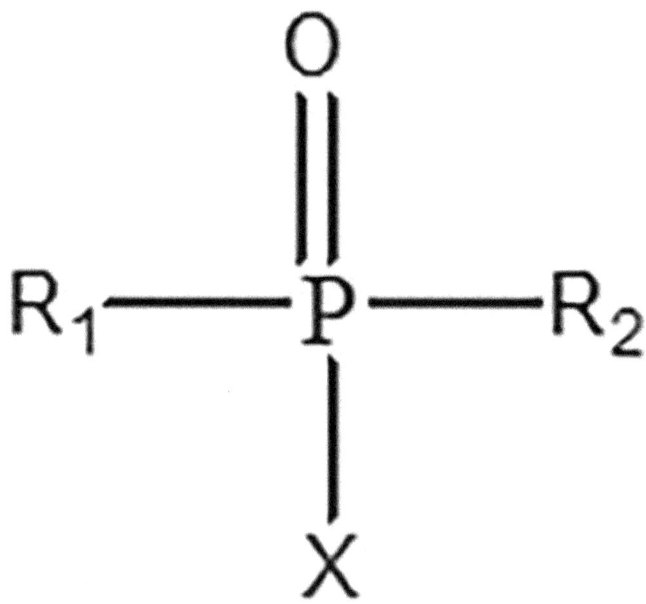

Numerous organophosphates (OPs), such as dursban, malathion, and chlorpyrifos, are used extensively around the world (Fenik et al., 2011). Many detection technologies, from chromatographic to enzymatic approaches, have been developed to control and track exposure to these pesticides (Kaushal et al., 2021). One of the most often utilized OPs is parathion, which has a significant adsorption to soil particles and is mildly hazardous. When parathion is exposed to sunlight, it photodegrades to produce its active metabolite, paraoxon (Figure 2), and then breaks down at about 200 °C. Pesticides like parathion are frequently persistent in the environment due to their relatively low solubility; consequently, surfactants, commercial adsorbents, and organic co-solvents are frequently used in increasing solubility and make removal easier.

Figure 2. Oxidation of Parathion to Paraoxon

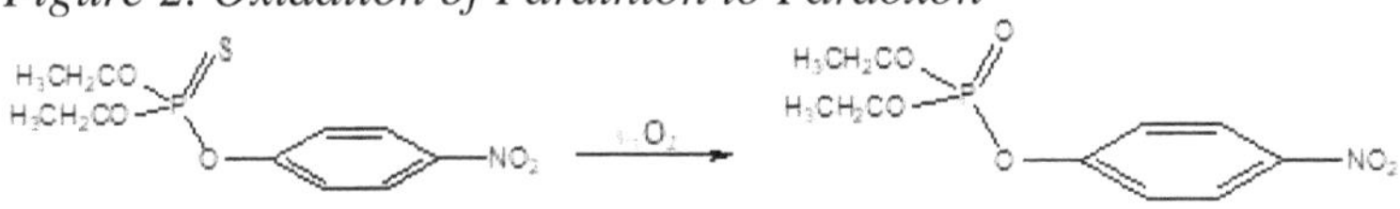

2.2. Organochlorine Pesticides

Historically, organochlorine pesticides (OCPs) have been employed due to their great effectiveness and affordability. They pose a serious risk to human health and the environment because of their high toxicity, persistence, long-distance movement, and propensity to bioaccumulate in the food chain. About 40% of all frequently used pesticides are OCPs, a class of chemical molecules made up of carbon, hydrogen, and chlorine. OCPs have been divided into three primary groups based on similarities in their chemical structures, namely, Chlorinated cyclodienes, Dichlorodiphenylethanes and Chlorinated benzenes and cyclohexanes (Ren et al., 2025). Among the most prevalent OCPs, residues of dichloro-diphenyl-trichloroethane (DDT), chlordane, heptachlor hexachlorocyclohexane (HCH), have been found in agricultural soils worldwide (Kafaei et al., 2020).

Figure 3. Different categories of OCPs

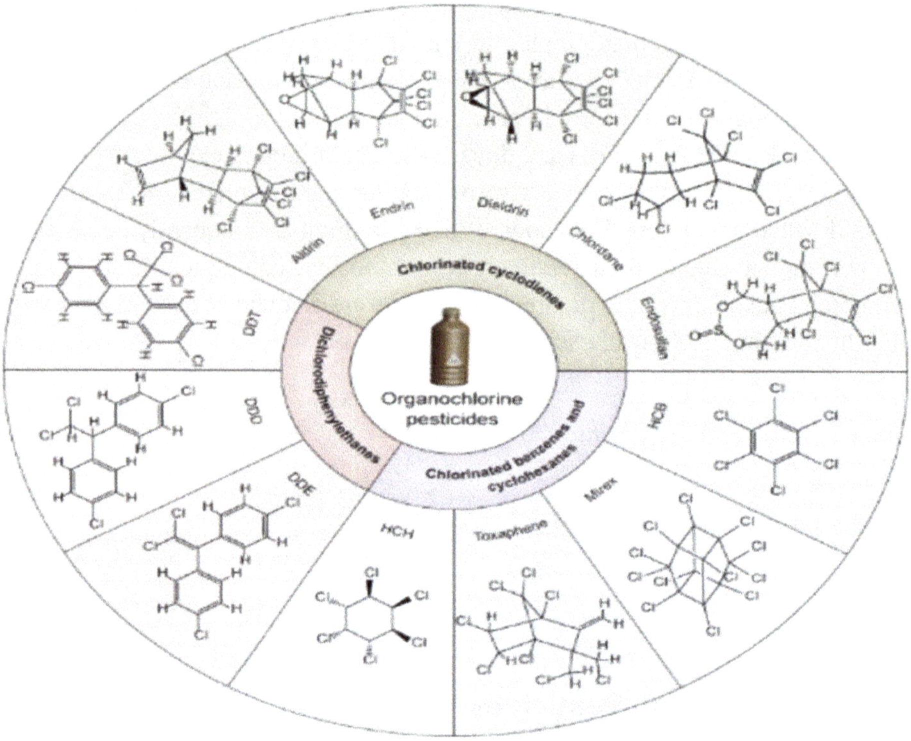

(Ren et al., 2025)

Since the 1970s, the majority of OCPs have been prohibited in more than 130 nations and included in the Stockholm Convention due to the severe toxicity they cause to humans and wildlife, which includes cancer, allergies, and neurological, reproductive, and immunological dysfunctions. OCPs are persistent and may be found in high concentrations in soil, water, sediment, atmosphere, and biota they can travel distances in the atmosphere, water, and migratory birds, even to areas where no pesticides are used, causing global pollution; and because they are lipophilic, OCPs in soil or water can build up in livestock and plants and ultimately endanger human health through the food chain, it is because of these reasons that their population is still a concern (Gerber et al., 2016; Chen et al., 2021).

Adsorbed onto soil colloids, organochlorine pesticides (OCPs) can go into aquifers with restorative microbes, where they may repartition due to microbial interactions. Through vital processes like organic matter breakdown and nutrient cycling, microbial communities play an essential role in preserving balance in soil ecosystems, which is a significant indicator of soil health. As a result, microbes that react quickly to observable lithosphere disruptions are frequently considered bioindicators of soil health (Egbe et al., 2021; Jia et al., 2025). Changes in the makeup of soil microbial communities can have a significant impact on soil fertility, which is essential for plant development and agricultural output (Jia et al., 2025).

2.3. Carbamate Pesticides

For many years, carbamate pesticides have been broadly utilized in agricultural production. Due to their being hazardous to living systems, their degradation is essential even if they are not as persistent in the environment as other pesticides like organophosphate. The carbamate insecticide was classified as dangerous, harmful, and restricted for use by the World Health Organization (WHO). These pesticides can contaminate land and water, and so there is currently a great deal of awareness about their detrimental impacts (Mustapha and Halimoon, 2019). Microbes can easily break down or metabolize most insecticides. The breakdown of carbamate pesticides by microorganisms depends on several ecological conditions in addition to the presence of bacteria with the right biodegradative enzymes (Cai et al., 2015; Mustapha and Halimoon, 2019).

2.4. Chlorophenols

Chlorophenols are a significant class of chemical compounds that have wide-ranging uses in the agricultural and pharmaceutical industries. To increase crop yields and guard against diseases and pests that frequently result in significant losses, especially in cereal crops, they are frequently used in agriculture as fungicides, herbicides, rodenticides, germicides, and other pesticide formulations. Depending on the number of chlorine atoms bonded to the benzene ring at particular locations to the hydroxyl group, chlorophenols are divided into five categories, which are monochlorophenols (CPs), dichlorophenols (DCPs), trichlorophenol (TCPs), tetrachlorophenols (TCPs) and pentachlorophenols (PCPs) respectively containing 1, 2, 3, 4 and 5 chlorine atoms respectively (Figure 4) (Yadav et al., 2023).

Figure 4. Structures ofmono and polycholophenols

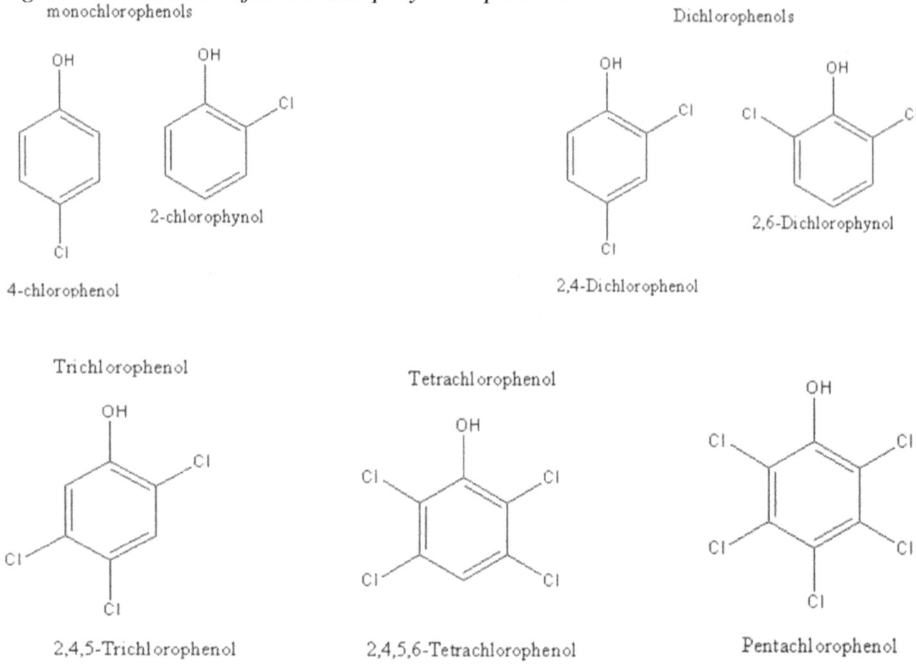

3. HYDROCARBONS

3.1. Petroleum Hydrocarbons

The demand for petroleum production has increased as a result of the world's population growth and the quick expansion of industrial, agricultural, and municipal activities. However, the main causes of hydrocarbon spills are activities related to the petroleum industry, such as drilling, storage, transportation, refining, and processing. As a result, hydrophobic compounds make up a large portion of the terrestrial oil pollution (Mekonnen et al., 2024; Singha and Pandey, 2021). Hydrocarbons, which are hydrophobic and extremely persistent hazardous substances obstruct soil pores and create anaerobic zones that impede biodegradation, this is because they prefer to form complexes and bind with soil particles and organic matter (Mayans et al., 2024). Total petroleum hydrocarbons (TPH) are derived from crude oil and are mostly composed of hydrogen and carbon, with lesser amounts of nitrogen, sulfur, and oxygen. They consist of asphaltenes, aromatics, resins, and saturated hydrocarbons/paraffins (Varjani, 2017).

Petroleum extractable hydrocarbons and volatile petroleum hydrocarbons are the two primary types of chemicals (Figure 4) (Alotaibi et al., 2021). Small-chain hydrocarbons (C6–C10), such as benzene, toluene, ethylbenzene, and xylene (BTEX), are included in the first group. Polycyclic aromatic hydrocarbons (PAH) and long-chain (C10–C40) aliphatic hydrocarbons are included in the second group (Hazaimeh et al.,2024). Because an impermeable layer of petroleum hydrocarbons forms over contaminated soils, gas exchange between the soil and air occurs, and the soil's capacity to sorb water and minerals is diminished. Anaerobic soil damages plant roots and reduces bacterial populations and metabolic activity in untreated soil (Kwon et al., 2023).

Figure 5. Schematic diagram showing classification of petroleum hydrocarbons (PHCs)

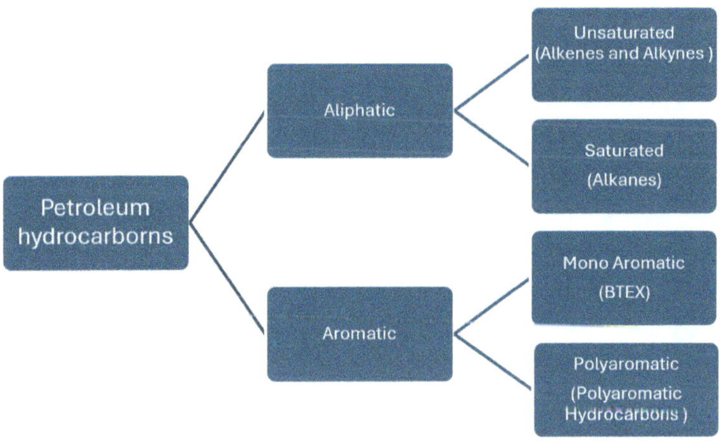

3.2. Polyaromatic Hydrocarbons

Polyaromatic hydrocarbons (PAHs) are aromatic hydrocarbons with benzene rings fused together in various structural configurations with no substituents and they are significant environmental concern. They are categorised as heavy or high molecular weight (HMW) and light or low molecular weight (LMW) according to the number of benzene rings they have. PAHs that contain four or less benzene rings are considered light and those that contain more than four rings are heavy, some examples of light and heavy PAHs include Naphthalene which has two benzene rings and Benzo(a)Pyrene which is the most commonly studied and measured

PAH, respectively (Figure 6) (Lawal, 2017). There are hundreds of possible PAH combinations, but the USEPA (United States Environmental Protection Agency) only designated roughly 28 of them as harmful in 2008 (Kuppusamy et al., 2017).

Figure 6. Structures of (A) light and (B) heavy PAH

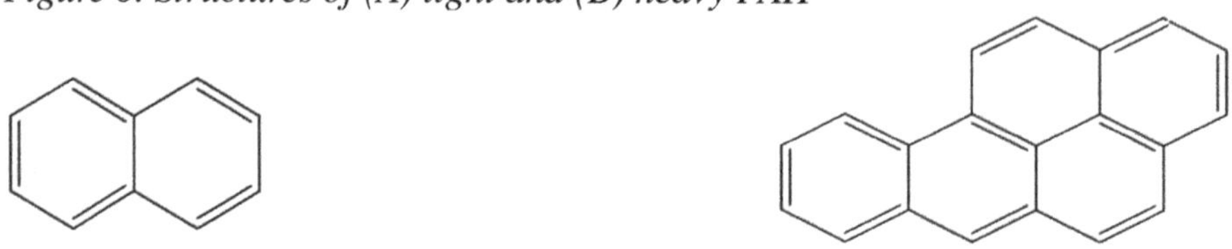

The greatest source of PAHs in the atmosphere, despite the fact that there are several possible sources, is human activity due to the rapid urbanization and industrialization. There are hundreds of possible PAH combinations, but the USEPA has classified sixteen of the more than 100 PAHs as priority contaminants because of their toxicity (Kuppusamy et al., 2017). These PAHs include naphthalene, acenaphthylene, acenaphthene and benzo[a]pyrene. The incomplete combustion of organic materials is the main source of PAHs (Patel et al., 2020; Mallah et al., 2020). The manufacture of metals, aluminium, motor vehicle exhaust, coal and gas-fired power plants, tobacco smoke, food processing, and home heating are some of the main occupational and non-occupational sources. Because of variations in their ring configurations, PAHs have distinctive absorption spectra under ultraviolet (UV) light, with each isomer exhibiting a distinct UV profile (Hong et al., 2023; Sahoo et al., 2020). Be consistent with your ref style.

Chemically, two or more benzene rings fused in angular, clustered, or linear configurations make up polycyclic aromatic hydrocarbons (PAHs), which are categorized as either cata-condensed or peri-condensed according to their structural diversity. The fusing of rings in peri-condensed PAHs creates cyclic connections between ring centers by utilizing common carbon atoms across numerous rings. Two groups of peri-condensed PAHs, such as alternants and non-alternants, can be distinguished. Six-membered rings predominately create alternants, whereas five-membered rings form non-alternants. One definition of cata-condensed PAHs is systems in which the lines do not form cycles (Sahoo et al., 2020; Cataldo et al., 2018). Examples of peri-condensed PAHs include pyrene and fluoranthene, while chrysene and triphenylene are cata-condensed PAHs (Figure 7).

Figure 7. Main structural diversity of PAHs

Peri-condensed		Cata-condensed	
Pyrene	Fluoranthene	Chrysene	Triphenylene
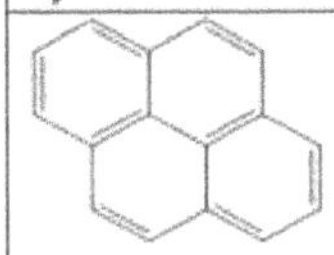	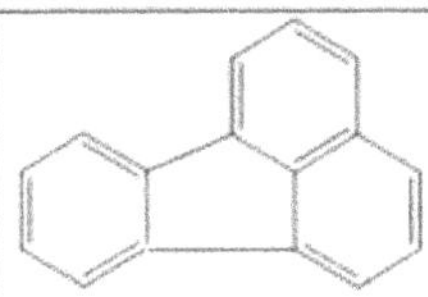	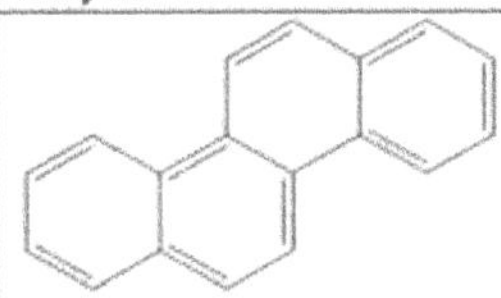	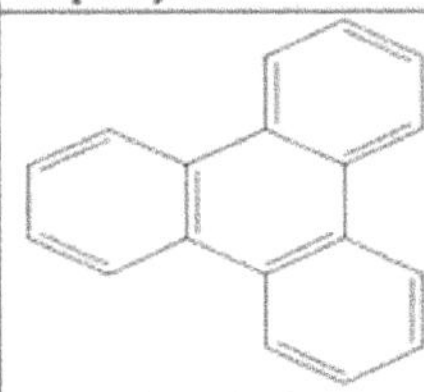
cycle-alternant	cycle non-alternant	Linear non-branched alternant	Linear branched alternant

Remediation of PAHs from the environment has been an international concern because of their inherent features, which make them persistent pollutants with a wide range of biological toxicity. According to Patel et al. (2020), PAHs, are pervasive contaminants that can be found in atmospheric, aquatic, and terrestrial settings. The U.S. EPA reports PAH contamination at nearly 600 of the 1,408 hazardous waste sites designated for long-term governmental treatment, making them a particularly important class of organic contaminants. These substances are quickly mobilized in the environment, lipophilic, semi-volatile, and only sporadically soluble in water. They also have modest vapour pressures. When exposed to ultraviolet (UV) light from solar radiation, they can decompose. They can be absorbed in air, water, sediment, and soil particles (Abdel-Shafy et al., 2016).

4. HEAVY METALS

Given their high toxicity and persistence, heavy metal poisoning has become a significant environmental concern and a primary cause of soil pollution. Rapid industrialization, the widespread use of synthetic fertilizers, atmospheric deposition, the application of sewage sludge and farmyard manure, and other factors all contribute to their buildup in soils (Angon et al., 2024). HMs, such as zinc (Zn), lead (Pb), arsenic (As), mercury (Hg), copper (Cu), and chromium (Cr), are thought to impact about 20 million hectares of land worldwide. Among these, the Agency for Toxic Substances and Disease Registry (ATSDR) has identified arsenic (As), lead (Pb), cadmium (Cd), and mercury (Hg) as the most dangerous to plants and people. They can infiltrate plant systems and contaminate the food chain, which is quite risky for both food quality and human health, therefore One of the most significant environmental concerns is the removal of these heavy metals from the soil (Liu et al., 2018; Pouresmaieli et al., 2022).

The group of proteins known as soil enzymes, which are often accumulated in soil, are essential for preserving the ecology, physical and chemical characteristics, fertility, and general health of soil. In addition to reducing soil organic pollution and influencing soil and food chain chemical processes, soil enzymes are also sensitive to ecological shifts. The balance of soil enzymes is altered by soil contamination, particularly by heavy metals (Wahsha et al., 2017). Chemical leaching, Adsorption, advanced oxidation processes, photodegradation, electro-remediation, biodegradation, and phytoremediation are only a few of the many soils heavy metal remediation methods that have been thoroughly studied (Chen et al.,2020). The various remediation techniques are primarily separated into two groups based on the remediation process: HMs in situ immobilization and HMs mobilization. The process of improving the removal of metals from soil is known as heavy metal (HM) mobilization. This can be achieved by adding mobilizing agents such as desorbers, surfactants and chelators, or by combining bioremediation and phytoremediation to encourage HM uptake by plants. By altering HMs' chemical forms, these amendments make them more soluble and make it easier for them to drain from the soil (Gao et al., 2023). In contrast, heavy metals in situ immobilization reduces HMs' mobility and bioavailability to sequester them in the soil, as well as their solubility in the soil (Chávez-Vergara et al., 2025).

5. BIOREMEDIATION: MICROBIAL RESPONSE TO POLLUTION

Contaminated land is now widely acknowledged to pose a risk to human health, international efforts have been made to remediate many of these sites, either to reduce the possibility of harmful environmental or health effects from contamination or to allow for the site's redevelopment for human use. The traditional methods of remediation have involved either capping and containing the polluted sections of a site or excavating contaminated soil and disposing of it in a landfill and these techniques have certain shortcomings (Kensa, 2011). Microorganisms and their byproducts are used in the bioremediation process to eliminate pollutants from soil. The process of improving the removal of metals from soil is known as heavy metal (HM) mobilization. Altering HMs' chemical forms, makes them more soluble and make it easier for them to drain from the soil. This process is called mineralization; the process of ionic exchange allows the bacteria to adhere to soil particles.

Soil particles are negatively charged, and an ionic connection including polyvalent cations can hold soil and bacteria together (Yang et al., 2021; Adams et al., 2015).

As a new and innovative technique, bioremediation technology has been established to eliminate and minimize persistent pollutants from contaminated waste flows (such as soil and water) (Saeed et al 2022). By lowering contamination in environmental compartments and getting rid of persistent or resistant substances, bioremediation has been shown to be a successful method for decreasing the negative effects of pollution. It is considered a reasonably priced, safe, and dependable environmental remediation technique. Microbial biomass-based environment compartment decontamination has been emphasized as a substitute strategy because of its special qualities that can be used to human advantage (Hlihor and Cozma, 2023).

Boyd et *al.* (2022) reported that crude oil and many medications contain thiophenes, which are enzymatically oxidized by enzymes such as, mono- and dioxygenases to produce metabolites including sulfoxides, epoxides, and dihydrodiols. Microorganisms enzymatically oxidize thiophenes and PAHs, a process essential to drug metabolism, biodesulfurization, and bioremediation. While mono- and dioxygenases that oxidize thiophene yield unstable metabolites with complex stereochemistry, enzymes including cytochrome P450s, epoxide hydrolases, and glutathione S-transferases are involved in the breakdown of PAHs, especially by yeasts. The ability of microorganisms to change these dangerous and persistent substances is being further enhanced by developments in genetic engineering and omics techniques (Padilla-Garfias et al 2024).

According to recent research, biosorption, bioleaching, biomineralization, biotransformation are economical and environmentally friendly methods of removing HMs from various environmental matrices (Pande et al., 2022). In biosorption, environmental factors, biofilm formation, and physicochemical processes as ion exchange, complexation, precipitation, and adsorption influence how well microorganisms work as effective biosorbents. The temperature range of 20 to 35 °C is usually where optimal biosorption occurs, indicating the potential of microbial systems to reduce HM contamination while simultaneously recognizing their vulnerability to metal toxicity (Priya et al., 2022). Microbial redox systems are essential for industrial biotechnology and bioremediation. *Dehalococcoides mccartyi* may reductively dehalogenate hazardous organics, and in bioelectrochemical systems, its electron transfer can be facilitated by cobalt chelates, providing safer and more environmentally friendly applications. The adaptability of microbial redox processes in environmental and industrial settings is further demonstrated by the fact that fungal laccases and related multicopper oxidases oxidize a broad variety of aromatic chemicals, with their activity being increased by redox mediators and protein engineering (Aza and Camarero, 2023; Eberwein et al., 2024).

5.1. Bioremediation Approaches for Pesticide Degradation

In bioremediation pesticides along with other nutrients are employed as co-substrates that are metabolised by microorganisms thus removing them from the environment. Pesticide characteristics including dispersion, bioavailability, and soil persistence affect how well this procedure works. Enhancing the pesticides' accessibility to bacteria is essential for efficient breakdown, but this can be hampered by their limited water solubility and high binding to soil particles (Raffa and Chiampo, 2021). In this process enzymes including hydrolases, peroxidases, and oxygenases are essential to various biotransformation systems because they catalyse and regulate the underlying biochemical events (Khajezadeh et al., 2020). Pesticides often go through three different stages of breakdown, including oxidation, reduction, or hydrolysis processes in the first phase, which results in metabolites that are usually less toxic and more water-soluble. The metabolites produced are further transformed into substances like sugars and amino acids in the second phase, which show improved solubility and decreased toxicity. These metabolites are converted into secondary conjugates with little toxicity in the third phase (Cheng et al., 2022).

Bacteria and fungi, which can release internal or extracellular enzymes, are the main microorganisms in charge of these processes. The degradation rate is a crucial factor in bioremediation planning and is often calculated using a first-order kinetic model that takes the starting and ultimate pollutant concentrations into account. This strategy, however, has significant drawbacks because a variety of biological and environmental parameters, such as soil temperature, microbial activity, moisture content, pesticide availability, and leaching dynamics, affect the pace of degradation (Raffa and Chiampo, 2021; Chia et al., 2024).

5.2. Microbial Strategies for the Remediation of Heavy Metal- and Hydrocarbon-Contaminated Soils

Traditional methods for cleaning up heavy metal contaminated soils include chemical, biological, and physical techniques, which are occasionally used in combination. Even while these methods can be very successful, many of them have serious disadvantages, such as causing environmental disruption, requiring a lot of energy and resources, and being expensive (Ramezani et al., 2021). Through the action of their produced enzymes, which change target molecules into less toxic forms while producing energy or other advantageous metabolites, microbes are used in microbial remediation to detoxify hazardous chemicals. Numerous microbiological species, such as bacteria, protozoa, and fungi, have shown the ability to extract significant amounts of heavy metals from soil. Through a variety of metabolic and physical processes, bacteria improve the bioavailability and mobility of HMs in soils. These include lowering the pH of the soil, releasing chemicals that promote plant growth (like hormones), and producing metal-chelating agents like organic acids, siderophores, and biosurfactants. All these things work together to change the redox conditions of the soil and make remediation easier (Pehlivan et al., 2021; Borozan et al., 2021; Ramezani et al., 2021).

Although a variety of macro- and microorganisms can be used in hydrocarbon bioremediation, bacteria are especially useful because of their widespread presence and capacity to break down a variety of hydrocarbon contaminants (Kebede et al., 2021). Both in situ and ex situ cleanup techniques employ these microbes. Using methods including bioventing, biostimulation, bioaugmentation, and natural attenuation, in situ bioremediation treats pollutants on the spot with little disturbance. Ex situ options, on the other hand, call for the removal of contaminated material from the site for destruction or treatment using techniques including composting, incineration, landfilling, excavation, biopiles, and bioreactors. In general, the process of hydrocarbon bioremediation can be divided into three main approaches: Bioaugmentation, biostimulation, and natural attenuation (Chaudhary and Kim, 2019).

Natural attenuation, also known as natural remediation, is a passive, economical, but frequently slow process that uses the innate metabolic activity of local microbial communities to change dangerous hydrocarbons into less or non-toxic forms. By boosting growth-limiting elements like nutrients, vitamins, and oxygen and optimizing environmental factors like temperature, moisture, pH, redox potential, and terminal electron acceptors, biostimulation increases this innate ability and boosts the metabolic efficiency of naturally occurring hydrocarbon-degrading microbes. On the other hand, bioaugmentation speeds up the breakdown process by introducing particular microbial strains or consortia that have been isolated from other contaminated settings or genetically modified (Ławniczak et al., 2020; Das and Chandran, 2011). The limitations and advantages of each approach are listed in Table 1.

Table 1. Advantages and limitations of bioremediation strategies.

	Advantages	Limitations
Natural attenuation	Low-cost, Eco friendly, and minimal intervention. less invasive (little secondary waste; minimal surface infrastructure).	Slow and may be less effective in sites that are heavily contaminated. Not appropriate for all pollutants, such as some environments' resistant chlorinated solvents or situations in which exposure paths are uncontrollable.
Biostimulation	Effective when native microbes are present. Boost natural processes. Improves on current degraders without bringing in organisms; it is often scalable in situ. Can be tailored to soils or groundwater and is more affordable than physical or chemical treatments. Success with hydrocarbons and certain chlorinated solvents has been shown when geochemistry is optimized.	Monitoring and optimization of conditions is required. Geochemical regulation (pH, redox, electron donors/acceptors, nutrients) has a significant impact on performance; incorrect dosage can halt or reroute routes.
Bioaugmentation	Effective when native microbes are insufficient. Speeds up cleanup by introducing high-activity degraders to replace absent or inadequate native processes.	Costly, Potential ecological risks. Persistence and performance can be restricted by survival, establishment, and competition with natural microbiota; these factors are frequently site-specific. There are practical constraints for real-scale applications, and effectiveness is limited for metals and other non-biodegradable contaminants.

6. CONCLUSION

In summary, soil continues to be an essential component of terrestrial ecosystems, but contaminants like pesticides, hydrocarbons, and heavy metals are putting increasing strain on it. These pollutants endanger not only the quality of the soil but also the larger environment and human health, and they are frequently made worse by human activities. By bioremediation, nutrient cycling, and the decomposition of organic matter, microorganisms both sensitive responders and active players in soil processes play a crucial role in reducing these effects. The role of microbes as bio-indicators underscores the importance of preserving their diversity and functional capacity to sustain overall soil health. Safeguarding this delicate balance is vital for maintaining the ecological integrity of soils and their ability to support life. Microbes change or immobilize pollutants by enzymatic and physicochemical processes, including oxidation, hydrolysis, reductive dehalogenation, biosorption, and bioaccumulation, which lowers pollutants' toxicity and bioavailability. Degradation and detoxification are further improved by oxidative enzymes such as laccases, monooxygenases, and dioxygenases as well as bioelectrochemical techniques employing redox mediators. With continuous advancements enhancing effectiveness and usability, these microbial technologies collectively offer long-term solutions for industrial applications, pharmaceutical metabolism, and environmental cleanup.

CONFLICTS OF INTEREST

The authors declare that there are no conflicts of interest regarding the publication of this paper.

ACKNOWLEDGMENTS

Special appreciation to the South African Medical Research Council, the National Research foundation (NRF), the South African Coal, Oil and Gas Corporation (SASOL) and the University of Fort Hare, South Africa for financial support.

REFERENCES

Abdel-Shafy, H. I., & Mansour, M. S. (2016). A review on polycyclic aromatic hydrocarbons: source, environmental impact, effect on human health and remediation. *Egyptian journal of petroleum, 25*(1), 107-123.

Abdelhafeez, I. A., El-Tohamy, S. A., Abdul-Malik, M. A., Abdel-Raheem, S. A., & El-Dars, F. M. S. (2022). A review on green remediation techniques for hydrocarbons and heavy metals contaminated soil. *Curr. Chem. Lett, 11*(1), 43–62. DOI: 10.5267/j.ccl.2021.9.006

Adams, G. O., Fufeyin, P. T., Okoro, S. E., & Ehinomen, I. (2015). Bioremediation, biostimulation and bioaugmention: A review. *International Journal of Environmental Bioremediation & Biodegradation, 3*(1), 28–39. DOI: 10.12691/ijebb-3-1-5

Alotaibi, F., Hijri, M., & St-Arnaud, M. (2021). Overview of approaches to improve rhizoremediation of petroleum hydrocarbon-contaminated soils. *Applied Microbiology (Basel, Switzerland), 1*(2), 329–351. DOI: 10.3390/applmicrobiol1020023

Angon, P. B., Islam, M. S., Das, A., Anjum, N., Poudel, A., & Suchi, S. A. (2024). Sources, effects and present perspectives of heavy metals contamination: Soil, plants and human food chain. *Heliyon, 10*(7), e28357. DOI: 10.1016/j.heliyon.2024.e28357 PMID: 38590838

Arora, J., Kumari, A., Ranjan, A., Rajput, V. D., Shende, S., Prazdnova, E. V. E., & Jindal, T. (2024). Microbial adaptation in different extreme environmental conditions and its usefulness in differently polluted soil. In *Extremophiles for Sustainable Agriculture and Soil Health Improvement* (pp. 47–62). Springer Nature Switzerland. DOI: 10.1007/978-3-031-70203-7_3

Aza, P., & Camarero, S. (2023). Fungal laccases: Fundamentals, engineering and classification update. *Biomolecules, 13*(12), 1716. DOI: 10.3390/biom13121716 PMID: 38136587

Azhar, U., Ahmad, H., Shafqat, H., Babar, M., Munir, H. M. S., Sagir, M., & Khoo, K. S. (2022). Remediation techniques for elimination of heavy metal pollutants from soil: A review. *Environmental Research, 214*, 113918. DOI: 10.1016/j.envres.2022.113918 PMID: 35926577

Bokade, P., Gaur, V. K., Tripathi, V., Bobate, S., Manickam, N., & Bajaj, A. (2023). Bacterial remediation of pesticide polluted soils: Exploring the feasibility of site restoration. *Journal of Hazardous Materials, 441*, 129906. DOI: 10.1016/j.jhazmat.2022.129906 PMID: 36088882

Borozan, A. B., Misca, C. D., Morar, A., Obistioiu, D., Raba, D. N., Pirvulescu, L., Caba, I. L., Alexa, E., Poiana, M. A., Bordean, D. M., & Petcu, C. D. (2021). Soil pollution with heavy metals and bioremediation methods. *AgroLife Scientific Journal, 10*(1), 54–66. DOI: 10.17930/AGL202115

Boyd, D. R., Sharma, N. D., Stevenson, P. J., Hoering, P., Allen, C. C., & Dansette, P. M. (2022). Monooxygenase-and dioxygenase-catalyzed oxidative dearomatization of thiophenes by sulfoxidation, cis-dihydroxylation and epoxidation. *International Journal of Molecular Sciences, 23*(2), 909. DOI: 10.3390/ijms23020909 PMID: 35055091

Cai, Z., Wang, J., Ma, J., Zhu, X., Cai, J., & Yang, G. (2015). Anaerobic degradation pathway of the novel chiral insecticide paichongding and its impact on bacterial communities in soils. *Journal of Agricultural and Food Chemistry, 63*(32), 7151–7160. DOI: 10.1021/acs.jafc.5b02645 PMID: 26216485

Chaudhary, D. K., & Kim, J. (2019). New insights into bioremediation strategies for oil-contaminated soil in cold environments. *International Biodeterioration & Biodegradation, 142*, 58–72. DOI: 10.1016/j.ibiod.2019.05.001

Chávez-Vergara, B., Solleiro-Rebolledo, E., López-Martínez, R., Beltrán-Paz, O., Ceniceros-Gómez, Á. E., & Yañez-Mendoza, G. (2025). The release of arsenic is a hidden risk during the in-situ decomposition of landed sargassum litter. *Aquatic Botany, 199*, 103884. DOI: 10.1016/j.aquabot.2025.103884

Chen, L., Zhou, M., Wang, J., Zhang, Z., Duan, C., Wang, X., Zhao, S., Bai, X., Li, Z., Li, Z., & Fang, L. (2022). A global meta-analysis of heavy metal (loid) s pollution in soils near copper mines: Evaluation of pollution level and probabilistic health risks. *The Science of the Total Environment, 835*, 155441. DOI: 10.1016/j.scitotenv.2022.155441 PMID: 35469881

Chen, W., Zeng, F., Liu, W., Bu, J., Hu, G., Xie, S., Yao, H., Zhou, H., Qi, S., & Huang, H. (2021). Organochlorine pesticides in karst soil: Levels, distribution, and source diagnosis. *International Journal of Environmental Research and Public Health*, *18*(21), 11589. DOI: 10.3390/ijerph182111589 PMID: 34770109

Chen, X., Zhao, Y., Zhang, C., Zhang, D., Yao, C., Meng, Q., Zhao, R., & Wei, Z. (2020). Speciation, toxicity mechanism and remediation ways of heavy metals during composting: A novel theoretical microbial remediation method is proposed. *Journal of Environmental Management*, *272*, 111109. DOI: 10.1016/j.jenvman.2020.111109 PMID: 32854897

Cheng, M., Chen, D., Parales, R. E., & Jiang, J. (2022). Oxygenases as powerful weapons in the microbial degradation of pesticides. *Annual Review of Microbiology*, *76*(1), 325–348. DOI: 10.1146/annurev-micro-041320-091758 PMID: 35650666

Chia, X. K., Hadibarata, T., Kristanti, R. A., Jusoh, M. N. H., Tan, I. S., & Foo, H. C. Y. (2024). The function of microbial enzymes in breaking down soil contaminated with pesticides: A review. *Bioprocess and Biosystems Engineering*, *47*(5), 597–620. DOI: 10.1007/s00449-024-02978-6 PMID: 38456898

Das, N., & Chandran, P. (2011). Microbial degradation of petroleum hydrocarbon contaminants: An overview. *Biotechnology Research International*, *2011*(1), 941810. DOI: 10.4061/2011/941810 PMID: 21350672

Eberwein, M., Hellmold, N., Frank, R., Deobald, D., & Adrian, L. (2024). Reductive dehalogenase of Dehalococcoides mccartyi strain CBDB1 reduces cobalt-containing metal complexes enabling anodic respiration. *Frontiers in Microbiology*, *15*, 1457014. DOI: 10.3389/fmicb.2024.1457014 PMID: 39507334

Egbe, C. C., Oyetibo, G. O., & Ilori, M. O. (2021). Ecological impact of organochlorine pesticides consortium on autochthonous microbial community in agricultural soil. *Ecotoxicology and Environmental Safety*, *207*, 111319. DOI: 10.1016/j.ecoenv.2020.111319 PMID: 32947214

Fenik, J., Tankiewicz, M., & Biziuk, M. (2011). Properties and determination of pesticides in fruits and vegetables. *Trends in Analytical Chemistry*, *30*(6), 814–826. DOI: 10.1016/j.trac.2011.02.008

Gao, J., Han, H., Gao, C., Wang, Y., Dong, B., & Xu, Z. (2023). Organic amendments for in situ immobilization of heavy metals in soil: A review. *Chemosphere*, *335*, 139088. DOI: 10.1016/j.chemosphere.2023.139088 PMID: 37268229

Gerber, R., Smit, N. J., Van Vuren, J. H., Nakayama, S. M., Yohannes, Y. B., Ikenaka, Y., Ishizuka, M., & Wepener, V. (2016). Bioaccumulation and human health risk assessment of DDT and other organochlorine pesticides in an apex aquatic predator from a premier conservation area. *The Science of the Total Environment*, *550*, 522–533. DOI: 10.1016/j.scitotenv.2016.01.129 PMID: 26845188

Hazaimeh, M., Kanaan, B. M., AlFaleh, F. A., Elhaig, M. M., Khamaiseh, E. I., Zia, Q., Alaidarous, M., Seth, C. S., Alsowayeh, N., & Ahmad, F. (2024). Biodegradation of petroleum hydrocarbons using a novel bacterial strain isolated from hydrocarbons contaminated soil of Saudi Arabia. *Biocatalysis and Agricultural Biotechnology*, *57*, 103074. DOI: 10.1016/j.bcab.2024.103074

Hlihor, R. M., & Cozma, P. (2023). Microbial bioremediation of environmental pollution. *Processes (Basel, Switzerland)*, *11*(5), 1543. DOI: 10.3390/pr11051543

Iqbal, S., Xu, J., Allen, S. D., Khan, S., Nadir, S., Arif, M. S., & Yasmeen, T. (2020). Unraveling consequences of soil micro-and nano-plastic pollution on soil-plant system: Implications for nitrogen (N) cycling and soil microbial activity. *Chemosphere*, *260*, 127578. DOI: 10.1016/j.chemosphere.2020.127578 PMID: 32683024

Jaga, K., & Dharmani, C. (2003). Sources of exposure to and public health implications of organophosphate pesticides. *Revista Panamericana de Salud Pública = Pan American Journal of Public Health*, *14*(3), 171–185. DOI: 10.1590/S1020-49892003000800004 PMID: 14653904

Jaipieam, S., Visuthismajarn, P., Sutheravut, P., Siriwong, W., Thoumsang, S., Borjan, M., & Robson, M. (2009). Organophosphate pesticide residues in drinking water from artesian wells and health risk assessment of agricultural communities, Thailand. *Human and Ecological Risk Assessment*, *15*(6), 1304–1316. DOI: 10.1080/10807030903306984 PMID: 20485459

Jia, J., de Goede, R., Li, Y., Zhang, J., Wang, G., Zhang, J., & Creamer, R. (2025). Unlocking soil health: Are microbial functional genes effective indicators? *Soil Biology & Biochemistry*, *204*, 109768. DOI: 10.1016/j.soilbio.2025.109768

Joos, L., & De Tender, C. (2022). Soil under stress: The importance of soil life and how it is influenced by (micro) plastic pollution. *Computational and Structural Biotechnology Journal*, *20*, 1554–1566. DOI: 10.1016/j.csbj.2022.03.041 PMID: 35422972

Kafaei, R., Arfaeinia, H., Savari, A., Mahmoodi, M., Rezaei, M., Rayani, M., Sorial, G. A., Fattahi, N., & Ramavandi, B. (2020). Organochlorine pesticides contamination in agricultural soils of southern Iran. *Chemosphere*, *240*, 124983. DOI: 10.1016/j.chemosphere.2019.124983 PMID: 31726590

Kalia, A., & Gosal, S. K. (2011). Effect of pesticide application on soil microorganisms. *Archives of Agronomy and Soil Science*, *57*(6), 569–596. DOI: 10.1080/03650341003787582

Kaushal, J., Khatri, M., & Arya, S. K. (2021). A treatise on Organophosphate pesticide pollution: Current strategies and advancements in their environmental degradation and elimination. *Ecotoxicology and Environmental Safety*, *207*, 111483. DOI: 10.1016/j.ecoenv.2020.111483 PMID: 33120277

Kebede, G., Tafese, T., Abda, E. M., Kamaraj, M., & Assefa, F. (2021). Factors influencing the bacterial bioremediation of hydrocarbon contaminants in the soil: Mechanisms and impacts. *Journal of Chemistry*, *2021*(1), 9823362. DOI: 10.1155/2021/9823362

Kensa, V. M. (2011). Bioremediation-an overview. *Journal of Industrial Pollution Control*, *27*(2), 161–168.

Khajezadeh, M., Abbaszadeh-Goudarzi, K., Pourghadamyari, H., & Kafilzadeh, F. (2020). A newly isolated Streptomyces rimosus strain capable of degrading deltamethrin as a pesticide in agricultural soil. *Journal of Basic Microbiology*, *60*(5), 435–443. DOI: 10.1002/jobm.201900263 PMID: 32128846

Kwon, J. H., Ji, M. K., Kumar, R., Islam, M. M., Khan, M. A., Park, Y. K., Yadav, K. K., Vaziri, R., Hwang, J.-H., Lee, W. H., Ahn, Y.-T., & Jeon, B. H. (2023). Recent advancement in enhanced soil flushing for remediation of petroleum hydrocarbon-contaminated soil: A state-of-the-art review. *Reviews in Environmental Science and Biotechnology*, *22*(3), 679–714. DOI: 10.1007/s11157-023-09657-0

Lawal, A. T. (2017). Polycyclic aromatic hydrocarbons. A review. *Cogent Environmental Science*, *3*(1), 1339841. DOI: 10.1080/23311843.2017.1339841

Ławniczak, Ł., Woźniak-Karczewska, M., Loibner, A. P., Heipieper, H. J., & Chrzanowski, Ł. (2020). Microbial degradation of hydrocarbons—basic principles for bioremediation: A review. *Molecules (Basel, Switzerland)*, *25*(4), 856. DOI: 10.3390/molecules25040856 PMID: 32075198

Liu, L., Li, W., Song, W., & Guo, M. (2018). Remediation techniques for heavy metal-contaminated soils: Principles and applicability. *The Science of the Total Environment*, *633*, 206–219. DOI: 10.1016/j.scitotenv.2018.03.161 PMID: 29573687

Mayans, B., Antón-Herrero, R., García-Delgado, C., Delgado-Moreno, L., Guirado, M., Pérez-Esteban, J., Escolástico, C., & Eymar, E. (2024). Bioremediation of petroleum hydrocarbons polluted soil by spent mushroom substrates: Microbiological structure and functionality. *Journal of Hazardous Materials, 473*, 134650. DOI: 10.1016/j.jhazmat.2024.134650 PMID: 38776816

Mdeni, N. L., Adeniji, A. O., Okoh, A. I., & Okoh, O. O. (2022). Analytical evaluation of carbamate and organophosphate pesticides in human and environmental matrices: A review. *Molecules (Basel, Switzerland), 27*(3), 618. DOI: 10.3390/molecules27030618 PMID: 35163876

Mekonnen, B. A., Aragaw, T. A., & Genet, M. B. (2024). Bioremediation of petroleum hydrocarbon contaminated soil: A review on principles, degradation mechanisms, and advancements. *Frontiers in Environmental Science, 12*, 1354422. DOI: 10.3389/fenvs.2024.1354422

Mustapha, M. U., & Halimoon, N. (2019). An overview on biodegradation of carbamate pesticides by soil bacteria. *Pertanika Journal of Science & Technology, 27*(2).

Nasrabadi, T., Nabi Bidhendi, G., Karbassi, A., Grathwohl, P., & Mehrdadi, N. (2011). Impact of major organophosphate pesticides used in agriculture to surface water and sediment quality (Southern Caspian Sea basin, Haraz River). *Environmental Earth Sciences, 63*(4), 873–883. DOI: 10.1007/s12665-010-0757-2

Padilla-Garfias, F., Araiza-Villanueva, M., Calahorra, M., Sánchez, N. S., & Peña, A. (2024). Advances in the degradation of polycyclic aromatic hydrocarbons by yeasts: A review. *Microorganisms, 12*(12), 2484. DOI: 10.3390/microorganisms12122484 PMID: 39770687

Pande, V., Pandey, S. C., Sati, D., Bhatt, P., & Samant, M. (2022). Microbial interventions in bioremediation of heavy metal contaminants in agroecosystem. *Frontiers in Microbiology, 13*, 824084. DOI: 10.3389/fmicb.2022.824084 PMID: 35602036

Pehlivan, N., Gedik, K., Eltem, R., & Terzi, E. (2021). Dynamic interactions of Trichoderma harzianum TS 143 from an old mining site in Turkey for potent metal (oid) s phytoextraction and bioenergy crop farming. *Journal of Hazardous Materials, 403*, 123609. DOI: 10.1016/j.jhazmat.2020.123609 PMID: 32798794

Pouresmaieli, M., Ataei, M., Forouzandeh, P., Azizollahi, P., & Mahmoudifard, M. (2022). Recent progress on sustainable phytoremediation of heavy metals from soil. *Journal of Environmental Chemical Engineering, 10*(5), 108482. DOI: 10.1016/j.jece.2022.108482

Priya, A. K., Gnanasekaran, L., Dutta, K., Rajendran, S., Balakrishnan, D., & Soto-Moscoso, M. (2022). Biosorption of heavy metals by microorganisms: Evaluation of different underlying mechanisms. *Chemosphere, 307*, 135957. DOI: 10.1016/j.chemosphere.2022.135957 PMID: 35985378

Quinn, L. P., De Vos, B. J., Fernandes-Whaley, M., Roos, C., Bouwman, H., Kylin, H., & van den Berg, J. (2011). Pesticide use in South Africa: One of the largest importers of pesticides in Africa. *Pesticides in the Modern World-Pesticides Use and Management, 4*, 50–95. DOI: 10.5772/16995

Raffa, C. M., & Chiampo, F. (2021). Bioremediation of agricultural soils polluted with pesticides: A review. *Bioengineering (Basel, Switzerland), 8*(7), 92. DOI: 10.3390/bioengineering8070092 PMID: 34356199

Ramezani, M., Enayati, M., Ramezani, M., & Ghorbani, A. (2021). A study of different strategical views into heavy metal (oid) removal in the environment. *Arabian Journal of Geosciences, 14*(21), 2225. DOI: 10.1007/s12517-021-08572-4

Ren, L., Zhang, J., Geng, B., Zhao, J., Jia, W., & Cheng, L. (2025). Ecological Shifts and Functional Adaptations of Soil Microbial Communities Under Petroleum Hydrocarbon Contamination. *Water (Basel), 17*(8), 1216. DOI: 10.3390/w17081216

Saeed, M. U., Hussain, N., Sumrin, A., Shahbaz, A., Noor, S., Bilal, M., Aleya, L., & Iqbal, H. M. (2022). Microbial bioremediation strategies with wastewater treatment potentialities–A review. *The Science of the Total Environment, 818*, 151754. DOI: 10.1016/j.scitotenv.2021.151754 PMID: 34800451

Sakshi, S., Singh, S. K., & Haritash, A. K. (2019). Polycyclic aromatic hydrocarbons: Soil pollution and remediation. *International Journal of Environmental Science and Technology, 16*(10), 6489–6512. DOI: 10.1007/s13762-019-02414-3

Singha, L. P., & Pandey, P. (2021). Rhizosphere assisted bioengineering approaches for the mitigation of petroleum hydrocarbons contamination in soil. *Critical Reviews in Biotechnology, 41*(5), 749–766. DOI: 10.1080/07388551.2021.1888066 PMID: 33626996

Sun, S., Sidhu, V., Rong, Y., & Zheng, Y. (2018). Pesticide pollution in agricultural soils and sustainable remediation methods: A review. *Current Pollution Reports, 4*(3), 240–250. DOI: 10.1007/s40726-018-0092-x

Varjani, S. J. (2017). Microbial degradation of petroleum hydrocarbons. *Bioresource Technology, 223*, 277–286. DOI: 10.1016/j.biortech.2016.10.037 PMID: 27789112

Wahsha, M., Nadimi-Goki, M., Fornasier, F., Al-Jawasreh, R., Hussein, E. I., & Bini, C. (2017). Microbial enzymes as an early warning management tool for monitoring mining site soils. *Catena, 148*, 40-45.Wang, Z., Luo, P., Zha, X., Xu, C., Kang, S., Zhou, M., ... & Wang, Y. (2022). Overview assessment of risk evaluation and treatment technologies for heavy metal pollution of water and soil. *Journal of Cleaner Production, 379*, 134043.

Wołejko, E., Jabłońska-Trypuć, A., Wydro, U., Butarewicz, A., & Łozowicka, B. (2020). Soil biological activity as an indicator of soil pollution with pesticides–a review. *Applied Soil Ecology*, *147*, 103356. DOI: 10.1016/j.apsoil.2019.09.006

Yadav, S., Kumar, S., & Haritash, A. K. (2023). A comprehensive review of chlorophenols: Fate, toxicology and its treatment. *Journal of Environmental Management*, *342*, 118254. DOI: 10.1016/j.jenvman.2023.118254 PMID: 37295147

Yang, Y., Zhang, Z. W., Liu, R. X., Ju, H. Y., Bian, X. K., Zhang, W. Z., Zhang, C.-B., Yang, T., Guo, B., Xiao, C.-L., Bai, H., & Lu, W. Y. (2021). Research progress in bioremediation of petroleum pollution. *Environmental Science and Pollution Research International*, *28*(34), 46877–46893. DOI: 10.1007/s11356-021-15310-6 PMID: 34254241

Zhichkina, L., Nosov, V., Zhichkin, K., Zhenzhebir, V., Abramov, Y., & Alborova, M. (2020). Pesticide monitoring of agricultural soil pollution. In *E3S web of conferences* (Vol. 193, p. 01068). EDP Sciences.

Chapter 5

Microbial Responses to Aquatic Pollution: Biotransformation of Oil, Plastics, and Eutrophic Contaminants

Kingsley Ehi Ebomah
http://orcid.org/0000-0002-8192-0428
University of Fort Hare, South Africa

Anthony Ifeanyi Okoh
University of Fort Hare, South Africa

ABSTRACT

Aquatic pollution is characterized by contaminants in waterbodies mostly because of microorganisms, oil, plastics and eutrophic contaminants occurring in components with rapid dissemination of both soluble and insoluble agents. Microbial responses to pollution, the role they play in biotransformation and bioremediation are indeed paramount. The purpose of this chapter is to introduce the primary microbial responses to oil, plastic and eutrophic contaminants. We explore their potential effects on the environment, highlighting key microbes involved in biotransformation and pollutant degradation. Enzymatic mechanisms and biochemical pathways are also discussed. The global challenges faced by ecosystems are reviewed and potential solutions are examined. The region is vulnerable to environmental abuse raising red flags in the efforts to heal wounds inflicted on the ecosystems. Finally, we discussed possible applications of biotransformation processes, providing insights to develop new ways of combating this problem, shedding light on various research directions and challenges.

DOI: 10.4018/979-8-2600-2100-2.ch005

1 INTRODUCTION

According to Don-Pedro (1990), pollution is the creation and/or introduction of materials or energy into the environment by humans, either directly or indirectly, that has a negative impact on all living organisms, or obstructed facilities and other environmental uses. Environmental pollution is one of the biggest issues facing people worldwide, especially in developing nations, and environmental pollutants (EPs) can also be referred to as environmental contaminants. This negative impact on the environment and resources of humans can be due to the production and activities of our existence as living beings. Anthropogenic activities including agricultural runoff, improper waste disposal, industrial discharge and urban wastewater (WW) may lead to the discharge of contaminants into aquatic habitats, posing health risks and reducing water quality (Chawla et al., 2024). Generally, pollution can be categorized mainly into physical, chemical and biological (Sharma and Kashyap, 2022). Pollution in the aquatic environment has become one of the most pressing challenges in the world today. Contaminants or pollutants in the available water resources have negatively impacted animal husbandry and public health for decades (Ahmed et al., 2016; Osunla and Okoh, 2017).

Alternatively, in terms of water components, microbial pollutants or contaminants (bacteria, viruses and parasites) are predominantly from various sources including sewage and animal feces; chemical (pesticides, heavy metals) from industrial discharge and Agric run-off, or physical (sediment, trash and algae contaminants) from soil and organic materials. Water pollution can be categorized in several ways. Broadly, it can be classified by the type of pollutant, the source of pollution, and the affected water body. Other categories of pollutants include radiological and thermal.

Pollution can occur at several stages including industrial activities, agricultural runoff, sewage and wastewater treatment, and even from everyday household activities (Madhav et al., 2019). It can also happen through oil spills, mining operations, and the discharge of heated water from industrial processes. Additionally, climate change and the disposal of solid waste contribute to water contamination. There have been several reports about environmental pollution resulting in contamination of surface waters caused by artisanal mining, which in turn affects drinking water sources leading to serious public health issues (Arthur-Holmes et al., 2022). Artisanal miners use highly toxic substances including mercury and other harmful chemicals that are detrimental also to aquatic life. Furthermore, the build-up of excess sludge may cause difficulties for certain species to survive.

The purpose of this chapter is to introduce the primary microbial responses to oil, plastic and eutrophic pollutants. We explore their probable effects of pollution on the existing waterbodies, discussing the notable vulnerabilities, briefly highlighting key microbes involved in oil and plastic biotransformation, nutrient cycling and pollutant degradation. There have been numerous reports and articles investigating risks and consequences of water pollution (WP) for decades. Below are some relevant terminologies related to repercussions because of aquatic pollution.

- **Biodiversity Loss:** WP leads to the deterioration of various aquatic species, including amphibians, fish, and invertebrates, exhausting ecological resilience.
- **Habitation Destruction:** Pollutants alter the physical and chemical properties of aquatic environs, reducing the availability of suitable habitats for many species.
- **Food Chain Disruption:** Pollutants amass in the food chain, affecting organisms at all levels (including humans who consume contaminated seafoods).
- **Public Health Risks:** Polluted water can cause waterborne infectious diseases and long-term health problems for humans who rely on it for domestic purposes.
- **Economic Consequences:** Pollution negatively affects sectors like agriculture, fisheries, and tourism, resulting in reduced crop yields, economic losses and fishery declines.
- **Ecological Alteration:** WP can alter the balance of aquatic ecosystems and favor certain species thereby, possibly causing bellicose species outcompeting native populations.
- **Oceanic Acidification:** The absorption of carbon dioxide (CO_2) by the ocean can lead to amplified acidity, making it tougher for shellfish and coral to live.
- **Plastic Pollution:** Oceanic debris, including plastic, can starve, strangle, and suffocate sea animals, and much of it ends up in floating garbage patches.

Air, water, and soil are highly contaminated by chemicals from agriculture, use of fossil fuels, household and industrial waste, and mining. The value of nature, public health and safety may all be significantly impacted by contaminated substances. Gases including carbon dioxide (CO_2), methane (CH_4), nitrous oxide (N_2O), nitrogen oxides (NO_2), hydrofluorocarbons (HFCs), perfluorocarbons (PFCs) and sulfur hexafluoride (SF_6) contribute to climate change in increased temperature of the earth's atmosphere and oceans, thus, these can be categorized as examples of air pollutants.

Unpredictable disparities in precipitation, sea level rise, and extreme weather events are all caused by climate change. By enacting legislation prohibiting the production of dangerous chemicals, using specialized equipment that lowers greenhouse gas emissions, and utilizing alternative energy sources, air pollution can be managed. Sewage, petroleum products (such as polychlorinated biphenyls), nitrates, pesticides, sediments, and surplus organic matter are among the main pollutants of water. Industrial pipe outlets, storage tank or pipeline leaks, mining activities, incorrect fertilizer and pesticide application in agricultural fields, and some ship leaks can all release hazardous chemicals into water bodies.

Contaminants in water harm the neurological system, important body organs, and cause various malignancies and cardiovascular problems (Bashir et al., 2020). The spread of harmful substances, acidification, salinization, and sodification, increased soil erosion, chemical fertilizers, pesticides, fungicides, and the buildup of heavy metals and other inorganic pollutants are all factors that contribute to soil pollution. If soil pollutants get into our bodies, they can cause a variety of illnesses, much as pollutants in the air and water. There are numerous concrete methods for managing soil pollutants, including recycling, composting, incineration, and landfills.

2 COMMON MICROBES ASSOCIATED WITH ENVIRONMENTAL POLLUTANTS

Microorganisms such as bacteria (*Acinetobacter* spp., *Bacillus*, *Pseudomonas* species) and fungi (*Aspergillus* spp., *Fusarium*) are associated with EPs, either by indicating pollution, for instance the fecal coliforms and *E. coli* in aquatic environment, or by bioremediating them, using their metabolic proficiencies to degrade organic and inorganic pollutants (Sutar et al., 2018). Additionally, some microbes can be negatively affected by contaminants, which can lead to shifts in their ecosystem structure and functions, as seen in the pharyngeal microbiota of individuals in polluted areas (Cundell, 2018; Wypych et al., 2019).

Certain microbes respond to pollutants by degrading them into less harmful products or by using them as certain food sources through "Bioremediation", it has been widely reported that these responses can serve as bioindicators which can signal the well-being of the ecosystem (Sagova-Mareckova et al., 2021). Microorganisms respond to pollutants by degrading them into less harmful products or by using them as certain food sources through "Bioremediation", it has been widely reported that these responses can serve as bioindicators which can signal the well-being of the ecosystem (Sagova-Mareckova et al., 2021). Aquatic contamination (AC) is a vital component and an extension of environmental pollution (EP).

Several variations in microbial species have led to ecological migration and these discrepancies are not far-fetched from being factors leading to potentially harmful elements of contamination. **Table 1** shows a summary of common microorganisms linked to pollutants and the roles they play. It is rare to observe the absence of pathogenic microbes in every habitat; hence pathogens that cause series of infections always thrive predominantly in environments that are considered inhospitable (Alabbosh, 2025). Notwithstanding, many ecological niches benefit from the enzymatic products, natural mechanisms and mechanistic pathways produced during the short life cycles of these creatures. Many regions across the world have been harnessing the unique abilities of certain microorganisms and several studies have investigated various metabolic processes involved to enable humanity benefit while limiting the losses. Specifically, *Acinetobacter* species and *Bacillus subtilis* have shown high efficiency in degrading complex compounds.

Table 1. Common microbes associated with relevant pollutants and their respective roles

Microbial species	Roles in pollution	Types of pollutants	References
E. coli and other coliforms, *Streptococcus, Vibrio, Pseudomonas* can act as indicators of overall water and environmental quality	Indicators of pollution	Fecal	Sumampouw and Risjani, 2014
Acinetobacter, Pseudomonas, Arthrobacter, Burkholderia, and *Rhodococcus* can break down complex organic compounds	Bioremediation (Degrading Pollutants)	Organic Pollutants	Reineke and Schlömann, 2023
Alcanivorax species and *Pseudomonas* spp. are hydrocarbon-degrading bacteria	Oil Biotransformation and bioremediation	Oil pollutants	Harayama et al., 2004
Bacillus subtilis, Myrothecium roridum, and *Aspergillus tamarii* have shown effectiveness in degrading industrial dyes	Bioremediation (Degrading Pollutants)	Industrial Dyes	Maglione et al., 2024
Microbacterium profundis, for example, is used in cadmium bioremediation, and can help in removing metals from contaminated sites	Bioremediation (Degrading Pollutants)	Heavy Metals	Khan et al., 2025
Arthrobacter, Bacillus, Pseudomonas (Microbes living in the root zone of plants (rhizosphere))	Rhizoremediation	Pollutants from soil and air	Bhadrecha et al., 2023

Other notable microorganisms have shown tenacity in dealing with inorganic materials and many reports have outlined how microbes harbor genes for sustenance in harsh environments (Froelich and Noble, 2014; Arora et al., 2024). Once exhibited, there can be a trigger of rapid gene transfer to other non-pathogenic strains. The environment should be perceived as a vehicle as well as a channel, it can also act as a platform for all organisms to thrive in favorable conditions that suit their metabolism and enzymatic pathways. From the emergence of novel microbes leading to unprecedented outbreaks to the advent of antimicrobial resistance giving rise to potential epidemics. Contaminants in water can lead to a variety of infections and **Table 2** summarizes the common microorganisms and infections associated with various pollutants.

Studies on microorganisms' responses should be carried out more often as there is still a lot to learn from these amazing creatures that are consistently evolving into species and sub-species, variants and sub-variants subsequently harboring virulence genes (VGs) and antimicrobial resistance genes (ARGs). Existing and co-existing in even harsh environments and extreme climate conditions, the microbial ecosystem is tenaciously ubiquitous, found flourishing in hostile, extreme habitats for instance frozen desert, deep-sea vents, and extremely salinized seas, as a result of specific modifications to their molecular, cellular, and biochemical structures (Sarojini et al., 2025).

Table 2. Polluted water related infections/diseases and their causative agents.

Causative agents	Microbial Species	Infections/Diseases	Potential pollutants	References
Bacteria	*Aeruginosa* spp. *Bacillus* spp. *Campylobacter* spp. *E. coli* (incl. *E. coli* O157:H7) *Helicobacter* spp. *Salmonella* spp. *Vibrio* spp.	Gastroenteritis Bloody diarrhea Abdominal pain, peptic ulcer, gastric cancer Diarrhea, Salmonellosis, Typhoid fever, Cholera	Fecal contaminated food and water	Gerba, 2009
Viruses	*Adenovirus* *Astrovirus* *Calicivirus* *Coronavirus* *Coxsackievirus* *Enerovirus* *Echovirus* *Hepatitis virus* *Norovirus* *Parvovirus* *Poliovirus* *Rotavirus*	Eye and Respiratory Gastroenteritis, aseptic meningitis, respiratory illness, paralysis Fever, rash Hepatitis Aseptic meningitis, Paralysis Gastroenteritis	Contaminated food and water	Bosch et al., 2018
Protozoans	*Cryptosporidium parvum* *Entamoeba histolytica* *Giardia lamblia*	Cryptosporidiosis Amoebic dysentery (Amoebiasis) Giardiasis	Contaminated water sources (like drinking water or swimming areas) and food that has been in contact with human or animal feces.	Di Genova and Tonelli, 2016
Fungi	*Aspergillus flavus*, *Penicillium* species, *Fusarium* species, *Cladosporium* *Stachybotrys* *Trichoderma*	Cancer from aflatoxins, pulmonary hemorrhage, ringworm, candidiasis	Heavy metals, pesticides, dyes, and microplastics	Akhtar and Mannan, 2020

By proving that they can endure and even thrive in these harsh environments, these "extremophiles" help us better understand the boundaries of life on Earth and beyond. **Table 3** represents the routes and sources of AC. Microbial pathogens that originate mainly from human and animal waste frequently travel via domestic and recreational events notwithstanding, organic pollutants have similar origins and routes as industrial pollutants, pesticides and petroleum related products have varying routes and sources.

Table 3. Routes and sources of aquatic contamination (Adapted with modifications from Farrington, 2014).

S/N	Contaminants/Pollutants	Routes	Origin
1.	Microbial pathogens	Domestic and recreational activities (washing, bathing and swimming)	• Human waste • Animal waste
2.	Pesticides and related products	Run-off from farms	• Organic chemicals • Inorganic chemicals
3.	Organic contaminants	Discharged into surrounding waterbodies via drains and gutters	• Human waste • Animal waste
4.	Industrial pollutants	Effluent discharges	• Food and beverages industries • Tobacco, soap and detergents industries • Textile factories
5.	Petroleum related products	Refining, exploration of crude oil, broken pipelines and transportation	• Crude oil accidental discharges Effluents from refineries (grease and other suspended solids) • Drilling remnants
6.	Nutrients from crops (Phosphate and nitrate)	Run-offs	• Fertilizers from farmlands

3 MICROBIAL RESPONSES TO OIL POLLUTION

The discharge of liquid petroleum hydrocarbons into the environmental settings, primarily in the "marine ecosystem," because of human activities is known as "oil spill." Nevertheless, petroleum hydrocarbons harm both aquatic and terrestrial organisms and are thus considered toxic to all forms of life. Researchers and environmentalists have taken notice of the contamination of marine habitats for decades. Oil pollution (OP) or oil spill (OS) is the release of oil (particularly liquid petroleum hydrocarbon) into the environment, predominantly the marine ecosystem, which includes flora and fauna (Thakur and Koul, 2022). Depending on the size and location of the oil spill, the damage might range from minor to severe (Schwacke et al., 2014). Abundant oil spills have a significant negative impact on aquatic organisms' life activities including breathing, feeding, spawning, and thermoregulation (Takeshita et al., 2021). Oil spillage on land and water have been established to be fatal to humans, fish, birds, cattle, and wildlife generally owing to mainly direct contact with the spilled oil and intake of polluted foods. OP is fatal to fish and smaller organisms or reduce their quality which are critical links in the global food chain, thus, resulting in economic losses for farmers and the government (Osuagwu and Olaifa, 2018; Onyena and Sam, 2020).

Immediate spill response with strategies including booms, skimmers, and dispersants is one way to mitigate the effects of OP on aquatic life and the economy and in a recent study by Ejiohuo et al. (2025), it was discussed that long-term approaches such as cutting boosting oil recovery, energy use, and refining oil spill trajectory models have been implemented. It has been widely reported that oil-degrading microbes break down hydrocarbons involving numerous reactions and mechanisms in oil spillage aquatic environs and in a study by Kleindienst et al. (2015), they investigated the physiological capacities of the microbes and stated controversially that the hydrocarbon class are the primary determinants of the pace of hydrocarbon biodegradation. The authors further argued that aerobic processes are generally more rapid than anaerobic processes, and low-molecular-weight hydrocarbons are degraded at higher rates than high-molecular-weight hydrocarbons. **Figure 1** illustrates the link between community, industry and pollution of aquatic environments showing how plastic disposed improperly, and oil spilled subsequently seepage into the surface waterbody which serves as domestic and recreational use.

Figure 1. A flow diagram depicting the link between community, industry and oil/ microplastic pollution of the aquatic environment

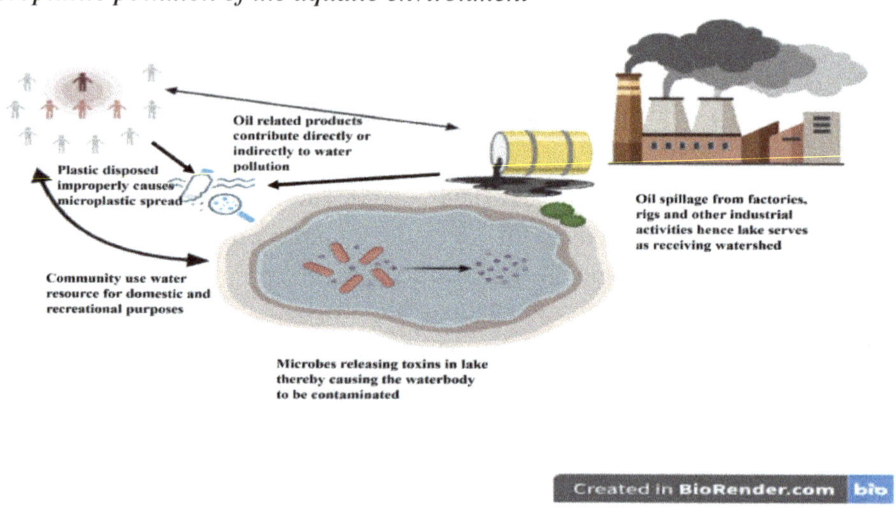

(created using BioRender)

Both *Pseudomonas* and *Alcanivorax* are genera of hydrocarbon-degrading bacteria that are crucial to oil biotransformation and bioremediation. *Pseudomonas* has a wider range of hydrocarbon metabolism capabilities, including the ability to metabolize aliphatic and aromatic compounds, while *Alcanivorax* specializes in the degradation of n-alkanes in marine oil spills. *Pseudomonas* occasionally produces biosurfactants that help with the emulsification and degradation of oil, and both are essential in the breakdown of petroleum components (Harayama et al., 2004).

The situation of oil spillage has become unbearable especially to marine animals. **Table 4** is a summary of the adverse effects due to oil spills. Many marine creatures are faced with lethal consequences, and several edible species contain these pollutants so therefore, this phenomenon can be regarded as an unfortunate situation, which goes from the contaminated water tables to our cups on the tables. **Figure 2** is an illustration of the general processes involved in the fate of crude oil spillage in the aquatic milieu.

Table 4. Oil spillage situation and adverse effect

S/N	Situation of oil spillage	Adverse effects of spilled oil	References
1.	Coating of marine birds	Direct toxicity Death by smothering in severe conditions	Farrington, 2014
2	Coating of marine mammals	Reduced resistance to environmental stress	Farrington, 2014
3	Oil ingested or dissolved oil in the body parts including gill, nostrils, mouth	Lethal effects to reproductive system, disruption of movement, chemotaxis failure	Farrington, 2014

Figure 2. Processes involved in the fate of spilled crude oil in the marine environment

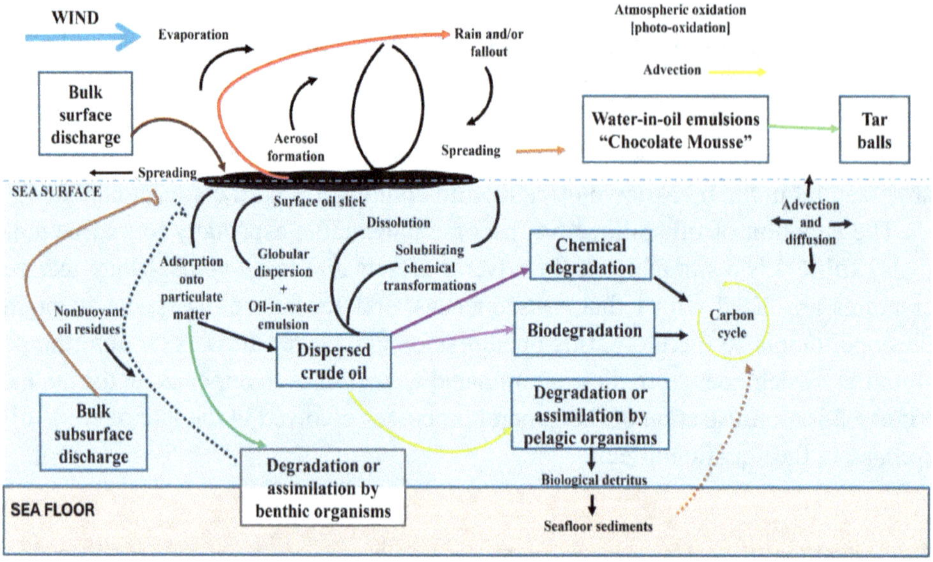

(Source: Adapted with little modifications from Farrington, 2014 that was previously adapted as a color version from the National Research Council Report in 1985 and from Burwood and Speers, 1974).

4 MICROBIAL RESPONSES TO PLASTIC POLLUTION

Natural gas and crude oil are the primary basic resources used to make plastics; these are converted into monomers, which are smaller hydrocarbon molecules. The basic structure of plastics is made up of lengthy chains known as "polymers", which are created when these monomers go through polymerization (Causin, 2015; Desidery and Lanotte, 2022). The raw ingredients for numerous plastics, including bioplastics, can also come from other natural resources such as cellulose, coal, waste products, and even CO_2 (Rajendran et al., 2025). Currently, plastics are mostly created from natural elements like cellulose, coal, crude oil, natural gas and salt via a "polymerization or polycondensation process". Plastics will increase from CO_2, waste, renewable materials. Plastics are lucrative and have become a part of our daily existence subsequently they are commonly used in agriculture, automative, construction, medical equipment, personal care goods, textiles and other industrial uses (Mazhandu et al., 2020; Nikiema et al., 2020). Studies about the recent surge in plastic pollution carries another twist. Reports suggested that plastic-degrading microorganisms produce enzymes to break plastics into smaller components that can be consumed (Sagova-Mareckova et al., 2021).

Unquestionably, plastics are pervasive and significant risks to ecosystems around the world. The aquatic environment undergoes physical and chemical disruption processes that inevitably convert plastic debris into microplastics (MPs). Unfortunately, MPs are globally transportable by wind or ocean currents, and they can be found in practically every niche area, including the most distant or extreme parts of the planet (Hale et al., 2020). The term "plastisphere" is conceptualized as MPs that are inhabited by microbial colonies that create biofilms (Zhai et al., 2023). Interest in the field of microbial ecology has increased when it was discovered that this special substrate can facilitate microbial spread. The growth, movement, persistence, and ecology of microorganisms are all impacted in concert by MPs (Metcalf et al., 2023).

There have been several reports of massive plastic particle accumulation in marine habitats worldwide, endangering the biota and probably ending up in our digestive systems. Although most of the plastic trash is produced onshore and carried into marine areas via rivers, freshwater ecosystems have gotten less attention (Dris et al., 2015). According to certain research, microplastics are present in freshwater ecosystems and the level of pollution is comparable to that seen in the oceans (Dris et al., 2015). Microplastics have been found in sediments (mostly lake margins but also riverbanks) and water samples (mostly lake and river surface water) in continental waters. Surprisingly, a recent study by Sarojini et al. (2025) even discussed microplastics in milk and its implications for human and animal health.

Low molecular weight, tiny molecules of dissolved organic materials seep from plastic and support aquatic metabolic activity and microbial adhesion to the plastic surface. For the polymers to experience intracellular oxidation and cross the microbial cell membrane, their molecular weight needs to be adequately low (Bher et al., 2022). While non-hydrolysable polymers (polypropylene, polyethylene, and expanded polystyrene) are commonly found in the pelagic zone of the marine environment, a recent report (Pal et al., 2025) has indicated that hydrolysable polymers (polyethylene terephthalate, polycarbonate, and polyurethane) in the photic zone serve as plastisphere substrates.

The process of microbial degradation of polyethylene (PE) and polypropylene (PP) is slow and challenging because of the high crystallinity of plastics, but it involves enzymes from bacteria and fungus breaking down polymeric polymers (Zhong et al., 2024). Microorganism strains like *Bacillus cereus* and *Pseudochrobactrum saccharolyticum* have been discovered in recent studies to have the ability of breaking down these microplastics, causing surface alterations and weight loss. Degradation rates are influenced by variables including pH, UV light, and salt concentration, and many studies have reported the breakdown is tracked using techniques such as scanning electron microscopy (SEM) and Fourier-transform infrared (FTIR) spectroscopy (Zhong et al., 2024).

Microorganisms that grow on the surface of polymers and enzymatic activities that cause the polymer to hydrolyze into oligomers and ultimately monomers are two examples of biological mechanisms for polymer degradation. There are several known enzymes that degrade the extremely stable backbones of non-hydrolysable polymers, including hydroxyquinone, alkB, laccase, oxygenase, peroxide, etc. Hydrolysable polymers are relatively susceptible to catalysis by enzymes such as cutinase, esterase, and PET hydrolase. Principally, *Ideonella sakaiensis* is accountable for the biotransformation of polyethylene terephthalate (PET) plastic. It does this by degrading PET into its component monomers, terephthalic acid (TPA) and ethylene glycol (EG), which it then uses as a source of carbon and energy. PETase and MHETase are two essential enzymes involved in this process. PET is hydrolyzed into mono (2-hydroxyethyl) terephthalate (MHET) by the bacterium's secreted PETase, which subsequently breaks down PET into TPA and EG.

According to several studies (Maity et al., 2021; Dey et al., 2024; Senko et al., 2024) on previously treated polyethylene (formerly exposed to thermal treatment, UV light, or containing additives) while measuring degradations, it is difficult to ascertain whether microbes act on high molecular weight polymers or smaller fragments produced by abiotic influences. Low molecular weight, tiny molecules of dissolved organic materials seep from plastic and support aquatic metabolic activity and microbial bond to the plastic surface. For the polymers to undergo intracellular oxidation and penetrate the microbial cell membrane, their molecular weight needs to be sufficiently low (Stewart et al., 2018).

According to recent reports (Khan and Faruque, 2023; Dey et al., 2024), non-hydrolyzable polymers (including polypropylene, polyethylene, and expanded polystyrene) are commonly found in the pelagic zone of the marine environment, while hydrolyzable polymers (such as polyethylene terephthalate, polycarbonate, and polyurethane) in the photic zone serve as plastisphere substrates. Polyethylene which are considered as obstinate polymers may be broken down by a class of alkane-degrading enzymes, including hexadecane and discussions have been carried out about the responses of *Aspergillus parasiticus* to polyethylene degradation (Siaperas et al., 2024). A study by Dhanraj et al. (2022) on mesophilic *Pseudomonas* sp. isolated from marine beach sediment and cultured with a low molecular weight polyethylene (LMWPE) as a sole carbon source demonstrated its ability to biodegrade polyethylene. Although hydrolysable plastics are vulnerable to microbial extracellular hydrolases (which are found in pre-existing degradation pathways) involved in cellulose and protein breakdown, the additional environmental requirements prevent the plastics from fully biodegrading. The recovery of PET hydrolytic enzyme PETase from the bacteria *Ideonella sakaiensis* (Oda and Wlodawer, 2024) and other subsequent enzymes from marine and terrestrial metagenomes included in public databases (Chen et al., 2024) demonstrate the ubiquity of PET degrading capacity of the plastisphere in those habitats. However, excessive oxidation via microbial biotransformation or partial hydrolysis of polymers results in the development of nanoplastic, which may pose a risk of ingestion through the food chain.

Although the effects of nano-plastic on human health have not been thoroughly studied, reports (Sun and Wang, 2023) have raised the possibility that microplastics (less than 150 μm) could move from the human gut epithelium into the lymphatic system, causing systemic exposure and ultimately affecting health after ingestion. The known terrestrial fungal representatives of plastic degraders include the ascomycete *Humicola insolens*, *Phanerochaete chrysosporium*, and the basiomycote *Trametes versicolor* which are soft-rot fungus and white-rot fungus respectively (Amaral-Zettler et al., 2020). There are numerous microbial species mentioned as possible plastic degraders. Most plastic biodegradation investigations are undertaken in vitro, while field degradation studies have revealed biodegradation rates in both marine (1.6-1.9% weight loss in one year) and terrestrial (0.1-0.4% weight loss in 800 days) environments.

The reported variations between field and laboratory-based investigations are largely due to a number of other factors, including polymer crystallinity, microbe preference for palatable carbon sources, and inadequate breakdown temperature. In a lab context, nylon and polyethylene have been shown to be degraded by a different species of fungi called Homobasidiomycetes, which contain esterase and other enzymes that break down polymers (Amaral-Zettler et al., 2020). Despite being uncommon marine residents, these creatures can be found in mangrove habitats.

In terms of biotransformation of plastic contaminants in the aquatic milieu, more studies are indeed needed. Recently, building bricks have been constructed from plastic waste and remnants which are being used as an alternative in construction of houses. **Figure 5.3** shows the microbial enzymatic breakdown of polymers.

Figure 3. Microbial enzymatic degradation of polymers

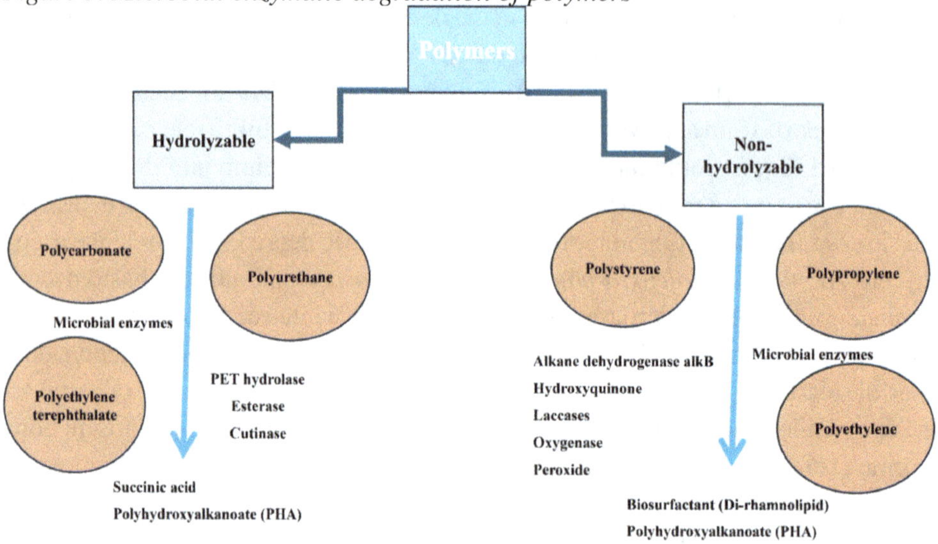

(Source: Modified from Fig 6b in Amaral-Zettler et al., 2020 (https://www.nature.com/articles/s41579-019-0308-0)).

5 MICROBIAL RESPONSES TO EUTROPHIC CONTAMINANTS

A major factor in the devastation of numerous freshwater and marine ecosystems worldwide is the naturally occurring phenomenon known as eutrophication. "Excessive plant and algal growth due to the augmented obtainability of one or more restrictive growth factors required for photosynthesis" (Schindler, 2006): sunlight, CO_2, and nutrients are some of its essential features. According to Harper (1992), as lakes mature and are filled with sediments over centuries, eutrophication occurs naturally. The process of eutrophication triggers changes in microbial community structures and function as they compete for available nutrients with certain microbial species thriving on the substantial resources including nitrogen (N) and phosphorus (P) (Sagova-Mareckova et al., 2021).

As stated by Mateo-Sagasta et al. (2017), aquaculture may be "a foremost contributor to eutrophication or biological loads in the receiving waterbodies." The primary sources of eutrophication include uneaten feed (predominantly from overfeeding) (Focardi et al., 2005), and decomposition of dead organisms and over fertilization (Feng et al., 2004). For instance, a study conducted in Scotland reviewed that the discharge of untreated organic waste from salmon production is equivalent to be over 75% of the pollution discharged by the human population (Tett et al., 2018).

Only about 20% to 50% of the total N supplemented to the cultured microbes is reserved as biomass by the farmed microorganisms, while the remainder is included in the sediment (Jackson et al., 2003) and ultimately released into the receiving ecosystems, intensifies the danger of algal blooms (such as poisonous microalgae-red tides) in marine areas, lakes, and aquatic reservoirs (Mateo-Sagasta et al., 2017; Bashir et al., 2020). Fish and other aquatic life are exposed to harsh conditions or perish because of hypoxia in water resources when organic contaminants consume dissolved oxygen (DO) in the surface water, which deteriorates the quality features of fresh water (Mateo-Sagasta et al., 2017). Microbial degradation transforms and cycles nitrogen (N) and phosphorus (P) in ecosystems through processes including mineralization and solubilization, which release orthophosphate and ammonia (NH_3) from organic compounds for plant uptake. N is processed via ammonification, nitrification, and denitrification, while P is released from organic matter through phosphatase enzymes and organic acid exudation. These microbial activities are essential for nutrient availability, soil fertility, and overall ecosystem health, with anthropogenic nutrient additions impacting microbial communities and their functions, leading to issues like algal blooms in water bodies.

6 BIOTRANSFORMATION MECHANISMS AND PATHWAYS

By means of enzymatic activities, microorganisms bio-transform oil, plastic, and excess nutritional pollutants, causing them to degrade or mineralize into less harmful substances like water and carbon dioxide. Activating bacteria that break down hydrocarbons, enzymes that break down plastic, and microorganisms that use nutrients are examples of microbial reactions. The rate of these activities is affected by environmental conditions such as temperature, nutrition, and the presence of other bacteria. Bioremediation techniques frequently use bioaugmentation or biostimulation to increase microbial activity.

When electrons are transported from an organic pollutant to oxygen, the terminal electron acceptor, natural biotransformation usually uses molecular oxygen (O_2):

Biomass + CO_2 + H_2O + other products = organic substrate + O_2.

Some bacteria use fermentation and anaerobic oxidation of organic carbon to generate energy when oxygen is scarce. Iron III salts, nitrates, and sulfate are useful substitutes for oxygen acceptors in a variety of anaerobic organisms (anaerobes).

Through physical and chemical processes like oxidation or evaporation, the components of petroleum and petroleum products undergo biotransformation, changing the crude oil's composition (Speight, 2014a; Taghvaei Ganjali et al., 2007). Petroleum compounds undergo a complex biotransformation process that is influenced by both nature and the concentration of hydrocarbons. According to Colwell et al. (1977), petroleum hydrocarbons fall into four classes: saturates, aromatics, resin ingredients, asphaltene constituents (phenols, fatty acids, ketones, esters, and porphyrins), and resins (pyridines, quinolines, carbazoles, sulfoxides, and amides).

Pollutants undergo biotransformation through Phase I oxidation, reduction, and hydrolysis reactions, which are mainly facilitated by enzymes such as cytochrome P450 (CYP) enzymes. Phase II conjugation reactions are then carried out, in which the modified compounds are joined to endogenous molecules such as glutathione, sulfate, or glucuronic acid. These enzymatic reactions, which can take place in living things or through microbial biodegradation, can make toxins more polar and water soluble, which makes it easier to excrete and detoxify them.

7 APPLICATIONS AND FUTURE DIRECTIONS

7.1 Revolutionary Process Involved in Transforming Plastic Waste Into Hydrogen Rich Gas Using Coal Catalysts

A groundbreaking approach is being pioneered by National Energy Technology Laboratory (NETL) to transform plastic waste into valuable hydrogen-rich syngas by co-gasifying it with coal. The method involves combining plastic waste with coal through a process called steam gasification. It aims to transform discarded plastics into valuable hydrogen-rich syngas, a potential alternative in energy production.

7.2 Integration of Recycled Plastic Waste Into Textile Industry

Plastics can be useful, converted to textiles and clothing. Recycled polyester and ocean plastics are being used by global brands to produce sneakers. Post-consumer plastic wastes (such as plastic bottles) are collected, cleaned, shred into flakes, melted down into pellets and then spin into polyester yarn, which is further processed, and footwears are manufactured. In a recent study by Solaja et al. (2025), the opportunities and challenges of using recycled polyethylene terephthalate (PET) plastics in the fashion industry in Nigeria were reviewed, however contrary to Sub-Saharan Africa, evidence from other studies show that producing polyester fibers from recycled PET can cut emission of carbon by over 50%. In addition, *Ideonella sakaiensis* is responsible for the biotransformation of polyethylene terephthalate (PET) plastic and provides a sustainable way to manage plastic garbage.

Using microbes and their enzymes, microbial biotransformation is used in remediation of pollutants thus, changing hazardous molecules into less dangerous ones. Heavy metals, hydrocarbons, pesticides, and dyes are just a few of the contaminants that can be removed with this method from contaminated groundwater, industrial waste and soil. Applications can be made in situ (treating the contamination at its original location) or ex situ (removing and treating the contaminated material elsewhere).

Hazardous contaminants, including hydrocarbons, dyes, heavy metals, and pesticides, are converted into less toxic byproducts by microbes for detoxification. Through the process of metal remediation, microorganisms lessen the environmental impact of hazardous metals such as arsenic, mercury, and chromium by changing them into forms that are more bioavailable or less soluble. This environmentally friendly process can enhance ecosystem health and soil fertility, resulting in safer food production. In Genetic Engineering, microorganisms are created and are genetically altered to improved capacities for more effectively breaking down contaminants. Microbial biotransformation can be used in conjunction with other remediation techniques, such phytoremediation, to produce synergistic effects in integrated systems. For real-time monitoring of microbial dynamics and contaminant transformation, the use of artificial intelligence (AI), Internet of Things (IoT), and omics-based techniques are being employed to enhance process control and scalability. To ensure substrate transfer and coordinated metabolic activity for strong pollutant breakdown, community engineering has focused on comprehending and manipulating complex microbial communities, however more studies are required in this area.

8 CONCLUSIONS

In order to manage sustainable land and water use and hold oil firms accountable, policy and regulation are essential. Furthermore, restoration initiatives including habitat construction and community involvement through stewardship and education initiatives aid in minimizing harm and promoting improved environmental habits. The importance of microbial biotransformation in addressing aquatic pollution cannot be overemphasized. There is a dire need to enhance our understanding of microbial metabolic networks and enzymes to create more efficient breakdown pathways for stubborn substances.

ACKNOWLEDGMENTS

The authors are grateful to the South African Medical Research Council and the National Research Foundation for financial support.

REFERENCES

Ahmed, T., Scholz, M., Al-Faraj, F., & Niaz, W. (2016). Water-related impacts of climate change on agriculture and subsequently on public health: A review for generalists with particular reference to Pakistan. *International Journal of Environmental Research and Public Health, 13*(11), 1051. DOI: 10.3390/ijerph13111051 PMID: 27801802

Akhtar, N., & Mannan, M. A. U. (2020). Mycoremediation: Expunging environmental pollutants. *Biotechnology Reports (Amsterdam, Netherlands), 26*, e00452. DOI: 10.1016/j.btre.2020.e00452 PMID: 32617263

Alabbosh, K. F. (2025). Leveraging the potential of environmental microorganisms: An extensive examination of their capacity in tackling global and local environmental and ecological challenges. *Polish Journal of Environmental Studies, 34*(1), 1–11. DOI: 10.15244/pjoes/185701

Amaral-Zettler, L. A., Zettler, E. R., & Mincer, T. J. (2020). Ecology of the plastisphere. *Nature Reviews. Microbiology, 18*(3), 139–151. DOI: 10.1038/s41579-019-0308-0 PMID: 31937947

Arora, J., Kumari, A., Ranjan, A., Rajput, V. D., Shende, S., Prazdnova, E. V., Mandzhieva, S. S., Sushkova, S., Minkina, T., Chauhan, A., & Srivastav, R. (2024). Microbial adaptation in different extreme environmental conditions and its usefulness in differently polluted soil. In *Extremophiles for sustainable agriculture and soil health improvement* (pp. 47–62). Springer Nature Switzerland. DOI: 10.1007/978-3-031-70203-7_3

Arthur-Holmes, F., Busia, K. A., Yakovleva, N., & Vazquez-Brust, D. A. (2022). Artisanal and small-scale mining methods and the Sustainable Development Goal 6: Perceived implications for clean water supply. *Environmental Science & Policy, 137*, 205–215. DOI: 10.1016/j.envsci.2022.08.017

Bashir, I., Lone, F. A., Bhat, R. A., Mir, S. A., Dar, Z. A., & Dar, S. A. (2020). Concerns and threats of contamination on aquatic ecosystems. In *Bioremediation and biotechnology: Sustainable approaches to pollution degradation* (pp. 1–26). Springer International Publishing. DOI: 10.1007/978-3-030-35691-0_1

Bhadrecha, P., Singh, S., & Dwibedi, V. (2023). 'A plant's major strength in rhizosphere': The plant growth promoting rhizobacteria. *Archives of Microbiology*, *205*(5), 165. DOI: 10.1007/s00203-023-03502-2 PMID: 37012531

Bher, A., Mayekar, P. C., Auras, R. A., & Schvezov, C. E. (2022). Biodegradation of biodegradable polymers in mesophilic aerobic environments. *International Journal of Molecular Sciences*, *23*(20), 12165. DOI: 10.3390/ijms232012165 PMID: 36293023

Bosch, A., Gkogka, E., Le Guyader, F. S., Loisy-Hamon, F., Lee, A., Van Lieshout, L., Marthi, B., Myrmel, M., Sansom, A., Schultz, A. C., Winkler, A., Zuber, S., & Phister, T. (2018). Foodborne viruses: Detection, risk assessment, and control options in food processing. *International Journal of Food Microbiology*, *285*, 110–128. DOI: 10.1016/j.ijfoodmicro.2018.06.001 PMID: 30075465

Causin, V. (2015). Polymers: An overview. In Polymers on the crime scene: Forensic analysis of polymeric trace evidence (pp. 9–52).

Chawla, H., Singh, S. K., & Haritash, A. K. (2024). Reversing the damage: Ecological restoration of polluted water bodies affected by pollutants due to anthropogenic activities. *Environmental Science and Pollution Research International*, *31*(1), 127–143. DOI: 10.1007/s11356-023-31295-w PMID: 38044406

Chen, J., Lin, L., Tu, Q., Peng, Q., Wang, X., Liang, C., Zhou, J., & Yu, X. (2024). Metagenomic-based discovery and comparison of the lignin degrading potential of microbiomes in aquatic and terrestrial ecosystems via the LCdb database. *Molecular Ecology Resources*, *24*(5), e13950. DOI: 10.1111/1755-0998.13950 PMID: 38567644

Cundell, A. M. (2018). Microbial ecology of the human skin. *Microbial Ecology*, *76*(1), 113–120. DOI: 10.1007/s00248-016-0789-6 PMID: 27245597

Desidery, L., & Lanotte, M. (2022). Polymers and plastics: Types, properties, and manufacturing. In *Plastic waste for sustainable asphalt roads* (pp. 3–28). Woodhead Publishing. DOI: 10.1016/B978-0-323-85789-5.00001-0

Dey, S., Rout, A. K., Ghosh, K., Jana, A. K., & Behera, B. K. (2024). Microbial ecology in microplastics: Impact on aquatic ecosystems and bioremediation. In *Current trends in fisheries biotechnology* (pp. 79–93). Springer Nature Singapore. DOI: 10.1007/978-981-97-3165-7_6

Dey, S., Veerendra, G. T. N., Babu, P. A., Manoj, A. P., & Nagarjuna, K. (2024). Degradation of plastics waste and its effects on biological ecosystems: A scientific analysis and comprehensive review. *Biomedical Materials & Devices*, *2*(1), 70–112. DOI: 10.1007/s44174-023-00085-w

Dhanraj, N. D., Hatha, A. M., & Jisha, M. S. (2022). Biodegradation of petroleum based and bio-based plastics: Approaches to increase the rate of biodegradation. *Archives of Microbiology*, *204*(5), 258. DOI: 10.1007/s00203-022-02883-0 PMID: 35419707

Di Genova, B. M., & Tonelli, R. R. (2016). Infection strategies of intestinal parasite pathogens and host cell responses. *Frontiers in Microbiology*, *7*, 256. DOI: 10.3389/fmicb.2016.00256 PMID: 26973630

Dris, R., Imhof, H., Sanchez, W., Gasperi, J., Galgani, F., Tassin, B., & Laforsch, C. (2015). Beyond the ocean: Contamination of freshwater ecosystems with (micro-) plastic particles. *Environmental Chemistry*, *12*(5), 539–550. DOI: 10.1071/EN14172

Ejiohuo, O., Onyeaka, H., Akinsemolu, A., Nwabor, O. F., Siyanbola, K. F., Tamasiga, P., & Al-Sharify, Z. T. (2025). Ensuring water purity: Mitigating environmental risks and safeguarding human health. Water Biology and Security, 4(2), Article 100341.

Farrington, J. W. (2014). Oil pollution in the marine environment II: Fates and effects of oil spills. *Environment*, *56*(4), 16–31. DOI: 10.1080/00139157.2014.922382

Feng, Y. Y., Hou, L. C., Ping, N. X., Ling, T. D., & Kyo, C. I. (2004). Development of mariculture and its impacts in Chinese coastal waters. *Reviews in Fish Biology and Fisheries*, *14*(1), 1–10. DOI: 10.1007/s11160-004-3539-7

Focardi, S., Corsi, I., & Franchi, E. (2005). Safety issues and sustainable development of European aquaculture: New tools for environmentally sound aquaculture. *Aquaculture International*, *13*(1), 3–17. DOI: 10.1007/s10499-004-9036-0

Froelich, B. A., & Noble, R. T. (2014). Factors affecting the uptake and retention of Vibrio vulnificus in oysters. *Applied and Environmental Microbiology*, *80*(24), 7454–7459. DOI: 10.1128/AEM.02042-14 PMID: 25261513

Gerba, C. P. (2009). Environmentally transmitted pathogens. In *Environmental microbiology* (pp. 445–484). Academic Press. DOI: 10.1016/B978-0-12-370519-8.00022-5

Hale, R. C., Seeley, M. E., La Guardia, M. J., Mai, L., & Zeng, E. Y. (2020). A global perspective on microplastics. Journal of Geophysical Research: Oceans, 125(1), Article e2018JC014719.

Harayama, S., Kasai, Y., & Hara, A. (2004). Microbial communities in oil-contaminated seawater. *Current Opinion in Biotechnology*, *15*(3), 205–214. DOI: 10.1016/j.copbio.2004.04.002 PMID: 15193328

Harper, D. M. (1992). Eutrophication of freshwaters.

Jackson, C. J., Preston, N., Burford, M. A., & Thompson, P. J. (2003). Managing the development of sustainable shrimp farming in Australia: The role of sedimentation ponds in treatment of farm discharge water. *Aquaculture (Amsterdam, Netherlands)*, *226*(1-4), 23–34. DOI: 10.1016/S0044-8486(03)00464-2

Khan, H. A., Sher, S., Bukhari, D. A., & Rehman, A. (2025). Bioremediation of heavy metals from electronic waste dumping sites with bacteria. *Current Research in Biotechnology*, *10*, 100309. DOI: 10.1016/j.crbiot.2025.100309

Khan, T. F., & Faruque, M. G. (2023). Degradation of polymer materials in environment. Ecology Journal, 237.

Kleindienst, S., Paul, J. H., & Joye, S. B. (2015). Using dispersants after oil spills: Impacts on the composition and activity of microbial communities. *Nature Reviews. Microbiology*, *13*(6), 388–396. DOI: 10.1038/nrmicro3452 PMID: 25944491

Madhav, S., Ahamad, A., Singh, A. K., Kushawaha, J., Chauhan, J. S., Sharma, S., & Singh, P. (2019). Water pollutants: Sources and impact on the environment and human health. In Sensors in water pollutants monitoring: Role of material (pp. 43–62).

Maglione, G., Zinno, P., Tropea, A., Mussagy, C. U., Dufossé, L., Giuffrida, D., & Mondello, A. (2024). Microbes' role in environmental pollution and remediation: A bioeconomy focus approach. *AIMS Microbiology*, *10*(3), 723–755. DOI: 10.3934/microbiol.2024033 PMID: 39219757

Maity, S., Banerjee, S., Biswas, C., Guchhait, R., Chatterjee, A., & Pramanick, K. (2021). Functional interplay between plastic polymers and microbes: A comprehensive review. *Biodegradation*, *32*(5), 487–510. DOI: 10.1007/s10532-021-09954-x PMID: 34086181

Mateo-Sagasta, J., Zadeh, S. M., Turral, H., & Burke, J. (2017). Water pollution from agriculture: A global review. Executive summary.

Nikiema, J., Mateo-Sagasta, J., Asiedu, Z., Saad, D., & Lamizana, B. (2020). Water pollution by plastics and microplastics: A review of technical solutions from source to sea.

Oda, K., & Wlodawer, A. (2024). Development of enzyme-based approaches for recycling PET on an industrial scale. *Biochemistry*, *63*(4), 369–401. DOI: 10.1021/acs.biochem.3c00554 PMID: 38285602

Onyena, A. P., & Sam, K. (2020). A review of the threat of oil exploitation to mangrove ecosystem: Insights from Niger Delta, Nigeria. *Global Ecology and Conservation*, *22*, e00961. DOI: 10.1016/j.gecco.2020.e00961

Osuagwu, E. S., & Olaifa, E. (2018). Effects of oil spills on fish production in the Niger Delta. *PLoS One*, *13*(10), e0205114. DOI: 10.1371/journal.pone.0205114 PMID: 30359365

Osunla, C. A., & Okoh, A. I. (2017). Vibrio pathogens: A public health concern in rural water resources in sub-Saharan Africa. *International Journal of Environmental Research and Public Health*, *14*(10), 1188. DOI: 10.3390/ijerph14101188 PMID: 28991153

Rajendran, S., Al-Samydai, A., Palani, G., Trilaksana, H., Sathish, T., Giri, J., Saravanan, R., Lalvani, J. I. J., & Nasri, F. (2025). Replacement of petroleum based products with plant-based materials, green and sustainable energy—A review. *Engineering Reports*, *7*(4), e70108. DOI: 10.1002/eng2.70108

Reineke, W., & Schlömann, M. (2023). Microbial degradation of pollutants. In *Environmental microbiology* (pp. 161–290). Springer. DOI: 10.1007/978-3-662-66547-3_6

Sagova-Mareckova, M., Boenigk, J., Bouchez, A., Cermakova, K., Chonova, T., Cordier, T., Eisendle, U., Elersek, T., Fazi, S., Fleituch, T., Frühe, L., Gajdosova, M., Graupner, N., Haegerbaeumer, A., Kelly, A.-M., Kopecky, J., Leese, F., Nõges, P., Orlic, S., & Stoeck, T. (2021). Expanding ecological assessment by integrating microorganisms into routine freshwater biomonitoring. *Water Research*, *191*, 116767. DOI: 10.1016/j.watres.2020.116767 PMID: 33418487

Sarojini, S., Balasubramanian, B., Jayaram, S., Nagaraj, V., Restine, A., Priyadarshini, P., Kaur, S., Rajeev, T. K., Pappuswamy, M., Kamyab, H., & Chelliapan, S. (2025). Invisible threats: Microplastics in milk and their implications for human and animal health. *Water, Air, and Soil Pollution*, *236*(10), 638. DOI: 10.1007/s11270-025-08267-6

Schindler, D. W. (2006). Recent advances in the understanding and management of eutrophication. *Limnology and Oceanography*, *51*(1, Part 2), 356–363. DOI: 10.4319/lo.2006.51.1_part_2.0356

Schwacke, L. H., Smith, C. R., Townsend, F. I., Wells, R. S., Hart, L. B., Balmer, B. C., Collier, T. K., De Guise, S., Fry, M. M., Guillette, L. J.Jr, Lamb, S. V., Lane, S. M., McFee, W. E., Place, N. J., Tumlin, M. C., Ylitalo, G. M., Zolman, E. S., & Rowles, T. K. (2014). Health of common bottlenose dolphins (Tursiops truncatus) in Barataria Bay, Louisiana, following the Deepwater Horizon oil spill. *Environmental Science & Technology*, *48*(1), 93–103. DOI: 10.1021/es403610f PMID: 24350796

Senko, O., Maslova, O., Stepanov, N., Aslanli, A., Lyagin, I., & Efremenko, E. (2024). Role of humic substances in the (bio)degradation of synthetic polymers under environmental conditions. *Microorganisms*, *12*(10), 2024. DOI: 10.3390/microorganisms12102024 PMID: 39458333

Sharma, M., & Kashyap, P. (2022). Pollution and its types. Journal Global Values, 13.

Siaperas, R., Taxeidis, G., Nikolaivits, E., & Topakas, E. (2024). Multi-omics insights into the response of Aspergillus parasiticus to long-chain alkanes in relation to polyethylene degradation [Preprint]. *bioRxiv*. https://doi.org/DOI: 10.1101/2024.12

Solaja, O. M., Osifo, O. K., & Adegoke, D. (2025). Integrating recycled PET plastics into Nigeria's textile industry: Environmental, ethical, and circular economy perspectives. *Circular Economy and Sustainability*, *5*(6), 1–29. DOI: 10.1007/s43615-025-00625-0

Sumampouw, O. J., & Risjani, Y. (2014). Bacteria as indicators of environmental pollution. *Environment*, *51*(52), 10–5923.

Sun, A., & Wang, W. X. (2023). Human exposure to microplastics and its associated health risks. *Environment & Health*, *1*(3), 139–149. DOI: 10.1021/envhealth.3c00053 PMID: 39473618

Sutar, R. S., Motghare, V. M., Kollur, S. C., Parikh, Y., & Asolekar, S. R. (2018). Significance of addressing persistence of pathogens and micropollutants to enhance reuse of treated sewages using constructed wetlands. In *Advances in waste management: Select proceedings of Recycle 2016* (pp. 355–367). Springer Singapore.

Takeshita, R., Bursian, S. J., Colegrove, K. M., Collier, T. K., Deak, K., Dean, K. M., De Guise, S., DiPinto, L. M., Elferink, C. J., Esbaugh, A. J., Griffitt, R. J., Grosell, M., Harr, K. E., Incardona, J. P., Kwok, R. K., Lipton, J., Mitchelmore, C. L., Morris, J. M., Peters, E. S., & Hall, A. J. (2021). A review of the toxicology of oil in vertebrates: What we have learned following the Deepwater Horizon oil spill. *Journal of Toxicology and Environmental Health. Part B, Critical Reviews*, *24*(8), 355–394. DOI: 10.1080/10937404.2021.1975182 PMID: 34542016

Tett, P., Benjamins, S., Black, K., Coulson, M., Davidson, K., Fernandes, T. F., Fox, C., Hart, M., Hicks, N., Hughes, A., & Hunter, D. C. (2018). Review of the environmental impacts of salmon farming in Scotland.

Thakur, A., & Koul, B. (2022). Impact of oil exploration and spillage on marine environments. In *Advances in oil-water separation* (pp. 115–135). Elsevier. DOI: 10.1016/B978-0-323-89978-9.00018-5

Wypych, T. P., Wickramasinghe, L. C., & Marsland, B. J. (2019). The influence of the microbiome on respiratory health. *Nature Immunology*, *20*(10), 1279–1290. DOI: 10.1038/s41590-019-0451-9 PMID: 31501577

Zhai, X., Zhang, X. H., & Yu, M. (2023). Microbial colonization and degradation of marine microplastics in the plastisphere: A review. *Frontiers in Microbiology*, *14*, 1127308. DOI: 10.3389/fmicb.2023.1127308 PMID: 36876073

Zhong, R., Li, M., Zhu, Z., Fang, X., Wang, X., Gong, H., & Yan, M. (2024). Bacterial degradation of polyethylene and polypropylene microplastics in a mangrove ecosystem. *Chemosphere*, *369*, 143908. DOI: 10.1016/j.chemosphere.2024.143908 PMID: 39643014

Chapter 6
Airborne Microorganisms and Their Role in the Degradation of Volatile Organic Compounds and Atmospheric Pollutants

Gift Ihotu Michael
Elizade University, Nigeria

Lateef Salam
http://orcid.org/0000-0001-8353-534X
Elizade University, Nigeria

ABSTRACT

This chapter examines the role of airborne microorganisms in the degradation of volatile organic compounds (VOCs) and atmospheric pollutants. It discusses the sources, diversity, and ecological functions of bioaerosols, including bacteria, fungi, archaea, and viruses, and highlights microbial survival strategies such as pigmentation, dormancy, DNA repair, and particle aggregation. Major pollutants, including NOx, SO_2, PAHs, ozone, and particulate matter, are reviewed alongside their environmental and health impacts. The chapter emphasizes microbial metabolic pathways involved in aerobic and anaerobic VOC degradation, including oxygenase-mediated oxidation, reductive dehalogenation, fumarate addition, and cometabolism. It also explores the role of airborne microbes in atmospheric chemistry, cloud formation, pollutant cycling, and climate regulation. Emerging biotechnologies such as biofiltration, engineered consortia, CRISPR-based optimization, and bioelectrochemical systems are highlighted as sustainable strategies for air-pollution control.

DOI: 10.4018/979-8-2600-2100-2.ch006

1. INTRODUCTION

Overview of Airborne Microorganisms

Airborne microbes form a significant part of bioaerosols, facilitating the movement of microorganisms throughout the atmosphere, biosphere, and human-influenced environments on Earth (Zhao et al., 2022). These bioaerosols contain a diverse and abundant range of bacterial and fungal species. Predominant microbial phyla identified within them include *Firmicutes, Proteobacteria, Bacteroidetes, Actinobacteria, Cyanobacteria* among bacteria, and *Ascomycota, Basidiomycota*, and *Chytridiomycota* among fungi (Chen et al., 2020; Zhao et al., 2022). Primarily, airborne microorganisms originate from environments and soils, followed by dispersal from human-related activities such as droplet release in composting facilities, waste management sites, and sewage treatment plants, where bacteria and fungi are predominantly utilized for degradation and waste processing (Ervens et al., 2025; Mahjadeen, 2023).

Since Louis Pasteur's pioneering studies of airborne microorganisms in 1860, these microbes have drawn scientific attention not merely for their occurrence in the atmosphere but also for their functional significance (Sebald & Hauser, 1995). Beyond chemical pollutants, there has been increasing interest in airborne microorganisms attached to atmospheric particles, as evidence highlights their influence on the atmospheric environment and their potential impact on ecosystem stability (Chen et al., 2020).

Importance of Atmospheric Pollutants and VOCs

Volatile Organic Compounds (VOCs) are a broad class of carbon-based chemicals that readily evaporate at ambient temperatures, allowing for their rapid transition into the atmosphere and making them key contributors to both indoor and outdoor air pollution. They originate from both natural processes, such as biogenic emissions from plants, microbial metabolism, forest fires, and anthropogenic activities, including industrial solvent use, fuel combustion, waste processing, paints, adhesives, and household cleaning agents (Kansal, 2008). VOCs play a significant role in atmospheric chemistry by participating in photochemical reactions that form ground-level ozone and secondary organic aerosols, which influence air quality and climate (Oliveira et al., 2024). Human exposure to VOCs has been associated with a range of adverse health outcomes, from mild irritation of the eyes and respiratory tract to long-term risks such as neurotoxicity and carcinogenicity, depending on the compound and exposure duration (Shuai et al., 2018; Alford and Kumar, 2021).

There is a clear consensus that global environmental pollution represents a complex international public health challenge. This issue is interconnected with social factors, economic activities, legislative frameworks, and prevailing lifestyle practices (Manisalidis et al., 2020) and is more severe in developing countries, where overpopulation, unchecked urban expansion, and the growth of industrial activities exacerbate environmental pollution challenges.

Significance of Microbial Degradation in Air Quality Improvement

The use of microbial consortia in bioremediation systems has proven more effective than traditional physicochemical approaches, as fungal–bacterial interactions exhibit improved degradation efficiency, particularly for hydrophobic pollutants (Ali et al., 2025). Microbial degradation processes have achieved removal efficiencies of 99.92% for acetic acid and 99.90% for ammonia under neutral pH conditions, whereas hydrogen sulfide and toluene recorded efficiencies of 99.24% and 99.90%, respectively, under acidic conditions (Haines et al., 2021). In addition to VOCs, microbial communities play a vital role in breaking down inorganic air pollutants like ammonia, hydrogen sulfide, and nitrogen oxides, key contributors to acid rain formation and particulate matter generation (Rahman et al., 2025).

Various bacterial genera, such as *Achromobacter*, *Bacillus*, *Mycobacterium*, *Burkholderia*, *Pseudomonas*, *Rhodococcus*, *Stenotrophomonas maltophilia*, and *Sphingomonas*, play key roles in degrading polycyclic aromatic hydrocarbons through assimilative pathways. These microorganisms utilize the pollutants as sources of carbon and energy to sustain their growth, while simultaneously transforming the toxic compounds into harmless mineral products (Yoshikawa et al., 2017). Research involving *Pseudomonas putida* has shown hydrocarbon reductions of up to 50% in smog-contaminated air samples, whereas *Acinetobacter* species, when used in specialized continuous-flow systems, have achieved up to 90% removal of naphthalene, a common organic air pollutant (David & Niculescu, 2021).

Furthermore, advancements in biofiltration and biotrickling technologies have effectively utilized microbial processes for large-scale purification of polluted air streams, attaining removal efficiencies of up to 99% for contaminants like toluene and hydrogen sulfide under optimized operational conditions (Yoshikawa et al., 2017).

This chapter examines the role of airborne microorganisms in the degradation of VOCs and other atmospheric pollutants, highlighting their potential contributions to natural air-cleaning processes. It also outlines the major sources and categories of air pollutants, such as NO_X, SO_2, PAHs, O_3, and VOCs, and discusses the mechanisms through which airborne microbes interact with, transform, and bioremediate these contaminants, ultimately contributing to improved air quality.

2. SOURCES AND TYPES OF AIRBORNE MICROORGANISMS

Natural Sources (Soil, Oceans, Plants, Animals)

A wide body of research demonstrates that vegetation is a major source of fungal particles released into the surrounding air, with plant surfaces consistently contributing substantial fungal biomass to local aerosols (Prussin & Marr, 2015; Qi et al., 2019; Al-Shaarani & Pecoraro, 2024). Soil and plant surfaces likewise act as important reservoirs for airborne bacteria, a pattern observed across diverse environments. In arid regions and deserts, dust-generating processes aerosolize bacteria that attach to mineral particles and can be transported across long distances by wind. Studies of aeolian dust in Southern Australia, for instance, have shown that airborne microbial communities predominantly originate from saline lake sediments and biological soil crusts (Chen et al., 2020; Malard & Pearce, 2022).

Marine environments also represent a major global source of airborne microorganisms. Oceans support vast microbial diversity, including bacteria, fungi, and some protozoans, and wave action and bubble bursting produce marine aerosols capable of transporting these organisms across the water–air interface (Chen et al., 2020; Al-Shaarani & Pecoraro, 2024). Many of these marine-derived microbes can later be detected in terrestrial, non-aquatic air masses, underscoring the long-range connectivity of microbial dispersal. Field studies further confirm that airborne community structures often closely resemble those found in soils and marine systems (Chen et al., 2020; Lang-Yona et al., 2021; He et al., 2024). For example, work conducted across mountainous and urban regions of Xi'an, China, reported that soil and leaf surfaces were the dominant sources of airborne bacteria during summer and autumn (Mu et al., 2019), consistent with broader evidence linking airborne bacterial assemblages to soil-origin communities (Al-Shaarani & Pecoraro, 2024).

Anthropogenic Sources (Industrial Emissions, Urban Activity)

Airborne bacteria are additionally derived from anthropogenic sources and human-influenced environments (Al-Shaarani & Pecoraro, 2024). In urban settings, these include hospital and household waste as well as pet excreta, whereas in rural areas they are primarily associated with agricultural practices, livestock rearing, and waste treatment processes such as wastewater management and composting (Al-Shaarani & Pecoraro, 2024). Composting relies on the breakdown of waste through microbial activity, during which a variety of bacteria and fungi are released at different stages, such as fresh waste delivery, shredding, pile turning, and compost screening (Chen et al., 2020). Additionally, research on bacterial community structure in a wastewater treatment plant indicated that the microbial composition of the water represents a major source of bacterial communities found in bioaerosols (Chen et al., 2020).

Under certain pathological conditions, humans and animals can serve as sources of fungi (Qi et al., 2019), which may be released into the air through skin shedding (Al-Shaarani & Pecoraro, 2024). The human body hosts about 10^{12} microorganisms on the skin and around 10^{14} in the digestive tract; they are considered a major source of bioaerosols, particularly in indoor environments (Prussin & Marr, 2015). Human activities like breathing (Morakinyo & Ana, 2025) and the daily shedding of millions of skin cells add significantly to bioaerosols in the environment. Both plants and domestic animals (e.g., dogs and cats) have been identified as sources of airborne microorganisms, contributing significantly to the composition of indoor bacterial communities (Al-Shaarani & Pecoraro, 2024). Gram-negative bacteria have been detected in metalworking fluids commonly utilized in engineering settings (Al-Shaarani & Pecoraro, 2024).

Classification of Airborne Microorganisms

Airborne microorganisms, or bioaerosols, comprise intact microbial cells, spores, fragments, metabolites, and other biological particles originating from soil, vegetation, water bodies, animals, humans, and decaying organic matter. Their continuous exchange between the biosphere and atmosphere highlights their ecological and public health significance. Classification frameworks include taxonomic, morphological, ecological, functional, and atmospheric behaviour–based approaches (Fig. 1).

Taxonomic Classification

The atmosphere contains a broad taxonomic diversity, including bacteria, fungi, viruses, archaea, microalgae, and protozoa. Airborne bacteria are often desiccation- and UV-resistant taxa, particularly Actinobacteria (e.g., *Micrococcus, Streptomyces*) and Firmicutes such as spore-forming *Bacillus* species. Gram-negative Proteobacteria (*Pseudomonas, Acinetobacter, Enterobacter*) are also common, especially in moist or human-impacted environments (Bowers et al., 2013). Fungi are major airborne constituents, mainly Ascomycota (*Aspergillus, Penicillium, Cladosporium, Alternaria*) and Basidiomycota basidiospores, with yeasts and Mucorales appearing in indoor or decomposing environments (Fröhlich-Nowoisky et al., 2016). Viruses occur as droplets or particle-bound entities, including influenza viruses, SARS-CoV-2, rhinoviruses, adenoviruses, and aerosolized enteric viruses such as norovirus and rotavirus (Prussin & Marr, 2015). Archaea (e.g., Euryarchaeota, Thaumarchaeota), microalgae (*Chlorella, Scenedesmus*), and protozoan cysts (*Acanthamoeba, Cryptosporidium*) are also present, mainly from soil, water, and wastewater sources.

Morphological Classification

Morphological traits, particularly particle size, determine atmospheric transport and deposition.

Ultrafine particles (<0.1 μm): viruses, cell fragments
Fine particles (0.1–2.5 μm): most bacteria, fungal fragments
Coarse particles (2.5–30 μm): fungal spores, pollen, protozoan cysts

Size, shape (cocci, rods, conidia), and protective structures influence survival under UV radiation, desiccation, and oxidative stress (Després et al., 2012).

Ecological Classification

Primary biological aerosols originate directly from natural and anthropogenic sources, including soil, marine spray, vegetation, animals, and humans. Secondary biological aerosols consist of non-viable biological derivatives such as endotoxins, mycotoxins, β-glucans, and cell wall fragments, which may cause strong immunological effects (Jones & Harrison, 2004). Source-based categories differentiate indoor, urban, rural, marine, and industrial bioaerosols. Indoor air tends to harbour human-associated bacteria (*Staphylococcus, Streptococcus, Bacteroides*), while outdoor air contains soil- and plant-derived taxa such as *Streptomyces* and *Cladosporium*.

Functional Classification

This framework highlights health and ecological roles:
Pathogens: *Mycobacterium tuberculosis, Legionella pneumophila, Aspergillus fumigatus*, respiratory viruses

Allergens: fungal spores (*Alternaria, Cladosporium*), endotoxins, β-glucans
Toxin producers: *Aspergillus* (aflatoxins), *Stachybotrys* (trichothecenes)
Extremotolerant taxa: *Deinococcus, Wallemia*, capable of long-distance atmospheric persistence (Womack et al., 2010)

These groups influence not only human health but also cloud formation, atmospheric chemistry, and ecosystem processes (Smets et al., 2016).

Atmospheric Behaviour-Based Classification

Microorganisms may remain suspended for hours to days (fine particles) or become settling particles that deposit quickly (coarse particles). They may be viable or non-viable, occurring as single cells or aggregates with dust, soot, or organic matter. Aggregation enhances protection against UV radiation and desiccation, promoting long-range transport.

Figure 1. Integrated classification framework for airborne microorganisms

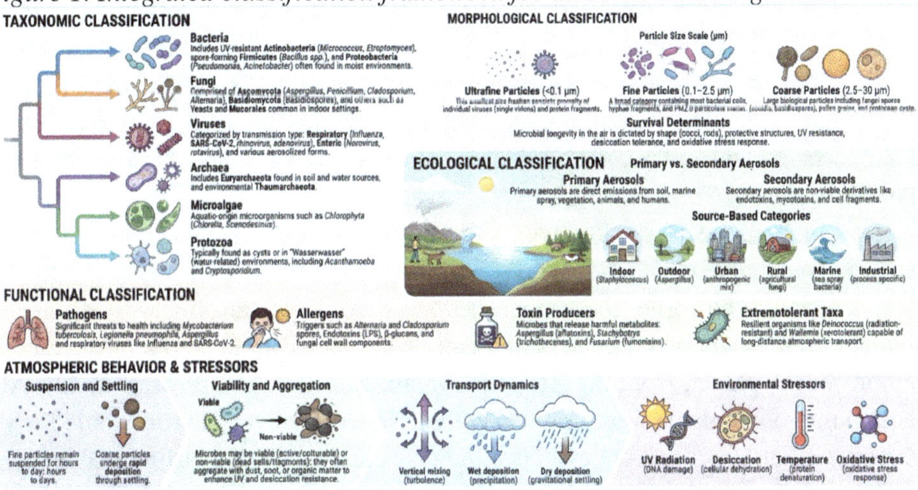

The schematic summarizes the major dimensions of bioaerosol classification, including taxonomic groups, particle-size–based morphology, ecological sources, functional categories, and key atmospheric behaviours governing suspension, viability, aggregation, and transport. Environmental stressors such as UV radiation, desiccation, temperature, oxidative stress, and humidity are also illustrated, providing an overview of factors shaping airborne microbial persistence and impacts.

Microbial Survival and Dispersion Mechanisms in the Atmosphere

Although the atmosphere is often considered a harsh environment for bacterial survival, diverse bacterial communities have also been discovered in other extreme habitats such as deep-sea environments, hot springs, and deserts (Smets et al., 2016). Life in the atmosphere is marked by limited nutrient and substrate availability, exposure to UV radiation, desiccation, fluctuations in temperature and pH, and the presence of reactive oxygen species (Smets et al., 2016). The survival of airborne microorganisms in the atmosphere is determined by their intrinsic adaptive traits, such as DNA repair mechanisms, pigment production, aggregation-promoting processes, and metabolic adaptations to nutrient limitation, such as the biosynthesis of cytochrome *bd* to cope with iron scarcity among others (Smets et al., 2016), and other surrounding environmental conditions (Ilori & Obayori, 2023).

Conversely, several atmospheric bacteria, such as *Geodermatophilus,* possess protective pigments that shield them from intense radiation exposure. Since most airborne microorganisms originate from soil environments, they generally exhibit similar resistance mechanisms (Aguilera et al., 2018). *Deinococcus* species are also commonly detected in atmospheric samples, and similar to soil-derived taxa, they possess multiple resistance traits linked to their exceptionally efficient DNA repair systems. In contrast, many non-spore-forming bacteria and some archaeal groups tend to have genomes with high G+C content, a feature that is thought to enhance UV resistance and improve overall survival under harsh atmospheric conditions (Ansari et al., 2015).

The atmosphere is predominantly populated by microbial aerosols such as bacterial vegetative cells, endospores, and fungal spores (Ilori & Obayori, 2023). After becoming airborne, these microbes are capable of traveling vast distances spanning thousands of kilometers and can be deposited in even the most remote regions of the planet, from the Arctic to Antarctica (Malard & Pearce, 2022). Although microorganisms were previously thought to be universally distributed with unrestricted dispersal, recent evidence indicates that ecological drift and dispersal constraints play a major role in shaping their biogeography (Malard & Pearce, 2022). Nevertheless, despite these harsh conditions, a significant proportion of airborne microorganisms survive and remain metabolically active, reaching altitudes as high as 20,000 meters (Bowers et al., 2009).

Airborne microorganisms that exist as single, free-floating cells tend to have a higher mortality rate compared to those associated with Particulate matter (PM) (Mahjadeen, 2023). A further bacterial survival mechanism is the induction of dormancy, a non-dividing state in which cells undergo morphological changes such as spore formation or cell wall modification, accompanied by a reduction or cessation of metabolic activity.

3. VOCS AND ATMOSPHERIC POLLUTANTS

Definition and Classification of VOCs

According to the United States Environmental Protection Agency (USEPA) (2023), VOCs are substances characterized by their tendency to evaporate easily due to high vapor pressure and their limited ability to dissolve in water. Due to their high volatility, mobility, and persistence, VOCs can be transported over extensive distances within environmental matrices following their emission (Pandey & Yadav, 2018). The USEPA has identified approximately 189 air pollutants, of which 97 are VOCs (Pandey & Yadav, 2018). Certain VOCs exhibit higher volatility than others; those that evaporate more rapidly tend to be more hazardous, posing greater risks to both the environment and human health (David & Niculescu, 2021).

Very Volatile Organic Compounds (VVOCs) are among the most hazardous classes of air pollutants, as they exhibit toxicity even at extremely low concentrations (Williams & Koppmann, 2007; Heeley-Hill et al., 2021; Yeoman & Lewis, 2021). This group includes compounds such as propane, butane, and methyl chloride. Propane (C_3H_8), a liquefied gas transported under vapor pressure, is commonly used for heating and cooking but poses significant safety risks. Similarly, butane (C_4H_{10}) is used for comparable purposes and is considered highly harmful when inhaled. Methyl chloride (CH_3Cl), also known as chloromethane, is a colorless, flammable, and toxic gas (Wang et al., 2007). Chloroform, or trichloromethane ($CHCl_3$), is a dense volatile organic compound (VOCs), colourless liquid with a strong smell that is produced on a large scale as a precursor to polytetrafluoroethylene (PTFE) and to various refrigerants (Williams & Koppmann, 2007; Yeoman & Lewis, 2021; Heeley-Hill et al., 2021).

Normal VOCs are equally hazardous as VVOCs and are commonly found in household products as well as in the environment. Examples include formaldehyde (CH_2O), vinyl chloride (C_2H_3Cl), carbon tetrachloride (CCl_4), toluene (C_7H_8), acetone, isopropyl alcohol, hexanal ($C_6H_{12}O$), and carbon disulfide (CS_2) (Yeoman & Lewis, 2021; Heeley-Hill et al., 2021). Semi-volatile organic compounds (SVOCs) have higher molecular weights and boiling points than VOCs, making them less prone to vaporize at room temperature. However, this reduced volatility does not make them any less harmful to human health or the environment (Heeley-Hill et al., 2021).

Among VOCs, aromatic hydrocarbons, such as benzene, toluene, ethylbenzene, and xylene, are most detected, along with halogenated hydrocarbons, including chloroethylene and trichloroethylene (Pandey & Yadav, 2018). VOCs are commonly found in products such as petroleum-based fuels, hydraulic fluids, paint thinners, and dry-cleaning solvents (USEPA, 2023). A summary of major atmospheric pollutants, their sources, key characteristics, and associated environmental and health impacts is provided in Table 1.

Table 1. Summary of major atmospheric pollutants, their sources, key properties, and associated environmental and health impacts.

Pollutant / Class	Type	Primary Sources	Key Characteristics	Environmental & Health Impacts	References
Ozone (O_3)	Secondary gaseous pollutant	Photochemical reactions involving NO_X and VOCs under sunlight	Strong oxidizing agent; major component of photochemical smog; formed mainly in the troposphere	Respiratory irritation, reduced lung function, aggravation of asthma; damages vegetation and contributes to climate forcing	Manisalidis et al., 2020; WHO, 2024
Carbon Monoxide (CO)	Primary gaseous pollutant	Incomplete combustion of fossil fuels (vehicles, generators, biomass burning)	Colorless, odorless gas; binds strongly to hemoglobin	Reduces oxygen delivery in the body; contributes indirectly to climate change through interactions with greenhouse gases	Manisalidis et al., 2020
Carbon Dioxide (CO_2)	Gaseous pollutant/ greenhouse gas	Fossil fuel combustion; human respiration (indoors)	Colorless, odorless, non-flammable; denser than air	Major driver of climate change; high indoor concentrations may cause oxygen deficiency	Maurya et al., 2023
Nitrogen Oxides (NO_X: NO, NO_2)	Primary & secondary gaseous pollutants	Vehicular emissions; fuel combustion; natural sources (lightning)	NO oxidizes to NO_2 in the atmosphere; NO_2 is acidic and corrosive	Formation of ozone and secondary PM; respiratory inflammation; ecosystem acidification	Manisalidis et al., 2020; Maurya et al., 2023
Sulfur Dioxide (SO_2)	Primary gaseous pollutant	Fossil fuel combustion; industrial processes	Toxic, acidic gas; regulatory annual limit ≈ 0.03 ppm	Causes respiratory disease; precursor of acid rain; damages vegetation and aquatic systems	Manisalidis et al., 2020
Polycyclic Aromatic Hydrocarbons (PAHs)	Semi-volatile organic compounds	Incomplete combustion (forest fires, engines, incineration); coal and tar deposits	Persistent, hydrophobic; widely dispersed via the atmosphere	Carcinogenic and mutagenic; bioaccumulation; long-term ecosystem contamination	Manisalidis et al., 2020

continued on following page

Table 1. Continued

Pollutant / Class	Type	Primary Sources	Key Characteristics	Environmental & Health Impacts	References
Volatile Organic Compounds (VOCs)	Gaseous organic pollutants	Industrial emissions; solvents; fuel evaporation; traffic	Highly reactive; precursors of ozone and secondary organic aerosols	Toxicity (e.g., benzene, formaldehyde); neurological, respiratory, cardiovascular effects; cancer risk	Yoshikawa et al., 2017; Pandey & Yadav, 2018; Cao et al., 2024; Oliveira et al., 2024
Particulate Matter ($PM_{2.5}$, PM_{10})	Particulate pollutant	Combustion processes; traffic; industrial emissions; natural dust	Fine particles penetrate deep into lungs; act as pollutant carriers	Cardiopulmonary disease, premature mortality; transports PAHs, VOCs, and metals	Manisalidis et al., 2020; Kuźma et al., 2020; Liu et al., 2022

Sources of VOCs and Pollutants (Natural vs. Anthropogenic)

VOCs arise from both natural and anthropogenic sources (Pandey & Yadav, 2018). Natural sources include emissions from vegetation and trees, forest fires of natural origin, and anaerobic decomposition in marshes and bogs. Plants release VOCs from fruits, flowers, leaves, stems, and roots in response to environmental stressors such as temperature shifts, light intensity, flooding, and drought. Microbial volatile organic compounds (mVOCs), produced by airborne microorganisms and fungal spores, also contribute to natural emissions (Fischer & Dott, 2003; Pandey & Yadav, 2018). Natural events such as volcanic eruptions, soil disturbances, and wildfires further introduce pollutants into the environment (Manisalidis et al., 2020).

Anthropogenic VOC emissions stem from domestic, commercial, and industrial activities. These include metal surface degreasing, textile cleaning, food processing, pesticide and fertilizer use, fumigation, chlorination, septic systems, landfills, hydrocarbon fuel evaporation, petroleum storage and distribution, printing processes, pharmaceutical manufacturing, and emissions from building materials (Pandey & Yadav, 2018). Vehicular emissions remain a major source, along with fuel storage, heating systems, and gas station operations. Agricultural practices contribute through fuel use in machinery, burning of crop residues, and the presence of VOCs in pesticide formulations (He et al., 2024). Composting is another significant emitter, especially during early aerobic decomposition (Pandey & Yadav, 2018). The use of ionic liquids as solvents represents an emerging VOC source (Wang et al., 2025).

Indoor environments are also affected by numerous VOC-emitting materials and activities. Perfumes, tobacco smoke, chlorinated water, adhesives, insulation foams, carpets, plastics, paint removers, new clothing, and kerosene heaters are key contributors to indoor air pollution (Spengler & Chen, 2000; Pandey & Yadav, 2018).

In suburban and metropolitan regions, industrial operations and vehicular traffic are major drivers of VOC emissions, accounting for over 28% of total TVOC concentrations. High levels of C8 and C9 aromatics from these activities contribute approximately 23% to total OH reactivity (Cao et al., 2024). Additionally, reliance on wood and solid fuels for heating and cooking, used by about three billion people globally, exposes households to high pollutant levels (Manisalidis et al., 2020). Although some human activities contribute at lower intensities, collectively they significantly influence environmental VOC loads.

4. MICROBIAL DEGRADATION OF VOCS AND AIR POLLUTANTS

Volatile organic compounds (VOCs) represent a chemically diverse group of low-molecular-weight organic chemicals defined by their substantial vapor pressures and capacity to volatilize at ambient temperatures. They include aliphatic hydrocarbons, aromatic hydrocarbons, oxygenated compounds, and numerous halogenated solvents. These compounds originate from both natural and anthropogenic sources. Biogenic emissions from vegetation, microbial metabolic processes, forest fires, and oceanic release contribute significantly to global VOC budgets (Rojo, 2009). Industrial processes, including solvent usage, petrochemical operations, refining activities, fuel combustion, waste management, and manufacturing, constitute major anthropogenic contributions (Lv et al., 2022).

Regardless of their origin, VOCs have significant impacts on atmospheric chemistry, human and ecological health, and global climate processes. They participate in ozone formation, secondary organic aerosol production, and diverse oxidative reactions in the troposphere (Šantl-Temkiv et al., 2022). Many VOCs, such as benzene, toluene, chloroform, trichloroethylene (TCE), dichloromethane (DCM), and vinyl chloride (VC), are recognized carcinogens, neurotoxins, or endocrine disruptors and are priority pollutants in air, soil, and groundwater systems (Wang et al., 2025).

Microorganisms play a central role in transforming VOCs in natural and engineered environments. Their metabolic versatility allows them to activate, oxidize, reduce, cleave, or mineralize structurally diverse VOCs under both aerobic and anaerobic conditions (Table 6.1). Their enzymatic repertoire includes oxygenases, dioxygenases, glycyl-radical enzymes, and reductive dehalogenases, each responsible for distinct transformation reactions (Harwood & Parales, 1996; Smidt & de Vos, 2004). As a result, microbial VOC degradation has emerged as a sustainable, efficient, and cost-effective remediation strategy with minimal secondary pollution (Lv et al., 2022; Isha et al., 2024).

4.1 Aerobic Microbial Degradation of VOCs

Oxygen as a Chemical Reactant and Electron Acceptor

Aerobic degradation of VOCs relies fundamentally on the availability of molecular oxygen (O_2). Oxygen plays two essential biochemical roles. First, it serves as the terminal electron acceptor in oxidative phosphorylation, enabling microorganisms to generate energy while metabolizing organic substrates. Second, and most critically in the context of VOC degradation, oxygen acts as a direct reactant incorporated into VOC molecules through the action of oxygenase enzymes (Hickey, 2021). This incorporation increases substrate polarity, reduces hydrophobicity, and thereby facilitates subsequent enzymatic transformations.

The primary enzymes responsible for aerobic VOC activation are monooxygenases and dioxygenases. Monooxygenases insert one oxygen atom into the substrate and reduce the second to water. Examples include methane monooxygenases, cytochrome P450 hydroxylases, toluene monooxygenases, and alkane hydroxylases (Sakai, 2023). Dioxygenases incorporate both oxygen atoms of O_2 into aromatic rings, yielding cis-dihydrodiols that serve as precursors for ring cleavage (Harwood & Parales, 1996). These initial oxidation reactions are considered rate-limiting steps in VOC biodegradation, as they fundamentally determine which substrates microorganisms can degrade.

Aerobic Biodegradation of Aromatic VOCs: The BTEX Paradigm

Aromatics such as benzene, toluene, ethylbenzene, and xylenes (collectively referred to as BTEX compounds) are ubiquitous environmental pollutants. Their aerobic degradation pathways have been extensively characterized, making them exemplary models for understanding microbial VOC transformation. The initial oxidation of BTEX compounds proceeds through the formation of distinct catechol derivatives. Benzene and toluene are converted to catechol, whereas o- and m-xylenes yield 3-methylcatechol, and p-xylene produces 4-methylcatechol (Wang et al., 2025). These intermediates then undergo ring cleavage via the ortho-cleavage (intradiol) or meta-cleavage (extradiol) pathway (Fig. 2).

Microorganisms capable of BTEX degradation include *Pseudomonas mendocina* KR1, which employs toluene-4-monooxygenase; *Ralstonia pickettii* PKO1, equipped with toluene-3-monooxygenase; and *Burkholderia vietnamiensis* G4, utilizing toluene-2-monooxygenase (Byrne et al., 1995; Yoshikawa et al., 2017). *Pseudomonas putida* F1 expresses toluene dioxygenase, enabling the degradation of benzene, toluene, and ethylbenzene (Yoshikawa et al., 2017). The diversity of monooxygenase and dioxygenase systems in these bacteria reflects the evolutionary adaptation of soil and water microorganisms to aromatic hydrocarbons.

Figure 2. Mechanistic overview of aerobic microbial pathways for VOC degradation, highlighting oxygen's dual role, key monooxygenase/dioxygenase reactions, BTEX catabolism, resistance mechanisms in chlorinated VOC oxidation, and cometabolic transformation processes

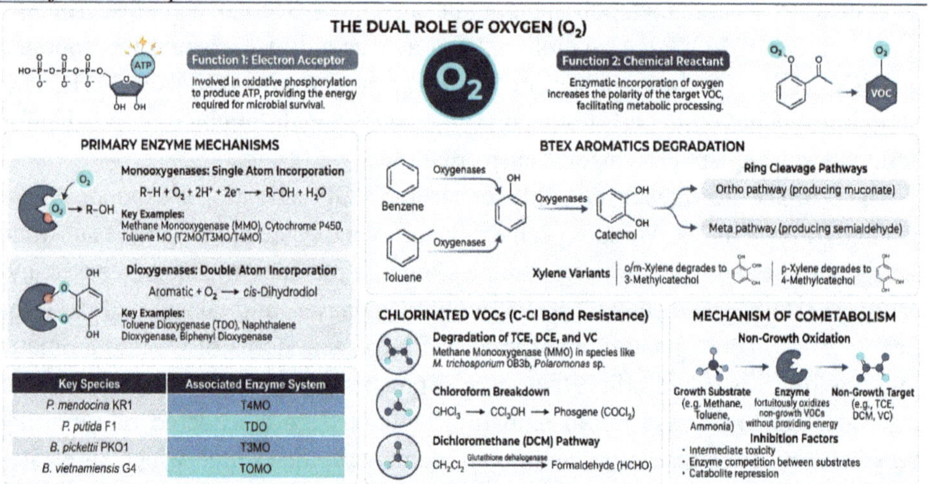

Aerobic Degradation of Chlorinated VOCs

Chlorinated VOCs, including TCE, DCEs, VC, chloroform, and DCM, pose unique challenges because carbon–chlorine bonds are relatively resistant to oxidation. Nevertheless, aerobic microorganisms degrade these contaminants through oxygenase-mediated pathways and cometabolic processes. Methanotrophic bacteria such as *Methylosinus trichosporium* OB3b oxidize TCE, DCE, and VC via methane monooxygenase (Yoshikawa et al., 2017; Sakai, 2023). Similarly, *Polaromonas sp.* JS666 and *Mycobacterium aurum* L1 oxidize VC and DCE during growth. Aerobic degradation of chloroform proceeds through oxidation to trichloromethanol, followed by a spontaneous conversion to phosgene (Wang et al., 2025). Dichloromethane is degraded primarily by glutathione-dependent dehalogenases that convert it into formaldehyde (Lin et al., 2002).

Cometabolism in Aerobic Environments

Cometabolism is a major determinant of VOC fates in aerobic systems. In cometabolic processes, enzymes induced by a growth substrate inadvertently oxidize non-growth substrates. Methane, toluene, ammonia, and propane often induce oxygenases capable of co-oxidizing chlorinated VOCs such as TCE, DCM, and VC (Alvarez-Cohen & Speitel, 2001). Although cometabolism does not yield metabolic energy from the co-oxidized compound, it can significantly accelerate the removal of recalcitrant VOCs in natural and engineered environments. However, cometabolism can also be inhibited by the toxicity of intermediates, enzyme competition, or catabolite repression (Toledo & Muñoz, 2025).

4.2 Anaerobic Microbial Degradation of VOCs

Metabolic Constraints and Energetics in Anoxic Systems

In the absence of oxygen, microorganisms rely on alternative terminal electron acceptors such as nitrate, sulfate, ferric iron, carbon dioxide, or organohalide compounds. Because oxygen-dependent activation cannot occur, anaerobic degradation requires chemical strategies that do not utilize molecular oxygen. These pathways are generally slower and more energetically constrained, yet they are crucial in saturated soils, sediments, aquifers, and engineered anaerobic reactors (Hickey, 2021).

Fumarate Addition: A Canonical Anaerobic Activation Pathway

Fumarate addition, catalyzed by glycyl-radical enzymes such as benzylsuccinate synthase (BSS), is one of the most important anaerobic activation mechanisms for aromatic and aliphatic hydrocarbons. These enzymes add hydrocarbons to fumarate, forming alkylsuccinate or benzylsuccinate intermediates that are further degraded via β-oxidation-like pathways (Callaghan, 2013). Genes in the *bssA* operon are widely used as biomarkers for anaerobic hydrocarbon degradation in contaminated sediments (Andorfer et al., 2023). Recent structural studies have revealed the radical-based catalytic mechanisms underpinning fumarate addition, offering opportunities for enzyme engineering (Mock et al., 2024).

Carboxylation and Hydroxylation Pathways in Anaerobic BTEX Degradation

Anaerobic degradation of BTEX compounds can also proceed through direct carboxylation of benzene to benzoate, or through anaerobic hydroxylation pathways in which water serves as the oxygen donor (Vogt et al., 2011). Although less widespread than fumarate addition, these pathways have been increasingly documented in oil-contaminated sediments, aquifers, and marine systems. Ethylbenzene degradation under nitrate-reducing conditions is initiated by anaerobic dehydrogenation, providing another example of oxygen-independent VOC activation (Yoshikawa et al., 2017).

Reductive Dehalogenation of Chlorinated VOCs

Halogenated VOCs undergo anaerobic degradation primarily through reductive dehalogenation. Organohalide-respiring bacteria, particularly *Dehalococcoides mccartyi*, use chlorinated compounds such as PCE, TCE, and VC as terminal electron acceptors. Corrinoid-dependent reductive dehalogenases (RDases) catalyze sequential chlorine removal, leading ultimately to non-toxic ethene (Smidt & de Vos, 2004). Key enzymes such as TceA, VcrA, and BvcA differ in substrate specificity and transformation efficiency (Eberwein et al., 2024). Carbon tetrachloride undergoes anaerobic reduction to carbon disulfide (CS_2) in the presence of vitamin B_{12} or to chloroform in its absence (Wang et al., 2025).

Syntrophy and Terminal Electron-Accepting Processes

Anaerobic degradation of many VOCs requires syntrophic interactions among bacteria and archaea. Fermentative microorganisms initiate hydrocarbon breakdown and produce intermediates such as hydrogen, formate, and volatile fatty acids. These intermediates are consumed by methanogens, sulfate-reducing bacteria, or iron-reducing bacteria, keeping hydrogen concentrations sufficiently low to drive otherwise thermodynamically unfavorable reactions (Gieg et al., 2014). Syntrophic interactions are especially important during the anaerobic degradation of BTEX compounds and higher hydrocarbons.

4.3 Enzymatic Systems Governing VOC Degradation

The enzymatic diversity underlying microbial VOC degradation is enormous. Oxygenases remain dominant under aerobic conditions, driving initial oxidation, dehalogenation, and hydroxylation reactions (Boyd et al., 2011). Monooxygenases catalyze diverse transformations, including hydroxylation, desulfurization, and dehalogenation, while dioxygenases catalyze aromatic ring cleavage (Karigar & Rao, 2011). Under anaerobic conditions, glycyl-radical enzymes mediate fumarate addition reactions, while reductive dehalogenases mediate sequential chlorine removal from halogenated VOCs (Smidt & de Vos, 2004). Fungal laccases, such as those produced by *Flavodon flavus* and *Coriolopsis gallica,* also contribute to the degradation of chlorinated phenolics and industrial effluents containing VOC-like contaminants (Isha et al., 2024). The combination of these enzymatic systems determines overall VOC degradation rates, pathways, and completeness.

4.4 Environmental and Ecological Factors Regulating VOC Degradation

Environmental parameters exert profound effects on microbial VOC degradation. Temperature influences enzyme kinetics, membrane fluidity, VOC volatility, and cellular stress responses. Humidity strongly affects airborne microorganisms, which may experience severe desiccation below 40% relative humidity, leading to reduced enzymatic activity and impaired degradation (Toledo & Muñoz, 2025). Excessive moisture, however, can create localized anoxic conditions that inhibit aerobic VOC degradation.

pH is a critical regulator, as extreme pH values denature enzymes and alter VOC chemistry, decreasing biodegradation efficiency (Mahjadeen, 2023). The presence of multiple co-occurring VOCs may enhance biodegradation through cometabolic stimulation or inhibit it due to toxicity, substrate competition, or catabolite repression. Atmospheric microorganisms also confront UV radiation, oxidative stress, and low nutrient availability, which act as environmental filters that select for radiation-tolerant and metabolically flexible taxa (Šantl-Temkiv et al., 2022). In soils and sediments, mass transfer limitations significantly restrict microbial access to hydrophobic substrates. Surfactants, biosurfactants, and biofilm formation can mitigate these limitations by increasing substrate solubility or enhancing partitioning into biologically accessible phases.

5 MICROBIAL COMMUNITIES INVOLVED IN VOC DEGRADATION

Microorganisms that degrade VOCs span multiple taxonomic groups, including bacteria, archaea, and fungi (Table 6.2). Atmospheric isolates of *Bacillus*, *Pseudomonas*, *Micrococcus*, *Frigoribacterium*, and *Sphingomonas* can degrade simple VOCs such as methanol and formaldehyde (Šantl-Temkiv et al., 2022). Methanotrophic bacteria play dual roles by oxidizing methane while cometabolizing chlorinated VOCs. Soil and aquatic systems harbor diverse BTEX degraders, reductive dechlorinators, nitrate reducers, sulfate reducers, iron reducers, and syntrophic bacteria. *Dehalococcoides mccartyi* remains the only known organism capable of complete reduction of DCE isomers and VC to ethene (Yoshikawa et al., 2017). Fungi contribute oxidative enzymes capable of transforming chlorinated phenolics and lignin-derived

VOCs in waste streams (Isha et al., 2024). The ecological organization, syntrophic partnerships, and metabolic specialization of these communities determine VOC degradation outcomes in natural and engineered environments.

6 DIAGNOSTIC TOOLS AND MULTI-OMICS APPROACHES FOR VOC DEGRADATION

Recent advances in molecular ecology have transformed the understanding of VOC degradation. Metagenomics allows the identification of functional genes and microbial taxa capable of VOC transformation, while metatranscriptomics and metaproteomics provide evidence of pathway expression under environmental conditions (Li et al., 2024). Stable isotope probing and compound-specific isotope analysis allow reconstruction of active degradation pathways and their thermodynamic properties. Activity-based protein profiling enables direct measurement of specific enzyme activities, including glycyl-radical enzymes and reductive dehalogenases (Andorfer et al., 2023; Mock et al., 2024). These multi-omics approaches offer powerful frameworks for diagnosing intrinsic biodegradation, designing biostimulation strategies, and assessing bioremediation performance.

7 ENGINEERING APPLICATIONS AND ADVANCES IN VOC BIOREMEDIATION

Microbial VOC degradation underpins multiple engineered bioremediation technologies. Aerobic biofilters, biotrickling filters, and bioscrubbers remain widely used for treating VOC-laden air emissions, with ongoing advances in support media, nutrient management, pH buffering, and real-time microbial monitoring (Silva et al., 2025). Anaerobic gas-phase bioreactors have emerged as alternatives for treating chlorinated VOCs without requiring aeration, thereby reducing operational energy demands (Chaudhary et al., 2025). Bioelectrochemical systems provide external electron donors or acceptors to stimulate reductive dehalogenation and syntrophic metabolism, representing a promising direction for in situ remediation of chlorinated VOC plumes (Eberwein et al., 2024). Synthetic ecology approaches enable the design of engineered consortia, combining methanotrophs, organohalide respirers, and complementary heterotrophs to support complex VOC degradation pathways (Li et al., 2024). These emerging technologies demonstrate the potential for integrating mechanistic understanding with engineering innovation to improve remediation outcomes.

Table 2. Some Microorganisms Involved in the Biodegradation of VOCs and Atmospheric Pollutants

Microorganism	Pollutant	Reference
Mycobacterium ethylenense NBB4	Vinyl Chloride	Yoshikawa et al., 2017
Bacillus spp.	Nitrogen oxides	Li et al., 2023; Yang et al., 2020
Pseudomonas spp., Achromobacter spp.	Ozone	Isha et al., 2024
Desulfitobacterium, Dehalococcoides	PCE	Yoshikawa et al., 2017
Acinetobacter	Sulfur compounds	Isha et al., 2024
Burkholderia sp.	Hydrocarbons	Lu et al., 1999
Sphingomonas sp.	VOCs	Karimi et al. 2015
Mycobacterium ethylenense NBB4	VC	Yoshikawa et al., 2017
Streptomyces sp.	Carbon monoxide	Ganguly et al., 2023
Pseudomonas putida	Hydrocarbons	Qamar et al., 2022
Pseudomonas putida	Nitrogen oxides (NOx)	Kour et al., 2022
Pseudomonas putida, Bacillus subtilis	Hydrocarbons, NO_x	Isha et al., 2024
Pseudoxanthomonas spadix BD-a59	BTEX	Yoshikawa et al., 2017
Acinetobacter sp.	Sulfur compounds	Ganguly et al., 2023
Pseudomonas fluorescens	BTEX	Lu et al., 1999
Nitrosomonas europaea	Benzene, Toluene, Ethylbenzene, and p-xylene	Yoshikawa et al., 2017
Rhodococcus jostii RHA1	Benzene, Toluene, Ethylbenzene, and o-xylene	Yoshikawa et al., 2017
Acinetobacter sp.	Ammonia	Lu et al., 1999
Flavobacterium sp.	Nitrogen oxides (NO_x)	Ganguly et al., 2023
Thauera sp. DNT-1	Toluene	Yoshikawa et al., 2017
Aromatoleum aromaticum EbN1, *Azoarcus* sp. T, *Thauera aromatica* K172	BTEX	Lin et al., 2002
Rhodococcus sp.	Nitrogen oxides (NO_x), catechol, and phenol	Hu et al., 2023; Jaber et al., 2020
Micrococcus luteus	Sulfur compounds	Ganguly et al., 2023
Azoarcus; Pseudomonas spp.	Benzene	Yoshikawa et al., 2017
Sphingomonas sp.	PAHs	Hu et al., 2023
Thiobacillus ferrooxidans	Inorganic sulfur compounds	Ganguly et al., 2023
Pseudomonas fluorescens	BTEX	Ganguly et al., 2023
Burkholderia sp.	Hydrocarbons	Karimi et al. 2015
Methylosinus trichosporium OB3b	Nitrous oxide (N_2O)	Hu et al., 2023
Sphingomonas sp.	PAHs	Yoshikawa et al., 2017

continued on following page

Table 2. Continued

Microorganism	Pollutant	Reference
Pseudomonas aeruginosa	Toluene	Chang et al., 2022
Methanobactin OB3b	Nitrous oxide	Muccee et al., 2019
Desulfobacterales, Coriobacteriaceae	Benzene	Yoshikawa et al., 2017
Desulfosarcina sp. PP31	p-xylene	Yoshikawa et al., 2017
Methylosinus trichosporium OB3b	DCM	Isha et al., 2024
Hyphomicrobium sp. DM2	DCM	Yoshikawa et al., 2017
Pseudomonas putida	PM	Isha et al., 2024
Aromatoleum aromaticum EbN1	Nitrate	(Lin et al., 2002).

8. ECOLOGICAL AND ENVIRONMENTAL ROLE OF AIRBORNE MICROBES

A key question in atmospheric microbiology is whether viable airborne bacteria simply persist in the harsh atmospheric environment by remaining metabolically inactive, or whether they actively contribute to modifying atmospheric conditions (Smets et al., 2016). Microbial communities found in cloud water have exhibited general indicators of metabolic activity under simulated atmospheric conditions, such as the presence of ATP, elevated ribosomal content, active RNA transcription, and measurable enzymatic and respiratory functions (Šantl-Temkiv et al., 2022).

8.1 Role in Atmospheric Chemistry and Pollutant Cycling

Multiple pathways associated with atmospheric trace gas metabolism, such as those involving hydrogen, methane, and carbon monoxide, have been widely detected in atmospheric metagenomes (Archer et al., 2020). Microbial participation in biophysical transformations can take place independently of metabolic processes, which are generally considered to be uncommon in the atmosphere (Šantl-Temkiv et al., 2022). Recent atmospheric modeling studies indicate that bacterial biodegradation within clouds can eliminate up to 20 parts per trillion per hour of formic acid and 5 parts per trillion per hour of acetic acid. These rates correspond to concentration reductions of approximately 20% and 3%, respectively, supplementing the losses already occurring through chemical oxidation (Yoshikawa et al., 2016; Roth et al., 2025).

Atmospheric microorganisms serve dual functions in atmospheric chemistry: they actively metabolize organic carbon and, at the same time, limit the availability of radical precursors. This dual activity reduces the oxidative capacity of cloud systems and affects the formation of secondary organic aerosols through the secretion of biosurfactants and extracellular polymeric substances (Kalalian et al., 2022).

Although photosynthesis in atmospheric microorganisms may be restricted by light inhibition, a variety of viable and stress-resistant microalgae and cyanobacteria have nonetheless been identified in cloud environments (Šantl-Temkiv et al., 2022). Photoautotrophic microorganisms serve as vital biological sinks for atmospheric CO_2, with microalgae accounting for about 50% of global photosynthetic activity and exhibiting CO_2 fixation efficiencies 10–50 times higher than those of terrestrial plants. Cyanobacteria contribute roughly 25% of the planet's annual carbon fixation and utilize specialized CO_2-concentrating mechanisms involving carboxysomes, protein-based microcompartments containing the carbon-fixing enzyme RuBisCO and carbonic anhydrase to maintain high intracellular CO_2 levels necessary for efficient carbon capture (Wang et al., 2024). Furthermore, methane oxidation at atmospheric concentrations has been demonstrated by *Methylocystis* and *Methylosinus* species enriched from the atmosphere (Šantl-Temkiv et al., 2022).

8.2 Interactions with Abiotic Factors and Atmospheric Particles

Recent metagenomic studies reveal that airborne microbial communities exhibit distinct biogeographical patterns influenced by both biotic and abiotic factors (Mahjadeen, 2023). Aerosol particles are constantly subjected to light, trace gases, oxidants, and various chemical pollutants, creating dynamic conditions that fluctuate rapidly and expose microorganisms to varying degrees of stress across different spatial and temporal scales.

Air pollutants and chemical constituents of particulate matter (PM) significantly influence airborne microbial communities by altering their abundance and diversity. Moderate PM levels may promote microbial growth, whereas high concentrations of toxic pollutants such as heavy metals, sulfates, and PAHs inhibit their survival (Precha et al., 2025). Also, certain chemical constituents of particulate matter (PM) serve as attachment sites for airborne microorganisms and create a favorable microenvironment that supports their growth and survival in the atmosphere (Chen et al., 2020). For example, the structure of microbial communities has been found to correlate strongly with water–soluble ions and metal elements (SO_4^{2-}, NO_3, NH_4^+, K^+, and Cl) concentrations, indicating that these aerosol particles are of great significance on bacterial compositions (Chen et al., 2020; Mahjadeen, 2023).

In some cases, suspended atmospheric plastics can also serve as potential carriers (fomites) for both pathogenic and non-pathogenic microorganisms (Mahjadeen, 2023). Additionally, seasonal variations and weather conditions significantly affect the composition of water-soluble ions in aerosols. It is therefore suggested that microbial activity in atmospheric aerosols may be influenced by the levels of water-soluble ions. Clouds and desert dust act as natural shields to atmospheric microorganisms, reducing exposure to harmful levels of direct ultraviolet (UV) radiation (Šantl-Temkiv et al., 2022).

8.3 Microbial Influence on Cloud Formation and Climate Regulation

Early evidence of microbial involvement in atmospheric processes emerged from the discovery that airborne bacteria can function as ice nuclei (IN) and cloud condensation nuclei (CCN) (Bauer et al., 2003). These biological particles facilitate cloud formation and precipitation (Smets et al., 2016) and have led researchers to describe clouds as "microbial oases," offering transient yet favorable conditions that sustain microbial activity (Ervens et al., 2025). Several airborne microorganisms act as efficient ice-nucleating and cloud-condensation particles (Bowers et al., 2009), thereby influencing cloud microphysics, atmospheric chemistry, and broader climate patterns (Aguilera et al., 2018; Mahjadeen, 2023). The biological nature of many atmospheric ice nuclei is supported by the loss of nucleation activity after exposure to 95°C heat treatment in snow samples (Sattler et al., 2001).

Ice nucleation is driven by specialized surface-bound or secreted proteins and operates largely as a biophysical mechanism, independent of microbial viability (Šantl-Temkiv et al., 2022). Several ice-nucleating proteins are associated with plant pathogens such as *Pseudomonas syringae*, which induces frost damage in crops (Smets et al., 2016; Vali and Schnell, 1976). *P. syringae* is one of the most effective ice-nucleating bacteria, capable of catalyzing ice formation at temperatures as high as –2°C (Vali and Schnell, 1976). Additional ice-nucleating genera include *Pseudomonas*, *Xanthomonas*, and *Erwinia*, in which genes encoding large, repetitive, surface-anchored ice nucleation proteins act as nucleating scaffolds (Bauer et al., 2003; Bowers et al., 2009).

Bioaerosols also alter atmospheric chemistry through metabolic activities, degradation of organic compounds, cell lysis, and desorption processes. As living primary biological aerosols (PBA), these microorganisms interact with various chemical species in the atmosphere and can modify the composition of organic aerosols.

Biological ice nucleation plays a significant role in cloud formation, precipitation, and climate regulation (Šantl-Temkiv et al., 2022). Ice crystals in clouds can form via the heterogeneous freezing of droplets, a process facilitated by microbial IN. When a bacterium initiates as a CCN and subsequently acts as an IN, it strongly influence cloud glaciation dynamics (Bauer et al., 2003). Among all biological nucleators identified, bacterial ice nucleation proteins (INpro) remain the most extensively studied. These >120 kDa membrane-associated proteins promote ice formation through mechanisms that are not yet fully understood (Lukas et al., 2020).

Ice-nucleating bacteria released from the plant phyllosphere are thought to participate in a biologically driven precipitation cycle. Once emitted into the atmosphere, these microorganisms enter cloud systems where their efficient ice-nucleation activity promotes droplet freezing and rainfall (Morris et al., 2013). The resulting precipitation enhances plant growth and supports epiphytic microbial communities, which in turn increase the release of additional ice-nucleating bacteria, reinforcing a positive feedback loop. This biologically mediated process operates alongside abiotic ice-nucleation mechanisms, indicating that atmospheric dynamics are shaped by both microbial and non-biological factors. Although evidence supports the existence of this feedback system, its overall influence on weather patterns and climate regulation remains insufficiently quantified (Šantl-Temkiv et al., 2022).

9 BIOTECHNOLOGICAL APPLICATIONS FOR AIR POLLUTANTS AND VOC DEGRADATION

9.1 Biofiltration and Bioreactors for Air Purification

Conventional air-pollutant removal methods such as chemical treatments and microfiltration are effective but often costly and require frequent maintenance (Kour et al., 2021). Biofiltration emerged as an alternative following the development of the first biofilter by Bach in 1923 and has since become a widely adopted pollution-control technology (Sheoran et al., 2022). It is particularly valuable for treating emissions from industries handling organic products, including paint and pharmaceutical manufacturing, vehicle exhaust, municipal waste and landfill operations, chemical and printing industries, fragrance and flavor production, coffee and cocoa roasting, wastewater odor control, composting, livestock farming, food processing, and foundries (Sheoran et al., 2022).

Modern biofiltration systems employ bacteria and fungi immobilized within biofilms to remove VOCs (Fig. 6.3). Pollutants diffuse into the active biofilm layers where aerobic microbial metabolism converts them into carbon dioxide, water, and microbial biomass (Haines et al., 2021). Bioactive coatings, composed of thin porous polymer matrices, further enhance this process by supporting microbial attachment or encapsulation, although the spatial separation of cells can limit natural biofilm formation (González-Martín et al., 2023). In bioreactors, VOC biodegradation typically begins with mass transfer across a moist biofilm, often the rate-limiting step for hydrophobic VOCs in multiple reactor designs (Wu et al., 2018a; Wu et al., 2018b).

Recent advances have significantly improved system performance, with biofilters achieving over 90% pollutant-removal efficiency and maintaining stable operation for more than eight months (Ali et al., 2025). Bioactive coatings enhance surface-to-volume ratios and support dense microbial populations (González-Martín et al., 2023). Comparative studies show fungal biofilters outperform bacterial systems in removing hydrophobic compounds such as toluene, achieving elimination capacities of 18.1 g m^3 h^{-1} versus 13.7 g m^3 h^{-1}, due to the absence of water barriers that favor rapid mass transfer (Sheoran et al., 2022).

Figure 3. Biofiltration systems employ bacterial and fungal biofilms for VOC removal

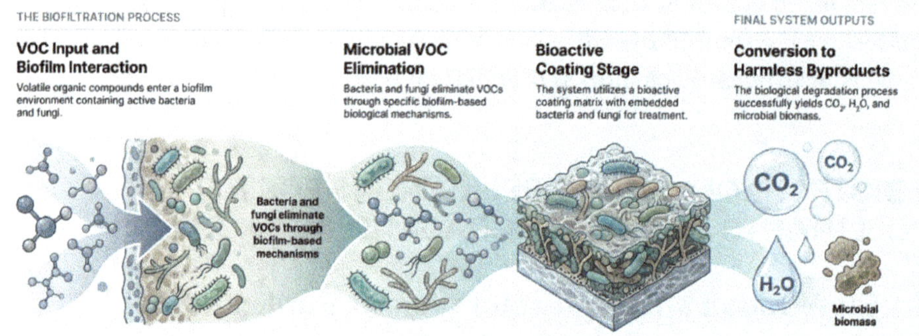

Volatile organic compounds (VOCs) enter the biofilm matrix, where microorganisms metabolize them into CO_2, H_2O, and microbial biomass. The bioactive coating enhances biodegradation efficiency through immobilized microbial communities.

Several factors significantly influence the performance of biofiltration systems. These include the type of packing material used, moisture content, residence time, temperature, and pH conditions. Other important considerations are the effects of shutdown periods, pressure drops across the bed (which largely determine operational costs), and the availability of essential nutrients. Additionally, the characteristics of the inlet pollutants, maintenance practices, and the empty bed residence time (EBRT) impact system efficiency. The composition and acclimation period of microorganisms, biofilm shredding or sloughing, and even rodent activity within the system can also affect overall biofilter performance (Sheoran et al., 2022).

9.2 Use of Microbial Consortia in VOC Remediation

Microbial consortia generally comprise diverse bacterial and fungal communities that work together to break down complex VOC mixtures through complementary metabolic interactions, such as co-metabolism, cross-feeding of intermediates, and stepwise degradation of persistent compounds (Prenafeta-Boldú et al., 2012). For example, during the biodegradation of aromatic hydrocarbons like benzene, toluene, ethylbenzene, and xylene (BTEX), some bacterial species carry out the initial ring hydroxylation, while others are responsible for the complete mineralization of the resulting intermediates. This metabolic division of labor enhances the overall efficiency of the degradation process (Lebrero et al., 2012).

While bioremediation has been extensively used to treat sites polluted with a single VOC or closely related VOCs, such as CE, its application becomes more challenging when dealing with locations contaminated by multiple different pollutants (Yoshikawa et al., 2016). For example, the environmental requirements for bioremediating CE and benzene vary, making it challenging to achieve complete cleanup of mixed VOC contamination with a single treatment strategy (Yoshikawa et al., 2016).

Molecular approaches such as 16S rRNA gene sequencing, metagenomics, and metatranscriptomics have shown that effective microbial consortia typically include primary degraders with specific catabolic genes for pollutant breakdown, secondary consumers that utilize degradation intermediates, and auxiliary populations that strengthen community stability by producing biosurfactants, exopolysaccharides, and stress-protective compounds (Lebrero et al., 2012). Employing single microbial species to eliminate difficult air pollutants is generally less effective than utilizing a microbial consortium (Nwankwegu et al., 2021).

Research has shown that bacterial genera such as *Pseudomonas*, *Bacillus*, *Rhodococcus*, and *Mycobacterium*, together with fungal genera like *Aspergillus*, *Penicillium*, and *Cladosporium*, commonly dominate VOC-degrading consortia in biofilters and biotrickling filters. The composition of these microbial communities' changes dynamically in response to variations in pollutant loading rates and operational conditions (Vergara-Fernández et al., 2017). In another study, Kleinheinz et al. (1999) investigated VOC removal by examining the degradation of α-pinene in a biofiltration column inoculated with *Pseudomonas fluorescens* and *Alcaligenes xylosoxidans*. Within 36 hours, the microbes effectively reduced α-pinene levels, initially averaging 295 ppm, highlighting the promise of biofiltration systems for VOC control and improved air quality.

9.3 Advances in Biotechnology for Enhancing Microbial Degradation

Recent progress in biotechnology has expanded the potential for enhancing microbial degradation of air pollutants, particularly through gene-editing tools such as CRISPR-Cas9, Transcription Activator-like Effector Nucleases (TALENs), and Zinc Finger Nucleases (ZFNs). These technologies enable precise modification of genes involved in degrading VOCs, NOx, SOx, and harmful hydrocarbons (Jaiswal et al., 2019; Isha et al., 2024). CRISPR-Cas9 functions by guiding engineered RNAs to specific genomic sequences, allowing targeted insertion or deletion of DNA fragments and facilitating the introduction or enhancement of pollutant-degrading enzymes (Martínez-García et al., 2014). Despite promising laboratory results, the environmental use of CRISPR-engineered microbes remains limited due to uncertainties surrounding efficiency, ecological stability, and biosafety once such organisms enter natural ecosystems (Minhas et al., 2024).

One major advancement includes engineering the fast-growing *Vibrio natriegens* with synthetic gene clusters enabling the simultaneous degradation of naphthalene, phenanthrene, phenol, and halogenated hydrocarbons, achieving up to 100% naphthalene removal and 93.4% toluene degradation (Su et al., 2025). However, current trends increasingly emphasize microbial consortia rather than single engineered strains. Mixed microbial communities distribute metabolic burdens across different species, reducing stress on individual cells and improving pollutant breakdown. Such division-of-labour systems, including cross-feeding, cooperative detoxification, and consolidated bioprocessing, have proven especially effective for VOC removal and are more resilient to environmental fluctuations such as pH changes, moisture stress, and sudden increases in pollutant loading (Rahmati et al., 2022; Isha et al., 2024; Hiroshi, 2024).

Bioaugmentation represents another biotechnology-driven strategy, involving the intentional introduction of specialized degraders into contaminated environments. These strains complement native microbial communities and accelerate pollutant removal where natural biodegradation is insufficient (Isha et al., 2024; Shah et al., 2024).

Advances in engineered biofilters, biotrickling filters, and two-phase partitioning bioreactors have further improved air purification efficiency. Innovations such as optimized packing materials (e.g., polyurethane foam, perlite–activated carbon composites) and the use of non-aqueous phase liquids (NAPLs) address mass-transfer limitations associated with hydrophobic VOCs by increasing surface area, improving water retention, and enhancing gas–biofilm contact (Pachaiappan et al., 2022). Two-phase reactors incorporate biocompatible organic phases that buffer toxic VOC concentrations, allowing higher inlet loads and more stable microbial performance. Additionally, computational fluid dynamics (CFD) modelling and real-time sensor technologies (e.g., O_2 and VOC sensors) now guide operational optimization under industrial-scale conditions (Irfana et al., 2016).

A promising emerging technology is the bio-electrochemical system for air pollution control, in which microorganisms such as *Shewanella oneidensis* grow on electrodes and use electrochemical reactions to degrade $PM_{2.5}$ and VOCs, offering a novel route for electrically enhanced bioremediation (Isha et al., 2024).

9.4 Case Studies of Successful Implementation

A genetically engineered *Pseudomonas putida* biofiltration system was developed in New Jersey, USA, to remediate benzene, a pollutant linked to brain tumors and leukemia. In this setup, the biofiltration unit was integrated into a ventilation system. Prior to the experiment, benzene concentrations were measured at 200 $\mu g/m^3$, with a leukemia incidence of 10 cases per 100,000 people. Following six months of operation, *P. putida* effectively reduced benzene levels to 75 $\mu g/m^3$, corresponding to a leukemia rate of seven cases per 100,000. This demonstrated nearly a 70% reduction in benzene concentration within the study period (Isha et al., 2024).

Qamar et al. (2022) demonstrated that *P. putida* could lower hydrocarbon levels in smog-contaminated air samples by up to 50%. Similarly, other studies have highlighted the effectiveness of bacterial bioremediation in reducing smog pollutants. For instance, Ganguly et al. (2023) reported that *Acinetobacter sp.* decreased sulfur compounds by 55%, while Kour et al. (2022) showed that *Bacillus subtilis* reduced nitrogen oxides by as much as 60%.

Industrial activities release large amounts of wastewater and flue gases into the environment. Research by Barla et al. (2024) using bacterial consortia in a bubble column bioreactor showed that wastewater-based growth media could be used to treat both simultaneously, achieving high biomass and removal efficiencies of about 81% for CO_2, 77% for SO_2, and 3.66 g/L for NO. Also, Sun et al. (2019) evaluated a biofilter for the simultaneous removal of NO and SO_2 under thermophilic, micro-oxygen conditions. The system achieved removal efficiencies above 90% for both pollutants, demonstrating stable performance in reducing key components of polluted air.

Karimi et al. (2015) demonstrated that *Pseudomonas aeruginosa* achieved about 90% removal of naphthalene through specialized continuous flow systems. Incorporating these systems into industrial operations further enhanced the effectiveness of pollutant reduction. In another study, Khan et al. (2024) applied a membrane bioreactor (MBR) system inoculated with mixed microbial consortia of *Pseudomonas* sp. and *Bacillus* sp. to treat phenolic contaminants in industrial effluents. The system showed high removal efficiency and proved more cost-effective than traditional physicochemical methods.

10. CHALLENGES AND KNOWLEDGE GAPS

Atmospheric VOC mixtures are chemically diverse and heterogeneous, comprising multiple compounds with varying structures and toxicities. These mixtures often generate synergistic or antagonistic interactions that influence microbial degradation efficiency (Yoshikawa, 2017). At extremely high concentrations, VOCs inhibit microbial enzymes and suppress degradation (Karimi et al., 2015). Hydrophobic VOCs, including BTEX, exhibit low solubility and poor bioavailability, reducing mass transfer into microbial cells (Isha et al., 2024). Spatial variability in VOC concentrations further demands site-specific bioremediation approaches (Bala et al., 2022).

Microorganisms typically show substrate specificity and depend on natural electron donors and acceptors, limiting the range of degradable VOCs. Industrial remediation often results in incomplete mineralization and the accumulation of toxic intermediates (Isha et al., 2024). Bioavailability remains a major constraint because VOCs must partition into aqueous biofilms before uptake, and hydrophobic compounds diffuse slowly, especially under fluctuating moisture, temperature, pH, and nutrient conditions (Cheng et al., 2016; Chaudhary et al., 2023).

Environmental stress further constrains biodegradation. Many bacteria function optimally only within narrow pH, temperature, or osmotic windows, complicating remediation under harsh conditions (Nwankwegu et al., 2021). VOC toxicity inhibits microbial growth, particularly at high concentrations (Yoshikawa et al., 2016). Mixed VOCs can generate toxic by-products that interfere with the degradation of co-occurring compounds. Epoxides, produced during aerobic degradation of chlorinated ethenes (CEs), strongly inhibit microbial activity (Yoshikawa, 2017). Detoxification enzymes such as epoxyalkane: coenzyme M transferase (EaCoMT), encoded by *etnE* and found in *Mycobacterium, Nocardioides*, and *Haliea*-like taxa, help mitigate epoxide toxicity (Liu & Mattes, 2016).

Co-existing VOCs also impede anaerobic pathways. For instance, chloroform inhibits anaerobic DCM degradation (Yoshikawa et al., 2016). In a toluene-degrading consortium of *Pseudomonas putida* strains and *Burkholderia vietnamiensis*, only *P. putida* mt-2 tolerated TCE, whereas others degraded TCE but were inhibited by toxic intermediates (Yoshikawa et al., 2017). Similarly, catechol derivatives generated during BTEX metabolism, such as those produced by *P. putida* PPO1 during p-xylene degradation, can suppress benzene and toluene degradation (Yoshikawa et al., 2017).

Catabolite repression further hinders VOC degradation; microorganisms preferentially consume easily metabolizable substrates such as succinate, suppressing toluene and xylene metabolism (Blázquez et al., 2018). A major knowledge gap persists regarding airborne microbial ecology, including the diversity, abundance, and metabolic potential of airborne microorganisms (Yoshikawa et al., 2017; Chaudhary et al., 2023). Greater mechanistic insight into metabolic pathways, gene regulation, stress tolerance, and resistance development is needed.

Effective remediation of mixed VOCs will likely require microbial consortia with complementary catabolic functions or single strains with broad metabolic capabilities (Yoshikawa et al., 2017). However, the scalability of biofilters and bioreactors remains limited by biomass accumulation, uneven nutrient distribution, and by-product management (Cheng et al., 2016). Advances in omics, molecular tools, and synthetic biology may enable the development of more resilient VOC-degrading systems.

11. FUTURE PERSPECTIVES

Metagenomics enables broad characterization of airborne microbial communities and their pollutant-degrading genes, offering a cultivation-independent approach for understanding microbial contributions to air quality. Existing air monitoring networks can be adapted for metagenomic sampling, enhancing scalability and cost-efficiency (Pozdniakova et al., 2025). Transcriptomics identifies actively expressed genes in airborne microbes, revealing real-time VOC degradation processes (Yang et al., 2022).

Artificial intelligence models that integrate multi-omics and environmental variables now allow prediction of microbial behavior and VOC degradation performance, supporting adaptable bioremediation strategies (Pozdniakova et al., 2025).

Synthetic biology and metabolic engineering can produce optimized microbial strains or consortia with enhanced stability and pollutant specificity suitable for airborne contexts (Mousavian et al., 2024). Integrating engineered microbes into biosensors and air treatment technologies may improve detection and mitigation of VOCs (Behzad et al., 2015).

Future pollutant management frameworks should incorporate climate dynamics, atmospheric chemistry, and airborne microbiome interactions. Policy enforcement is also crucial. For example, the U.S. Environmental Protection Agency (USEPA) issued an interim final revised compliance deadline for the National Volatile Organic Compound Emission Standards for Aerosol Coatings, extending the original compliance date from July 2025 to January, 2027. This regulatory action aims to reduce VOC emissions from both consumer-grade and industrial aerosol products in order to curb urban air pollution and limit tropospheric ozone formation. In contrast, bioaerosol regulation remains underdeveloped globally due to limited standardized protocols and risk assessment frameworks. Future policies should require monitoring systems that ensure safe use of microbial agents, preventing unintended pathogen dissemination (Loveniers et al., 2024). Therefore, close collaboration among microbiologists, environmental engineers, and regulatory authorities will be critical to establish scientifically grounded guidelines that achieve pollutant reduction while simultaneously safeguarding public health (Loveniers et al., 2024).

12 CONCLUSIONS

Airborne microorganisms form a diverse, metabolically adaptable community capable of degrading volatile organic compounds (VOCs) and other atmospheric pollutants. Originating from soil, marine, and vegetative environments, these microbes withstand harsh atmospheric conditions through structural and molecular adaptations. Their oxygenase-driven pathways enable the transformation of hazardous compounds such as BTEX and chlorinated ethenes into less toxic products, and microbial consortia and biofiltration systems demonstrate strong pollutant removal potential.

Growing molecular and ecological evidence confirms that airborne microbes remain metabolically active, influencing both pollutant degradation and broader atmospheric processes, including chemical transformations, cloud formation, and climate dynamics. This dual role underscores their significance in global biogeochemical cycles.

Advancing sustainable air pollution control will depend on engineered microbial consortia, bioaugmentation, and optimized biofiltration technologies. Addressing knowledge gaps through omics-based analyses, integrative modeling, and synthetic biology, supported by robust regulatory frameworks, will be essential for safely harnessing airborne microorganisms to improve air quality and protect human and environmental health.

REFERENCES

Aguilera, A., de Diego-Castilla, G., Osuna, S., Bardera, R., Sor Mendi, S., & Blanco, Y.. (2018). Microbial ecology in the atmosphere: the last extreme environment. In Najjari, A., Cherif, A., Sghaier, H., & Ouzari, H. I. (Eds.), *Extremophilic Microbes and Metabolites-Diversity, Bioprospecting and Biotechnological Applications*. IntechOpen., DOI: 10.5772/intechopen.81650

Al-Shaarani, A. A. Q. A., & Pecoraro, L. (2024). A review of pathogenic airborne fungi and bacteria: Unveiling occurrence, sources, and profound human health implication. *Frontiers in Microbiology*, *15*, 1428415. DOI: 10.3389/fmicb.2024.1428415 PMID: 39364169

Alford, K. L., & Kumar, N. (2021). Pulmonary Health Effects of Indoor Volatile Organic Compounds—A Meta-Analysis. *International Journal of Environmental Research and Public Health*, *18*(4), 1578. DOI: 10.3390/ijerph18041578 PMID: 33562372

Ali, Q., Khan, A. R., Raza, W., Bilal, M. S., Khalid, S., Ayaz, M., Khan, A., & Mundra, S. (2025). Mechanisms of Microbial VOC-Mediated communication in plant ecosystems and agricultural applications. *Journal of Sustainable Agriculture and Environment*, *4*(1), e70044. Advance online publication. DOI: 10.1002/sae2.70044

Alvarez-Cohen, L., & Speitel, G. E. (2001). Cometabolic biodegradation of chlorinated solvents. *Environmental Health Perspectives*, *109*(Suppl 2), 225–238.

Andorfer, M. C., King-Roberts, D. T., Imrich, C. N., Brotheridge, B. G., & Drennan, C. L. (2023). Development of an *in vitro* method for activation of X-succinate synthases for fumarate hydroalkylation. *iScience*, *26*(6), 106902. DOI: 10.1016/j.isci.2023.106902 PMID: 37283811

Ansari, T. U., Valsan, A. E., Ojha, N., Ravikrishna, R., Narasimhan, B., & Gunthe, S. S. (2015). Model simulations of fungal spore distribution over the Indian region. *Atmospheric Environment*, *122*, 552–560. DOI: 10.1016/j.atmosenv.2015.10.020

Archer, S. D. J., & Pointing, S. B. (2020). Publisher Correction: Anthropogenic impact on the atmospheric microbiome. *Nature Microbiology*, *5*(5), 776. DOI: 10.1038/s41564-020-0680-6 PMID: 32099094

Bala, S., Garg, D., Thirumalesh, B. V., Sharma, M., Sridhar, K., Inbaraj, B. S., & Tripathi, M. (2022). Recent Strategies for Bioremediation of Emerging Pollutants: A review for a Green and Sustainable environment. *Toxics*, *10*(8), 484. DOI: 10.3390/toxics10080484 PMID: 36006163

Barla, R. J., Gupta, S., & Raghuvanshi, S. (2024). Sustainable synergistic approach to chemolithotrophs—Supported bioremediation of wastewater and flue gas. *Scientific Reports, 14*(1), 16529. Advance online publication. DOI: 10.1038/s41598-024-67053-2 PMID: 39019921

Bauer, H., Giebl, H., Hitzenberger, R., Kasper-Giebl, A., Reischl, G., Zibuschka, F., & Puxbaum, H. (2003). Airborne bacteria as cloud condensation nuclei. *Journal of Geophysical Research, 108*(D21), 2003JD003545. Advance online publication. DOI: 10.1029/2003JD003545

Behzad, H., Gojobori, T., & Mineta, K. (2015). Challenges and opportunities of airborne metagenomics. *Genome Biology and Evolution, 7*(5), 1216–1226. DOI: 10.1093/gbe/evv064 PMID: 25953766

Bowers, R. M., Clements, N., Emerson, J. B., Wiedinmyer, C., Hannigan, M. P., & Fierer, N. (2013). Seasonal variability in bacterial and fungal diversity of the near-surface atmosphere. *Environmental Science & Technology, 47*(21), 12097–12106. DOI: 10.1021/es402970s PMID: 24083487

Bowers, R. M., Lauber, C. L., Wiedinmyer, C., Hamady, M., Hallar, A. G., Fall, R., Knight, R., & Fierer, N. (2009). Characterization of airborne microbial communities at a High-Elevation site and their potential to act as atmospheric ice nuclei. *Applied and Environmental Microbiology, 75*(15), 5121–5130. DOI: 10.1128/AEM.00447-09 PMID: 19502432

Boyd, D. R., Sharma, N. D., McMurray, B., Haughey, S. A., Allen, C. C. R., Hamilton, J. T. G., McRoberts, W. C., More O'Ferrall, R. A., Nikodinovic-Runic, J., Coulombel, L. A., & O'Connor, K. E. (2011). Bacterial dioxygenase- and monooxygenase-catalysed sulfoxidation of benzo[b]thiophenes. *Organic & Biomolecular Chemistry, 10*(4), 782–790. DOI: 10.1039/C1OB06678A PMID: 22134441

Byrne, A. M., Kukor, J. J., & Olsen, R. H. (1995). Sequence analysis of the gene cluster encoding toluene-3-monooxygenase from Pseudomonas pickettii PKO1. *Gene, 154*(1), 65–70. DOI: 10.1016/0378-1119(94)00844-I PMID: 7867951

Callaghan, A. V. (2013). Enzymes involved in the anaerobic biodegradation of hydrocarbons. *Frontiers in Microbiology, 4*, 89. DOI: 10.3389/fmicb.2013.00089 PMID: 23717304

Cao, L., Men, Q., Zhang, Z., Yue, H., Cui, S., Huang, X., Zhang, Y., Wang, J., Chen, M., & Li, H. (2024). Significance of Volatile Organic Compounds to Secondary Pollution Formation and Health Risks Observed during a Summer Campaign in an Industrial Urban Area. *Toxics, 12*(1), 34. DOI: 10.3390/toxics12010034 PMID: 38250990

Chang, J., Peng, P., DiSpirito, A. A., & Semrau, J. D. (2022). Variable inhibition of nitrous oxide reduction in denitrifying bacteria by different forms of methanobactin. *Applied and Environmental Microbiology*, *88*(7), e02346-21. Advance online publication. DOI: 10.1128/aem.02346-21 PMID: 35285718

Chaudhary, D. K., Park, J., Kim, P., Ok, Y. S., & Hong, Y. (2023). Enrichment cultivation of VOC-degrading bacteria using diffusion bioreactor and development of bacterial-immobilized biochar for VOC bioremediation. *Environmental Pollution*, *320*, 121089. DOI: 10.1016/j.envpol.2023.121089 PMID: 36669717

Chaudhary, P., Padhi, S. K., Pattanaik, L., & Dikshit, P. K. (2025). Recent advances in treating volatile organic compounds from waste gas streams: The potential of anoxic/anaerobic bioreactors, parameters affecting their performance and future perspectives. *Biomass Conversion and Biorefinery*, *15*(11), 16383–16408. DOI: 10.1007/s13399-025-06635-0

Chen, X., Kumari, D., & Achal, V. (2020). A review on Airborne Microbes: The Characteristics of Sources, Pathogenicity and Geography. *Atmosphere (Basel)*, *11*(9), 919. DOI: 10.3390/atmos11090919

Cheng, Y., He, H., Yang, C., Zeng, G., Li, X., Chen, H., & Yu, G. (2016). Challenges and solutions for biofiltration of hydrophobic volatile organic compounds. *Biotechnology Advances*, *34*(6), 1091–1102. DOI: 10.1016/j.biotechadv.2016.06.007 PMID: 27374790

David, E., & Niculescu, V. (2021). Volatile Organic compounds (VOCs) as environmental pollutants: Occurrence and mitigation using nanomaterials. *International Journal of Environmental Research and Public Health*, *18*(24), 13147. DOI: 10.3390/ijerph182413147 PMID: 34948756

Després, V. R., Huffman, J. A., Burrows, S. M., Hoose, C., Safatov, A., Buryak, G., Fröhlich-Nowoisky, J., Elbert, W., Andreae, M. O., Pöschl, U., & Jaenicke, R. (2012). Primary biological aerosol particles in the atmosphere: A review. Tellus B. *Tellus. Series B, Chemical and Physical Meteorology*, *64*(1), 15598. DOI: 10.3402/tellusb.v64i0.15598

Eberwein, M., Hellmold, N., Frank, R., Deobald, D., & Adrian, L. (2024). Reductive dehalogenase of *Dehalococcoides mccartyi* strain CBDB1 reduces cobalt- containing metal complexes enabling anodic respiration. *Frontiers in Microbiology*, *15*, 1457014. DOI: 10.3389/fmicb.2024.1457014 PMID: 39507334

Ervens, B., Amato, P., Aregahegn, K., Joly, M., Khaled, A., Labed-Veydert, T., Mathonat, F., Nuñez López, L., Péguilhan, R., & Zhang, M. (2025). Ideas and perspectives: Microorganisms in the air through the lenses of atmospheric chemistry and microphysics. *Biogeosciences, 22*(1), 243–256. DOI: 10.5194/bg-22-243-2025

Fröhlich-Nowoisky, J., Kampf, C. J., Weber, B., Huffman, J. A., Pöhlker, C., Andreae, M. O., Lang-Yona, N., Burrows, S. M., Gunthe, S. S., Elbert, W., Su, H., Hoor, P., Thines, E., Hoffmann, T., Després, V. R., & Pöschl, U. (2016). Bioaerosols in the Earth system: Climate, health, and ecosystem interactions. *Atmospheric Research, 182*, 346–376. DOI: 10.1016/j.atmosres.2016.07.018

Ganguly, P., Mandal, J., Sengupta, S., & Sinha, B. (2023). Basics of environment, pollution, and bioremediation techniques. In *Apple Academic Press eBooks* (pp. 1–17). https://doi.org/DOI: 10.1201/9781003412236-1

Gieg, L. M., Fowler, S. J., & Berdugo-Clavijo, C. (2014). Syntrophic biodegradation of hydrocarbon contaminants. *Current Opinion in Biotechnology, 27*, 21–29. DOI: 10.1016/j.copbio.2013.09.002 PMID: 24863893

González-Martín, J., Cantera, S., Muñoz, R., & Lebrero, R. (2023). Indoor air VOCs biofiltration by bioactive coating packed bed bioreactors. *Journal of Environmental Management, 349*, 119362. DOI: 10.1016/j.jenvman.2023.119362 PMID: 37897901

Haines, S. R., Hall, E. C., Marciniak, K., Misztal, P. K., Goldstein, A. H., Adams, R. I., & Dannemiller, K. C. (2021). Microbial growth and volatile organic compound (VOC) emissions from carpet and drywall under elevated relative humidity conditions. *Microbiome, 9*(1), 209. Advance online publication. DOI: 10.1186/s40168-021-01158-y PMID: 34666813

Harwood, C. S., & Parales, R. E. (1996). The β-ketoadipate pathway and the biology of aromatic compound degradation. *Annual Review of Microbiology, 50*, 553–590. DOI: 10.1146/annurev.micro.50.1.553 PMID: 8905091

He, Q., Wang, Z., Liu, H., Xu, P., Duan, R., Xu, C., Chen, J., & Wei, M. (2024). Influence of terrestrial and marine air mass on the constituents and intermixing of bioaerosols over a coastal atmosphere. *Atmospheric Chemistry and Physics, 24*(22), 12775–12792. DOI: 10.5194/acp-24-12775-2024

Heeley-Hill, A. C., Grange, S. K., Ward, M. W., Lewis, A. C., Owen, N., Jordan, C., Hodgson, G., & Adamson, G. (2021). Frequency of use of household products containing VOCs and indoor atmospheric concentrations in homes. *Environmental Science. Processes & Impacts, 23*(5), 699–713. DOI: 10.1039/D0EM00504E PMID: 34037627

Hickey, W. J. (2021). Biodegradation of environmental pollutants. In *Elsevier eBooks* (pp. 581–605). https://doi.org/DOI: 10.1016/B978-0-12-820202-9.00021-6

Hiroshi, S. (2024). Controlling genetic engineering in Environmental biotechnology: Potentials and challenges. *Advancements in Genetic Engineering*, *13*(02). Advance online publication. DOI: 10.35841/2169-0111.24.13.264

Hu, F., Wang, P., Li, Y., Ling, J., Ruan, Y., Yu, J., & Zhang, L. (2023). Bioremediation of environmental organic pollutants by Pseudomonas aeruginosa: Mechanisms, methods and challenges. *Environmental Research*, *239*, 117211. DOI: 10.1016/j.envres.2023.117211 PMID: 37778604

Ilori, M. O., & Obayori, O. S. (2023). Introduction to aeromicrobiology. Ilori MO, Obayori OS. Salam LB (eds), Aeromicrobiology. First Edition, Publisher: Academic Press, an imprint of Elsevier, London, United Kingdom, pp. 1–16. https://doi.org/DOI: 10.1016/B978-0-323-96122-6.00005-0

Irfana, S., Ahmad Farooq, L., Moieza, A., Aneesul Mohammad, M., & Asmat, R. (2016). Biofilters in Mitigation of Odour Pollution - A Review. *Nature Environment and Pollution Technology*, *15*(4), 1177–1185.

Isha, N., Ali, S., Khalid, A., Naseer, I. A., Raza, H., & Chang, Y. (2024). Bioremediation of smog: Current trends and future perspectives. *Processes (Basel, Switzerland)*, *12*(10), 2266. DOI: 10.3390/pr12102266

Jaiswal, S., Singh, D. K., & Shukla, P. (2019). Gene Editing and Systems Biology Tools for Pesticide Bioremediation: A review. *Frontiers in Microbiology*, *10*, 87. Advance online publication. DOI: 10.3389/fmicb.2019.00087 PMID: 30853940

Jones, A. M., & Harrison, R. M. (2004). The effects of meteorological factors on atmospheric bioaerosol concentrations—A review. *The Science of the Total Environment*, *326*(1-3), 151–180. DOI: 10.1016/j.scitotenv.2003.11.021 PMID: 15142773

Kalalian, C., Grira, A., Illmann, J. N., Patroescu-Klotz, I., Dib, G. E., & Coddeville, P.. (2022). Experimental and theoretical studies of Trans-2-Pentenal atmospheric ozonolysis. *Atmosphere (Basel)*, *13*(2), 291. DOI: 10.3390/atmos13020291

Kansal, A. (2008). Sources and reactivity of NMHCs and VOCs in the atmosphere: A review. *Journal of Hazardous Materials*, *166*(1), 17–26. DOI: 10.1016/j.jhazmat.2008.11.048 PMID: 19136203

Karigar, C. S., & Rao, S. S. (2011). Role of microbial enzymes in the bioremediation of pollutants: A review. *Enzyme Research*, *2011*, 1–11. DOI: 10.4061/2011/805187 PMID: 21912739

Karimi, B., Habibi, M., & Esvand, M. (2015). Biodegradation of naphthalene using Pseudomonas aeruginosa by up-flow anoxic–aerobic continuous flow combined bioreactor. *Journal of Environmental Health Science & Engineering*, *13*(1), 26. Advance online publication. DOI: 10.1186/s40201-015-0175-1 PMID: 25859393

Khan, M., Wibowo, A., Karim, Z., Posoknistakul, P., Matsagar, B., Wu, K., & Sakdaronnarong, C. (2024). Wastewater Treatment Using Membrane Bioreactor Technologies: Removal of Phenolic Contaminants from Oil and Coal Refineries and Pharmaceutical Industries. *Polymers*, *16*(3), 443. DOI: 10.3390/polym16030443 PMID: 38337332

Kleinheinz, G. T., Bagley, S. T., St John, W. P., Rughani, J. R., & McGinnis, G. D. (1999). Characterization of Alpha-Pinene-Degrading microorganisms and application to a Bench-Scale biofiltration system for VOC degradation. *Archives of Environmental Contamination and Toxicology*, *37*(2), 151–157. DOI: 10.1007/s002449900500 PMID: 10398764

Kour, D., Kaur, T., Devi, R., Yadav, A., Singh, M., Joshi, D., Singh, J., Suyal, D. C., Kumar, A., Rajput, V. D., Yadav, A. N., Singh, K., Singh, J., Sayyed, R. Z., Arora, N. K., & Saxena, A. K. (2021). Beneficial microbiomes for bioremediation of diverse contaminated environments for environmental sustainability: Present status and future challenges. *Environmental Science and Pollution Research International*, *28*(20), 24917–24939. DOI: 10.1007/s11356-021-13252-7 PMID: 33768457

Kour, D., Khan, S. S., Kour, H., Kaur, T., Devi, R., Judy, C., Rai, P. K., McQuestion, C., Spells, S., Bianchi, A., Mohan, R., Rai, A. K., & Yadav, A. N. (2022). Microbe-mediated bioremediation: Current research and future challenges. *Journal of Applied Biology and Biotechnology*, •••, 6–24. DOI: 10.7324/JABB.2022.10s202

Kuźma, Ł., Dąbrowski, E. J., Kurasz, A., Bachórzewska-Gajewska, H., & Dobrzycki, S. (2020). The 10-Year study of the impact of particulate matters on mortality in two transit cities in North-Eastern Poland (PL-PARTICLES). *Journal of Clinical Medicine*, *9*(11), 3445. DOI: 10.3390/jcm9113445 PMID: 33120927

Lang-Yona, N., Flores, J. M., Haviv, R., Alberti, A., Poulain, J., Belser, C., ... (2021). A cross-biomes bacterial diversity shed light on ocean-atmosphere microbial transmission. bioRxiv *(Cold Spring Harbor Laboratory)*. https://doi.org/DOI: 10.1101/2021.06.06.445733

Lebrero, R., Rodríguez, E., Estrada, J. M., García-Encina, P. A., & Muñoz, R. (2012). Odor abatement in biotrickling filters: Effect of the EBRT on methyl mercaptan and hydrophobic VOCs removal. *Bioresource Technology*, *109*, 38–45. DOI: 10.1016/j.biortech.2012.01.052 PMID: 22310214

Li, B., Mao, Z., Xue, J., Xing, P., & Wu, Q. L. (2024). Metabolic versatility of aerobic methane-oxidizing bacteria under anoxia in aquatic ecosystems. *Environmental Microbiology Reports, 16*(5), e70002. DOI: 10.1111/1758-2229.70002 PMID: 39232853

Li, Z., Gao, Y., Lu, Z., Xie, J., Liu, Y., Wang, G., & Cheng, X. (2023). Strengthening and microbial regulation mechanism of *Bacillus* on purification device for grass carp culture wastewater. *Frontiers in Environmental Science, 11*, 1128329. Advance online publication. DOI: 10.3389/fenvs.2023.1128329

Lin, B., Van Verseveld, H. W., & Röling, W. F. M. (2002). Microbial aspects of anaerobic BTEX degradation. *PubMed, 15*(2), 130–144. https://pubmed.ncbi.nlm.nih.gov/12244754

Liu, C., Cai, J., Chen, R., Sera, F., Guo, Y., Tong, S., Li, S., Lavigne, E., Correa, P. M., Ortega, N. V., Orru, H., Maasikmets, M., Jaakkola, J. J. K., Ryti, N., Breitner, S., Schneider, A., Katsouyanni, K., Samoli, E., Hashizume, M., & Kan, H. (2022). Coarse particulate air pollution and daily mortality: A global study in 205 cities. *American Journal of Respiratory and Critical Care Medicine, 206*(8), 999–1007. DOI: 10.1164/rccm.202111-2657OC PMID: 35671471

Liu, X., & Mattes, T. E. (2016). Epoxyalkane:Coenzyme M Transferase Gene Diversity and Distribution in Groundwater Samples from Chlorinated-Ethene-Contaminated Sites. *Applied and Environmental Microbiology, 82*(11), 3269–3279. DOI: 10.1128/AEM.00673-16 PMID: 27016563

Loveniers, P., Devlieghere, F., & Sampers, I. (2024). Towards tailored guidelines for microbial air quality in the food industry. *International Journal of Food Microbiology, 421*, 110779. DOI: 10.1016/j.ijfoodmicro.2024.110779 PMID: 38852216

Lukas, M., Schwidetzky, R., Kunert, A. T., Pöschl, U., Fröhlich-Nowoisky, J., Bonn, M., & Meister, K. (2020). Electrostatic interactions control the functionality of bacterial ICE nucleators. *Journal of the American Chemical Society, 142*(15), 6842–6846. DOI: 10.1021/jacs.9b13069 PMID: 32223131

Lv, Y., Wang, L., Liu, X., Chen, B., & Zhang, M. (2022). Degradation kinetics of aromatic VOCs polluted wastewater by functional bacteria at laboratory scale. *Scientific Reports, 12*(1), 19053. Advance online publication. DOI: 10.1038/s41598-022-21356-4 PMID: 36351963

Mahjadeen, I. (2023). Airborne Microorganisms in the Environment. *Research Gate.* DOI: DOI: 10.13140/RG.2.2.27961.58729

Malard, L. A., & Pearce, D. A. (2022). Bacterial colonisation: From airborne dispersal to integration within the soil community. *Frontiers in Microbiology*, *13*, 782789. Advance online publication. DOI: 10.3389/fmicb.2022.782789 PMID: 35615521

Manisalidis, I., Stavropoulou, E., Stavropoulos, A., & Bezirtzoglou, E. (2020). Environmental and Health Impacts of Air Pollution: A review. *Frontiers in Public Health*, *8*, 14. Advance online publication. DOI: 10.3389/fpubh.2020.00014 PMID: 32154200

Martínez-García, E., Nikel, P. I., Aparicio, T., & De Lorenzo, V. (2014). Pseudomonas 2.0: Genetic upgrading of *P. putida* KT2440 as an enhanced host for heterologous gene expression. *Microbial Cell Factories*, *13*(1), 159. Advance online publication. DOI: 10.1186/s12934-014-0159-3 PMID: 25384394

Maurya, A., Sharma, D., Partap, M., Kumar, R., & Bhargava, B. (2023). Microbially-assisted phytoremediation toward air pollutants: Current trends and future directions. *Environmental Technology & Innovation*, *31*, 103140. DOI: 10.1016/j.eti.2023.103140

Minhas, W. R., Saleemi, H., Raheem, M. F., Khalid, S. S., Aslam, H., & Ahmad, W. (2024). CRISPR-Engineered Microbes for Enhanced biodegradation of recalcitrant pollutants a genomic approach to environmental remediation. *Frontiers in Health Informatics*, •••, 8281–8304. DOI: 10.63682/fhi787

Mock, J., Schühle, K., Linne, U., Mock, M., & Heider, J. (2024). A Synthetic Pathway for the Production of Benzylsuccinate in *Escherichia coli. Molecules (Basel, Switzerland)*, *29*(2), 415. DOI: 10.3390/molecules29020415 PMID: 38257328

Morakinyo, O. M., & Ana, G. R. E. E. (2025). Seasonal levels and size fractions of indoor air bioaerosols as predictors of respiratory morbidities among school pupils in Ibadan, Nigeria. *Frontiers in Public Health*, *13*, 1611167. DOI: 10.3389/fpubh.2025.1611167 PMID: 41112650

Morris, C. E., Sands, D. C., Glaux, C., Samsatly, J., Asaad, S., Moukahel, A. R., Gonçalves, F. L. T., & Bigg, E. K. (2013). Urediospores of rust fungi are ice nucleation active at > −10 °C and harbor ice nucleation active bacteria. *Atmospheric Chemistry and Physics*, *13*(8), 4223–4233. DOI: 10.5194/acp-13-4223-2013

Mu, F., Li, Y., Lu, R., Qi, Y., Xie, W., & Bai, W. (2019). Source identification of airborne bacteria in the mountainous area and the urban areas. *Atmospheric Research*, *231*, 104676. DOI: 10.1016/j.atmosres.2019.104676

National Volatile Organic Compound Emission Standards for Aerosol Coatings: Interim Final rule. (2025, July 2). Federal Register. https://www.federalregister.gov/documents/2025/07/02/2025-12323/national-volatile-organic-compound-emission-standards-for-aerosol-coatings-interim-final-rule

Nwankwegu, A. S., Zhang, L., Xie, D., Onwosi, C. O., Muhammad, W. I., Odoh, C. K., Sam, K., & Idenyi, J. N. (2021). Bioaugmentation as a green technology for hydrocarbon pollution remediation. Problems and prospects. *Journal of Environmental Management*, *304*, 114313. DOI: 10.1016/j.jenvman.2021.114313 PMID: 34942548

Oliveira, K., Guevara, M., Jorba, O., Petetin, H., Bowdalo, D., Tena, C., Montané Pinto, G., López, F., & Pérez García-Pando, C. (2024). On the uncertainty of anthropogenic aromatic volatile organic compound emissions: Model evaluation and sensitivity analysis. *Atmospheric Chemistry and Physics*, *24*(12), 7137–7177. DOI: 10.5194/acp-24-7137-2024

Pachaiappan, R., Cornejo-Ponce, L., Rajendran, R., Manavalan, K., Rajan, V. F., & Awad, F. (2022). A review on biofiltration techniques: Recent advancements in the removal of volatile organic compounds and heavy metals in the treatment of polluted water. *Bioengineered*, *13*(4), 8432–8477. DOI: 10.1080/21655979.2022.2050538 PMID: 35260028

Pandey, P., & Yadav, R. (2018). A review on Volatile Organic Compounds (VOCs) as Environmental Pollutants: Fate and Distribution. *International Journal of Plant and Environment*, *4*(02), 14–26. DOI: 10.18811/ijpen.v4i02.2

Pozdniakova, S., Uchida, A., Fontal, A., Cañas, L., Santamaria, S., Hui, L. Y., Luhung, I., Schuster, S. C., Rodó, X., & Borràs, S. (2025). Integrating air microbiome for comprehensive air quality analysis. *iScience*, *28*(7), 113015. DOI: 10.1016/j.isci.2025.113015 PMID: 40808715

Precha, N., Chaisiri, K., Worakhunpiset, S., Limpanont, Y., Yamamoto, N., Suksong, W., Kliengchuay, W., & Tantrakarnapa, K. (2025). Comparison of airborne bacterial communities in PM2.5 between a dry-season haze period and a wet-season non-haze period in thailand. *Scientific Reports*, *15*(1), 12918. Advance online publication. DOI: 10.1038/s41598-025-97966-5 PMID: 40234570

Prenafeta-Boldú, F. X., Guivernau, M., Gallastegui, G., Viñas, M., Hoog, G. S., & Elías, A. (2012). Fungal/bacterial interactions during the biodegradation of TEX hydrocarbons (toluene, ethylbenzene and p-xylene) in gas biofilters operated under xerophilic conditions. *FEMS Microbiology Ecology*, *80*(3), 722–734. DOI: 10.1111/j.1574-6941.2012.01344.x PMID: 22380456

Prussin, A. J.II, & Marr, L. C. (2015). Sources of airborne microorganisms in the built environment. *Microbiome*, *3*(1), 78. Advance online publication. DOI: 10.1186/s40168-015-0144-z PMID: 26694197

Qamar, S. A., Bhatt, P., Ghotekar, S., & Bilal, M. (2022). Nanomaterials for bioremediation of air pollution. In *Elsevier eBooks* (pp. 243–261). https://doi.org/DOI: 10.1016/B978-0-12-823962-9.00008-8

Qi, Y., Li, Y., Xie, W., Lu, R., Mu, F., Bai, W., & Du, S. (2019). Temporal-spatial variations of fungal composition in PM2.5 and source tracking of airborne fungi in mountainous and urban regions. *The Science of the Total Environment*, *708*, 135027. DOI: 10.1016/j.scitotenv.2019.135027 PMID: 31787277

Rahman, M. A., Dey, B., Halim, M. A., & Ahmed, R. (2025). Mobilizing Microbes for Bioremediation Strategies in the Context of Climate Change. In *Sustainable Remediation for Pollution and Climate Resilience* (pp. 315–346). https://doi.org/DOI: 10.1007/978-981-96-5674-5_13

Rahmati, F., Lajayer, B. A., Shadfar, N., Van Bodegom, P. M., & Van Hullebusch, E. D. (2022). A review on biotechnological approaches applied for marine hydrocarbon spills remediation. *Microorganisms*, *10*(7), 1289. DOI: 10.3390/microorganisms10071289 PMID: 35889007

Rojo, F. (2009). Degradation of alkanes by bacteria. *Environmental Microbiology*, *11*(10), 2477–2490. DOI: 10.1111/j.1462-2920.2009.01948.x PMID: 19807712

Roth, E., Tape, C., Michonneau, P., & Chakir, A. (2025). Time-dependent fluxes and emission factors of volatile organic compounds from pig manure and solid/liquid digestates. *The Science of the Total Environment*, *996*, 180147. DOI: 10.1016/j.scitotenv.2025.180147 PMID: 40743754

Sakai, Y. (2023). Methane monooxygenases: Physiology, biochemistry and structural insights. *Chemical Reviews*, *123*(8), 5432–5468. DOI: 10.1021/acs.chemrev.2c00712

Šantl-Temkiv, T., Amato, P., Casamayor, E. O., Lee, P. K. H., & Pointing, S. B. (2022). Microbial ecology of the atmosphere. *FEMS Microbiology Reviews*, *46*(4), fuac009. Advance online publication. DOI: 10.1093/femsre/fuac009 PMID: 35137064

Sattler, B., Puxbaum, H., & Psenner, R. (2001). Bacterial growth in supercooled cloud droplets. *Geophysical Research Letters*, *28*(2), 239–242. DOI: 10.1029/2000GL011684

Sebald, M., & Hauser, D. (1995). Pasteur, oxygen and the anaerobes revisited. *Anaerobe*, *1*(1), 11–16. DOI: 10.1016/S1075-9964(95)80353-X PMID: 16887501

Shah, B. A., Malhotra, H., Papade, S. E., Dhamale, T., Ingale, O. P., Kasarlawar, S. T., & Phale, P. S. (2024). Microbial degradation of contaminants of emerging concern: Metabolic, genetic and omics insights for enhanced bioremediation. *Frontiers in Bioengineering and Biotechnology, 12*, 1470522. Advance online publication. DOI: 10.3389/fbioe.2024.1470522 PMID: 39364263

Sheoran, K., Siwal, S. S., Kapoor, D., Singh, N., Saini, A. K., Alsanie, W. F., & Thakur, V. K. (2022). Air pollutants removal using biofiltration technique: A challenge at the frontiers of sustainable environment. *ACS Engineering Au, 2*(5), 378–396. DOI: 10.1021/acsengineeringau.2c00020 PMID: 36281334

Shuai, J., Kim, S., Ryu, H., Park, J., Lee, C. K., Kim, G., Ultra, V. U.Jr, & Yang, W. (2018). Health risk assessment of volatile organic compounds exposure near Daegu dyeing industrial complex in South Korea. *BMC Public Health, 18*(1), 528. Advance online publication. DOI: 10.1186/s12889-018-5454-1 PMID: 29678134

Silva, J. R., Quinta-Ferreira, R. M., & Castro, L. M. (2025). Biological Treatments for VOC-Contaminated Off-Gas: Advances, Challenges, and Energetic Valorization Opportunities. *Sustainability (Basel), 17*(11), 4802. DOI: 10.3390/su17114802

Smets, W., Moretti, S., Denys, S., & Lebeer, S. (2016). Airborne bacteria in the atmosphere: Presence, purpose, and potential. *Atmospheric Environment, 139*, 214–221. DOI: 10.1016/j.atmosenv.2016.05.038

Smidt, H., & de Vos, W. M. (2004). Anaerobic microbial dehalogenation. *Annual Review of Microbiology, 58*(1), 43–73. DOI: 10.1146/annurev.micro.58.030603.123600 PMID: 15487929

Su, C., Jiang, S., & Dai, J. (2025). Development of a versatile synthetic microbe for the concurrent degradation of organic pollutants. *Advanced Biotechnology, 3*(2), 16. Advance online publication. DOI: 10.1007/s44307-025-00072-w PMID: 40408009

Sun, C., Yuan, J., Xu, H., Huang, S., Wen, X., Tong, N., & Zhang, Y. (2019). Simultaneous removal of nitric oxide and sulfur dioxide in a biofilter under micro-oxygen thermophilic conditions: Removal performance, competitive relationship and bacterial community structure. *Bioresource Technology, 290*, 121768. DOI: 10.1016/j.biortech.2019.121768 PMID: 31323510

Toledo, M., & Muñoz, R. (2025). Odour prevention strategies in wastewater treatment and composting plants: A review. *Journal of Environmental Management, 375*, 124402. DOI: 10.1016/j.jenvman.2025.124402 PMID: 39908607

Vali, G., & Schnell, R. (1976). Biogenic Ice Nuclei: Part I. Terrestrial and Marine Sources. *Journal of the Atmospheric Sciences*, *33*(8), 1554–1564. DOI: 10.1175/1520-0469(1976)033<1565:BINPIB>2.0.CO;2

Vergara-Fernández, A., Yánez, D., Morales, P., Scott, F., Aroca, G., Diaz-Robles, L., & Moreno-Casas, P. (2017). Biofiltration of benzo[α]pyrene, toluene and formaldehyde in air by a consortium of Rhodococcus erythropolis and Fusarium solani: Effect of inlet loads, gas flow and temperature. *Chemical Engineering Journal*, *332*, 702–710. DOI: 10.1016/j.cej.2017.09.095

Vogt, C., Kleinsteuber, S., & Richnow, H. H. (2011). Anaerobic benzene degradation by bacteria. *Microbial Biotechnology*, *4*(6), 710–724. DOI: 10.1111/j.1751-7915.2011.00260.x PMID: 21450012

Wang, H., Liu, X., Wu, C., & Lin, G. (2024). Regional to global distributions, trends, and drivers of biogenic volatile organic compound emission from 2001 to 2020. *Atmospheric Chemistry and Physics*, *24*(5), 3309–3328. DOI: 10.5194/acp-24-3309-2024

Wang, S., Ang, H., & Tade, M. O. (2007). Volatile organic compounds in indoor environment and photocatalytic oxidation: State of the art. *Environment International*, *33*(5), 694–705. DOI: 10.1016/j.envint.2007.02.011 PMID: 17376530

Wang, Y., Bi, R., Ge, J., & Xu, P. (2025). VOCs absorption using ionic liquids. *Separation and Purification Technology*, *132265*, 132265. Advance online publication. DOI: 10.1016/j.seppur.2025.132265

Williams, J., & Koppmann, R. (2007). Volatile Organic Compounds in the atmosphere: An overview. In *Volatile Organic Compounds in the Atmosphere* (pp. 1–32). https://doi.org/DOI: 10.1002/9780470988657.ch1

Womack, A. M., Bohannan, B. J. M., & Green, J. L. (2010). Biodiversity and biogeography of the atmosphere. *Philosophical Transactions of the Royal Society of London. Series B, Biological Sciences*, *365*(1558), 3645–3653. DOI: 10.1098/rstb.2010.0283 PMID: 20980313

Wu, H., Guo, C., Yin, Z., Quan, Y., & Yin, C. (2018a). Performance and bacterial diversity of biotrickling filters filled with conductive packing material for the treatment of toluene. *Bioresource Technology*, *257*, 201–209. DOI: 10.1016/j.biortech.2018.02.108 PMID: 29501953

Wu, H., Yan, H., Quan, Y., Zhao, H., Jiang, N., & Yin, C. (2018b). Recent progress and perspectives in biotrickling filters for VOCs and odorous gases treatment. *Journal of Environmental Management*, *222*, 409–419. DOI: 10.1016/j.jenvman.2018.06.001 PMID: 29883876

Yang, H., He, Y., Liao, J., Li, X., Zhang, J., Liebl, W., & Chen, F. (2022). RNA-Seq transcriptomic analysis reveals gene expression profiles of acetic acid bacteria under high-acidity submerged industrial fermentation process. *Frontiers in Microbiology*, *13*, 956729. Advance online publication. DOI: 10.3389/fmicb.2022.956729 PMID: 36246236

Yang, T., Xin, Y., Zhang, L., Gu, Z., Li, Y., Ding, Z., & Shi, G. (2020). Characterization on the aerobic denitrification process of Bacillus strains. *Biomass and Bioenergy*, *140*, 105677. DOI: 10.1016/j.biombioe.2020.105677

Yeoman, A. M., & Lewis, A. C. (2021). Global emissions of VOCs from compressed aerosol products. *Elementa*, *9*(1), 00177. Advance online publication. DOI: 10.1525/elementa.2020.20.00177

Yoshikawa, M., Zhang, M., & Toyota, K. (2016). Integrated Anaerobic-Aerobic biodegradation of multiple contaminants including chlorinated ethylenes, benzene, toluene, and dichloromethane. *Water, Air, and Soil Pollution*, *228*(1), 25. Advance online publication. DOI: 10.1007/s11270-016-3216-1 PMID: 28042183

Yoshikawa, M., Zhang, M., & Toyota, K. (2017). Biodegradation of Volatile Organic Compounds and Their Effects on Biodegradability under Co-Existing Conditions. *Microbes and Environments*, *32*(3), 188–200. DOI: 10.1264/jsme2.ME16188 PMID: 28904262

Zhao, J., Jin, L., Wu, D., Xie, J., Li, J., Fu, X., Cong, Z., Fu, P., Zhang, Y., Luo, X., Feng, X., Zhang, G., Tiedje, J. M., & Li, X. (2022). Global airborne bacterial community—Interactions with Earth's microbiomes and anthropogenic activities. *Proceedings of the National Academy of Sciences of the United States of America*, *119*(42), e2204465119. Advance online publication. DOI: 10.1073/pnas.2204465119 PMID: 36215495

Chapter 7
Sustainable Restoration of Urban and Agricultural Ecosystems Through Microbial Interventions

Ayodele Elizabeth Omotayo
http://orcid.org/0000-0003-1271-517X
University of Lagos, Nigeria

Olufunke O. Oyebamiji
http://orcid.org/0000-0002-6044-2084
Yaba College of Technology, Nigeria

Chinenye P. Anozie
Christopher University, Mowe, Nigeria

Olanike M. Buraimoh
University of Lagos, Nigeria

Adewale K. Ogunyemi
Trinity University, Nigeria

Olukayode O. Amund
University of Lagos, Nigeria

ABSTRACT

The accelerating degradation of urban and agricultural ecosystems due to anthropogenic pressures poses significant threats to biodiversity, soil health, and water quality. In response, there is a growing emphasis on nature-based and sustainable restoration strategies that align with global environmental goals. This chapter explores microbial interventions as a central paradigm for sustainable ecosystem restoration, highlighting their functional roles, mechanistic pathways, and applicability across diverse environmental systems. Their synergistic interactions within soils, wastewater systems, and green infrastructure significantly enhance ecosystem services, including soil fertility, pollutant removal, and climate regulation. Nonetheless, practical deployment is constrained by challenges related to environmental variability,

DOI: 10.4018/979-8-2600-2100-2.ch007

scalability, and microbial survival in situ. By positioning microbes as pivotal agents of ecological recovery, this chapter contributes to the conceptual and practical advancement of sustainable restoration strategies.

1 INTRODUCTION

1.1 Ecosystem Degradation: Impact of Human Activities on Urban and Agricultural Ecosystems

The 21st century has introduced unprecedented environmental changes, characterized by rapid and simultaneous evolution of human cultures and the ecosystems they rely on. Despite this dynamic nature presenting significant challenges, the diversity inherent to ecosystems acts as a crucial buffer, preventing catastrophic failures by mitigating extreme disruptions and inherited ecological faults. Early in this century, ecosystem degradation became a key issue in human ecological research (Singh and Gupta, 2018), demanding immediate action from scientists and policymakers.

The scale of human impact on the biosphere cannot be overstated. Activities such as urban sprawl, intensive agriculture, deforestation, oil exploration, industrial operations, mining, and tidal flat farming have led to vast biodiversity loss, habitat destruction, marine pollution, and alien species invasions, all severely affecting human well-being (Xie et al., 2018). The Millennium Ecosystem Assessment (2005) reported that over 60% of global ecosystem services have been seriously degraded, primarily driven by human activity. Future projections foresee escalating threats to biodiversity (Sala et al., 2000), making the creation of effective restoration strategies an existential necessity. In this context, ecological restoration has become a central focus in global environmental efforts (Jacobs et al., 2015). The urgency is reflected in global policies such as the UN Convention on Biological Diversity, which targeted restoration of at least 15% of degraded ecosystems between 2011 and 2020 (CBD, 2010). Practical restoration ecology offers a potentially cost-effective and resilient path forward, emphasizing critical interactions between microorganisms and ecosystems (Suding et al., 2015).

2 URBAN AND AGRICULTURAL WASTE MANAGEMENT PRACTICES

Figure 1. Circulation of materials in urban agroecosystem

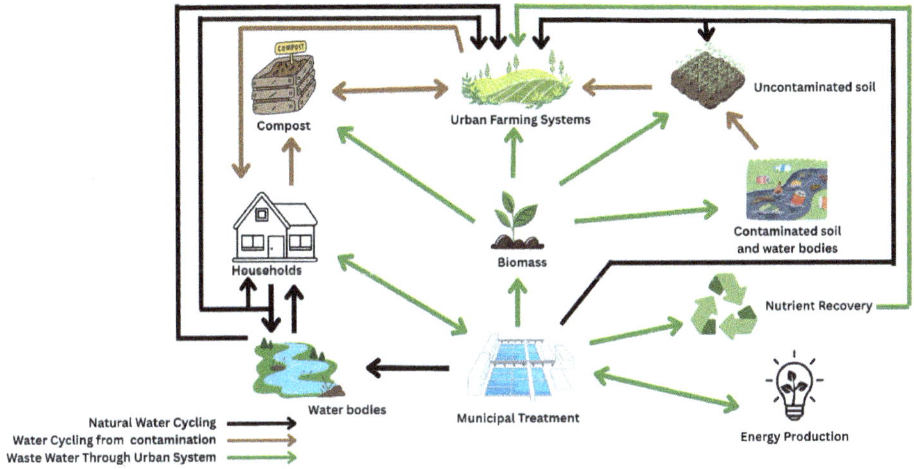

(Omotayo et al., 2025)

2.1. Microbial Interventions in Urban Ecosystem Restoration

Materials are continuously circulated in the environment (Figure 1) and major groups of microorganisms play indispensable and interconnected roles in maintaining ecosystem stability and driving biogeochemical cycles. Photoautotrophic bacteria such as cyanobacteria form the foundation of aquatic food webs through primary production and oxygen generation via photosynthesis (Falkowski, 2015). Chemoautotrophic bacteria further sustain ecosystems by driving nutrient cycling processes, including nitrification, sulfur oxidation, and iron oxidation, thereby linking inorganic compounds to biological productivity (Madigan et al., 2018). Heterotrophic bacteria and saprophytic fungi dominate organic matter decomposition, using extracellular enzymes to mineralize complex organic compounds and pollutants, which is critical for nutrient recycling and soil health (Rastogi and Madamwar, 2021). Methanogenic archaea and methanotrophic bacteria jointly influence the global carbon cycle by producing and oxidizing methane in anaerobic and aerobic environments, respectively, thereby modulating greenhouse gas emissions (Thauer et al., 2008). Mycorrhizal and lichenized fungi contribute to plant nutrition, soil formation, and rock weathering, supporting ecosystem establishment and long-term resilience (Smith and Read, 2008). Finally, viruses, including DNA and RNA viruses, regulate microbial population dynamics, facilitate horizontal gene transfer, and drive evolutionary processes that shape ecosystem structure and function (Koonin et al., 2015).

2.1.1 Urban Soil Remediation: Microbial Degradation of Pollutants in Urban Soils

Remediation preserves soil health, biodiversity, and vital ecosystem functions, essential for urban revitalization (Vannucchi et al., 2021). Contemporary approaches include physical, chemical, biological, and integrated techniques (Hou et al., 2023; Yaashikaa et al., 2022). Among these, bioremediation stands out for sustainability and for its ancillary benefits such as carbon sequestration and public health improvement (Landa-Acuña et al., 2020; Guidi Nissim et al., 2023).

Soil bioremediation, using microorganisms or plants to detoxify contaminants, offers a cost-efficient, nature-based solution (Megharaj and Naidu, 2017). Microbial mechanisms involve organic pollutant degradation through metabolism into non-toxic forms (Ali et al., 2022; Sharma et al., 2022) and heavy metal immobilization that reduces bioavailability and limits groundwater contamination (Cornu et al., 2017; Derakhshan Nejad et al., 2018). Phytoremediation offers a sustainable remediation approach that utilizes pollution-tolerant plant species working in concert with microbial partners and plant-derived compounds in the root zone to accelerate contaminant degradation (Petrova et al., 2022; Fernández-Braña et al., 2023). This biological remediation technique capitalizes on the inherent ability of select plant species to extract, transform, or immobilize environmental pollutants present in contaminated matrices. The effectiveness of this technology is fundamentally rooted in the cooperative interactions between vegetation and their rhizospheric microbial communities, a phenomenon known as rhizoremediation (Kuiper et al., 2004). Field investigations using *Paspalum vaginatum* exemplify the potential of this methodology, where the combined action of the grass species and its associated hydrocarbon-metabolizing bacteria resulted in 87.7% removal of crude oil from polluted soil over a 45-day period (Omotayo et al., 2014). The grass species possesses a highly branched fibrous root architecture that creates an extensive interfacial area suitable for microbial establishment, while plant-secreted exudates promote the proliferation and metabolic performance of beneficial microorganisms, culminating in enhanced contaminant removal efficiency (Kuiper et al., 2004; Paul et al., 2006). Rhizospheric bacterial strains, specifically *Arthrobacter* sp., *Bacillus pumilus*, *Bacillus sphaericus*, and *Serratia marcescens*, exhibited remarkable biodegradative potential, consuming between 95.6% and 98.3% of petroleum hydrocarbons when cultivated as sole carbon sources over a three-week incubation period (Omotayo et al., 2014). This combined biological system demonstrates exceptional efficacy for addressing petroleum-derived contamination while providing an economically viable and ecologically sound option compared to traditional remediation technologies.

Biotechnological advances on the other hand have augmented remediation options with genetically modified microbes and plants for targeted contaminant removal (Rafeeq et al., 2023), fungal enzymatic degradation pathways, notably from white-rot fungi (Qattan, 2025), and amendments like biochar and nanomaterials to boost microbial activity (Dong et al., 2024). Integrated systems combining plants, microbes, and these amendments hold promise for complex urban contamination under circular economy and green infrastructure frameworks (Xiang et al., 2022).

2.1.2 Wastewater Treatment: Microbial Treatment of Urban Wastewater for Improved Water Quality

Wastewater harbors numerous biological contaminants including pathogenic bacteria (fecal coliforms, *Escherichia coli*, *Salmonella*, *Shigella*, *Vibrio cholerae*), parasites, viruses, and fungi, causing diseases like cholera, typhoid, hepatitis, and dysentery. Pathogens often adhere to suspended solids, complicating removal and posing risks to human and aquatic health if untreated wastewater is released (Okoh et al., 2010).

Biological treatment systems, such as activated sludge, lagoons, constructed wetlands, and rotating biological contactors, effectively reduce pathogens, often achieving removal efficiencies close to or exceeding 99% (Sharafi et al., 2016). Microorganisms enhance wastewater treatment by metabolizing organic pollutants, reducing biochemical oxygen demand (BOD), and removing nutrients such as nitrogen and phosphorus. Bacteria like *Pseudomonas*, *Bacillus*, and *Nitrosomonas* degrade organic matter and facilitate nitrification, while *Nitrobacter* converts nitrite to nitrate. Fungi such as *Aspergillus* aid in breaking down complex compounds, and protozoa (e.g., *Vorticella*) help clarify water by consuming suspended bacteria. Algae like *Chlorella* contribute to oxygen production and nutrient uptake, improving overall water quality (Mishra et al., 2025). Ensuring microbial safety in treated wastewater is critical to prevent disease transmission and protect agricultural workers.

2.1.3 Green Infrastructure: Microbial Enhancement of Green Roofs, Walls, and Urban Gardens

Green infrastructure, defined by the European Commission (2013), consists of natural and semi-natural urban spaces planned to deliver ecosystem services (Hansen and Pauleit, 2014; Dorst et al., 2019). Urban parks, forests, gardens, green roofs, and wetlands provide habitats for microbial communities and multiple environmental benefits. These benefits include flood control and improved water management by enhancing infiltration and reducing runoff (Bonoli et al., 2013; Cristiano et al., 2020; Valois et al., 2023), mitigation of urban heat island effects aiding climate adaptation (la Cecilia et al., 2020; Bouzouidja et al., 2021; Bochenek et al., 2022), and promotion of sustainable urban environments (Russo et al., 2017; Silvestrini et al., 2021; Mihalakakou et al., 2023). These effects balance anthropogenic heat and alter urban microclimates (Wu et al., 2024).

Green infrastructure also enhances biodiversity (Donati et al., 2022; Bretzel et al., 2024), reduces pollutant concentrations (Calfapietra, 2020; Alpaidze and Pace, 2021), limits soil sealing to foster gas exchange (Aimar, 2023), sequesters carbon in plants and soils (Alpaidze and Pace, 2021), supports urban agriculture (Walters and Midden, 2018; Zareba et al., 2021), and improves human well-being through recreational and aesthetic benefits (Orta-Ortiz and Geneletti, 2023). Microbial communities underpin these functions by cycling nutrients, degrading organic matter, and sustaining soil structure.

2.2 Microbial Interventions in Agricultural Ecosystem Restoration

2.2.1 Microbial Diversity of Agroecosystems

Agricultural soils harbor immense microbial diversity. Nematodes alone represent about 80% of all animal abundance on earth (van den Hoogen et al., 2019). A teaspoon of soil contains thousands of bacterial species and millions of bacterial cells, surpassing the global human population (Bardgett and van der Putten, 2014). Yet, a few dominant bacterial taxa constitute most microbial biomass (Delgado-Baquerizo et al., 2018). Understanding these taxa and their ecological roles is fundamental for restoration strategies.

2.2.2 Soil Fertility/Health Management: Microbial Promotion of Soil Fertility and Plant Growth

Rhizosphere microbial communities, influenced by root exudates, include mycorrhizal fungi, nitrogen-fixers, biocontrol agents, and plant growth-promoting rhizobacteria (PGPR), all essential for plant health (Mendes et al., 2013). While beneficial microbes facilitate nutrient uptake and pathogen defense, harmful pathogens also persist, requiring nuanced management to optimize outcomes. Common PGPR genera include *Acinetobacter*, *Agrobacterium*, *Arthrobacter*, *Azospirillum*, *Bacillus*, *Bradyrhizobium*, *Frankia*, *Pseudomonas*, *Rhizobium*, *Serratia* and *Thiobacillus*. These bacteria play a critical role in nutrient mobilization, phytohormone production, pathogen suppression, fertilizer efficiency and pollutant degradation in soil (Kloepper et al., 2004).

Microbial activity also supports soil formation and fertility by mediating nutrient cycling of carbon, nitrogen, and other elements, sustaining agricultural productivity (Schulz et al., 2013). The activities of theses bacteria promote sustainable agriculture.

2.2.3 Pest and Disease Management: Microbial Biocontrol of Pests and Diseases in Crops

Beneficial microorganisms, including bacteria like *Bacillus* and *Pseudomonas*, and fungi such as *Trichoderma*, are increasingly applied in contemporary agriculture as an eco-friendly and effective approach to controlling plant diseases by protecting against various microbial pathogens (Boro et al., 2022). These microorganisms utilize multiple strategies to suppress pathogens, including direct antagonism via the secretion of antibiotics and lytic enzymes, competition for vital nutrients and habitat, and indirect mechanisms like triggering systemic resistance in the host plant, thereby boosting its innate defense systems (Gawai, 2018; Köhl et al., 2019).

2.2.4 Nutrient Cycling: Microbial Enhancement of Nutrient Cycling in Agricultural Soils

Plant Growth Promoting Bacteria (PGPB), a group of beneficial microorganisms, are essential for nutrient cycling by increasing the availability of key elements to host plants (Boro et al., 2022). These bacteria support nutrient cycling through mechanisms such as biologically fixing atmospheric nitrogen to supply plants with accessible nitrogen and producing siderophores that capture iron, thereby improving the availability of these vital nutrients in the rhizosphere (Singh et al., 2019; Boro et al., 2022). Additionally, microbial inoculants applied as biofertilizers can stimulate microbial activity in the soil, leading to enhanced nutrient absorption by plants (Suyal et al., 2016).

2.3 Biocatalytic Removal of Environmental Pollutants

2.3.1 Biocatalysis

Biocatalysis is the use of enzymes or whole microorganisms to catalyze chemical reactions that would otherwise proceed slowly or require harsh chemical treatments. In the context of environmental restoration, biocatalytic processes offer an efficient, eco-friendly, and sustainable approach to mitigating pollution. Unlike the conventional physicochemical treatments that may generate secondary contaminants, biocatalytic removal harnesses natural biological systems to transform pollutants into less toxic, biodegradable, or even beneficial products. A typical example is the use of laccase (an oxidoreductase enzyme) immobilized on biochar (to increase reusability and stability) to degrade bisphenol (Dong et al., 2023). Broad studies show that laccases can degrade dyes, PAHs, phenols, etc. This makes biocatalysis particularly valuable in urban and agricultural ecosystems, where environmental degradation from industrial effluents, agrochemicals, and urban waste poses significant challenges (Janeeshma et al., 2024).

2.3.2 Types of Biocatalysts

Biocatalysts can be broadly categorized into enzymes and microorganisms, as described by Madhuri et al. (2019).

Enzymes: These are specialized protein molecules that accelerate specific biochemical reactions. Enzymes such as laccase, peroxidase, and oxygenase have been widely studied for their ability to degrade pesticides, dyes, hydrocarbons, and phenolic compounds. Their high specificity ensures targeted degradation, minimizing harm to non-polluting compounds (Ahsan et al., 2021).

Microorganisms: Bacteria, fungi, and microalgae play a central role in natural bioremediation processes. Many microbes are capable of metabolizing pollutants as primary carbon or nitrogen sources, while others engage in co-metabolism, transforming pollutants alongside other substrates. For instance, *Pseudomonas* spp. can degrade hydrocarbons (Syuhadah and Raikhan, 2022), while white-rot fungi such as *Phanerochaete chrysosporium* employ ligninolytic enzymes to break down complex aromatic pollutants (Kato et al., 2024). Microalgae also contribute by removing heavy metals and absorbing nutrients from wastewater (Oyebamiji et al., 2019).

2.3.3 Mechanisms of Biocatalytic Removal

The efficiency of biocatalysis lies in its mechanisms, which can be enzymatic or microbial and their enhanced forms.

Enzymatic degradation: Enzymes act directly on pollutants, catalyzing their breakdown into smaller, less harmful compounds. For example, organophosphate hydrolases cleave organophosphate pesticides, reducing their toxicity, while laccases oxidize phenolic pollutants into less reactive forms (Chia et al., 2024).

Microbial degradation: Microorganisms use pollutants either as sole energy and carbon sources or co-metabolize them in the presence of other substrates. Through metabolic pathways, pollutants can be mineralized into harmless end products such as carbon dioxide, water, and biomass. Some microbes also possess plasmid-borne genes that enhance their adaptability in degrading synthetic compounds, making them promising tools in engineered bioremediation.

Immobilized Enzyme Systems: Free (soluble) enzymes are often subject to rapid deactivation under industrial or municipal wastewater conditions and they are difficult to recover and reuse. In contrast, immobilized enzymes (e.g., adsorbed, covalently bound, entrapped on/within a solid support) offer significant advantages: enhanced stability, better operational lifetime, ease of separation from the reaction mixture, and the ability to be reused across multiple cycles or integrated into continuous-flow bioreactors (Zdarta et al., 2021).

Metagenomics-Driven Analysis: Using environmental metagenomic analyses, researchers are now mining the vast uncultured microbial "dark matter" in soils and sediments to discover novel enzymes (for example, ring-hydroxylating dioxygenases and peroxidases) that are capable of degrading persistent organic pollutants such as polycyclic aromatic hydrocarbons (PAHs). These newly discovered enzymes - borne by previously unknown or uncultured microbes - expand our enzymatic toolkit for soil remediation by providing diverse catalytic mechanisms, broader substrate specificities, and adaptations to real-world contaminated settings. Because metagenomics bypasses the need for culturing, it enables targeted discovery of "hidden" biocatalysts directly from contamination-adapted environments (Zdarta et al., 2021).

Biofilms and Enhanced Pollutant Capture: The use of microbial biofilms in pollutant-removal systems offers several important operational advantages. Biofilm matrices (microbial communities embedded within an extracellular polymeric substance (EPS) matrix attached to surfaces or carriers) can facilitate enhanced trapping, sorption and local concentration of pollutants, while also enabling *in situ* degradation via microbial metabolic processes. This provides greater microbial stability and can lead to reduced sludge (biomass) generation compared to suspended-growth systems. Biofilm matrices facilitate pollutant trapping and degradation while maintaining microbial stability and reducing sludge generation—though fine-tuning operational parameters remains an active area of research (Saini et al., 2023).

2.3.4 Applications

The applications of biocatalytic interventions spans multiple environmental domains:

i. **Water treatment:** Wastewater from urban and agricultural sources often contains pesticides, pharmaceuticals, dyes, and heavy metals. Enzyme-based bioreactors and microbial consortia are increasingly used to degrade these pollutants, offering cost-effective and sustainable alternatives to chemical treatments (Sesay et al., 2025).

ii. **Soil remediation:** Contaminated soils, especially in agricultural lands exposed to pesticides and fertilizers, can be restored through microbial inoculation or enzyme amendments. These interventions accelerate the breakdown of persistent organic pollutants and improve soil health for crop production (Oro et al., 2024).
iii. **Air purification:** Although less explored than water and soil applications, biocatalytic removal is emerging in air pollution control. Biofilters and bioreactors employing microorganisms have shown promise in degrading volatile organic compounds (VOCs), hydrogen sulfide, and other airborne pollutants in industrial and urban settings (Marycz et al., 2022).

Biocatalytic removal of environmental pollutants represents a cornerstone of sustainable ecosystem restoration. By leveraging the catalytic power of enzymes and microorganisms, this approach provides an environmentally sound, cost-effective, and adaptable solution for urban and agricultural ecosystems alike. Future research focusing on genetic engineering, microbial consortia optimization, and large-scale deployment will further enhance the efficiency and scalability of biocatalytic technologies in addressing global environmental challenges.

2.3.5 Microbial Biocatalytic Degradation and Detoxification of Environmental Pollutants

Environmental pollutants including insecticides, herbicides, lignocellulosic wastes, cyanogenic compounds, oils and petrochemicals, and heavy metals pose serious risks to ecosystems and human health. These pollutants differ in chemical composition and origin, but commonly exert toxicity through enzyme inhibition, oxidative stress, metabolic disruption, and interference with cellular respiration. Conventional remediation methods are often expensive and environmentally disruptive. Microbial biocatalytic remediation offers a sustainable alternative, utilizing bacteria and fungi capable of transforming or detoxifying pollutants through enzymatic processes such as hydrolysis, oxidation–reduction, biosorption, bioaccumulation, and mineralization. Enzymes including oxygenase, hydrolase, ligninolytic enzymes, and metal-reducing proteins play key roles.

A. Pesticides

Pesticides encompass several major chemical classes, including organochlorine compounds like Dichlorodiphenyltrichloroethane (DDT) and dieldrin, which persist in the environment despite widespread bans due to their neural toxicity (Parte et al., 2017). Organophosphate pesticides such as chlorpyrifos and malathion were developed as replacements for organochlorines but present their own distinct environmental and health risks requiring biodegradation strategies (Singh and Walker,

2006). Additional pesticide categories include carbamates like carbofuran, which inhibit acetylcholinesterase (Parte et al., 2017), Pyrethroid pesticides and triazine herbicides such as atrazine, widely used in crops like maize and sugarcane, which contaminate soil and groundwater while disrupting endocrine function in animals and humans (Omotayo et al., 2011, 2013, 2016; Singh et al., 2004, 2018).

Case study - Atrazine:

Atrazine (2-chloro-4-ethylamino-6-isopropylamino-1,3,5-triazine) is a widely used s-triazine herbicide in crops like maize and sugarcane, contaminating soil and groundwater and posing significant risks (Omotayo et al., 2011, 2013, 2016; Singh et al., 2004, 2018). It is moderately persistent with half-lives from days to months and greater soil mobility than other herbicides (Ramanathan et al., 2022). Atrazine disrupts endocrine function causing reproductive and developmental harm in animals and humans and reduces soil microbial populations (Hong et al., 2022). Atrazine is degraded by bacteria general such as *Pseudomonas, Nocardioides, Agrobacterium, Arthrobacter, Acinetobacter* and *Bacillus* (Siripattanakul et al., 2010; Omotayo et al., 2013, 2016; Li et al., 2020).

Atrazine is sequentially broken down by bacterial specialized enzymes (AtzA, B, C/TrzN, AtzB, C) encoded by *atzA, B*, and *C* or *trzN-atzB* and *C* genes to cyanuric acid, a central metabolite in its degradation pathway and requires further bacterial enzymes (AtzD, E, F) encoded by *atzD, E* and *F* genes, for complete mineralization to carbon dioxide and ammonia (Figure 2) (De Souza et al., 1998; Topp et al., 2000; Martinez et al., 2001; Omotayo et al., 2011, 2013). Many microbes degrade intermediates but rarely mineralize atrazine alone (Douglass et al., 2017). Factors like atrazine application history, temperature, pH, soil depth, oxygen, carbon and nitrogen impact degradation, microbial growth and enzyme activity (Morán et al., 2010; Omotayo et al., 2013, 2016; Douglass et al., 2017; Yale et al., 2017).

Figure 2. Biocatalytic degradation pathway of atrazine (Omotayo and Kukoyi, 2025)

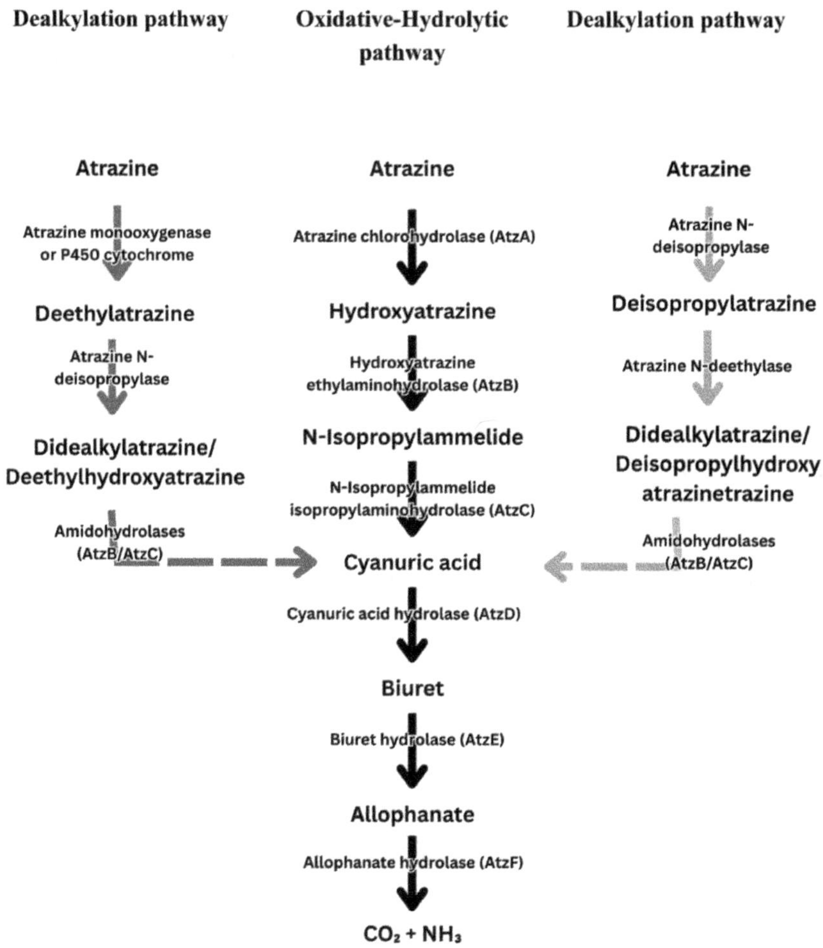

Source:
Boundy-Mills *et al.* (1997); De Souza *et al.* (1998); Govindaswamy and Gilbride (2010); Martinez *et al.* (2001); Omotayo *et al.* (2013); Strong *et al.* (2002); Topp *et al.* (2000).

B. Lignocellulose

One of the most bountiful renewable resources globally is lignocellulosic biomass. Cellulose, hemicellulose, and lignin are the structural components of plant cell walls of lignocellulosics. Among the three, cellulose is the most abundant polysaccharide, an organic component made up of long, linear β-1, 4-glucan chains with crystalline and amorphous regions (Bautista-Cruz et al., 2024). Hemicellulose is a branched heterogeneous polysaccharide including xylans, mannans, arabinoxylans that links with cellulose and lignin. Lignin, is an uneven, cross-linked phenylpropanoid polymer made from aromatic components. These are coniferyl alcohol (guaiacy), sinapyl alcohol (syringyl), and p-Coumaryl (p-hydroxyphenyl) monomers that fills in the space between polysaccharide fibrils and imparts rigidity, hydrophobicity, and resistance to microbial attack. Coniferyl Alcohol (G unit) are found in softwood lignin, sinapyl alcohol (S unit) are common in hardwood lignin, and p-Coumaryl Alcohol (H unit) found in grass lignin along with G and S units (Chen et al., 2025).

Enzymes, Active Components and Mechanisms of Microbial Degradation of Lignocelluloses

The breakdown of lignocellulose relies on multiplex arrays of microbial hydrolytic and oxidative enzymes. The predominant ones include:

Cellulases, which are composed of endoglucanases, exoglucanases, and β-glucosidases that collaborate to hydrolyze cellulose into glucose (Gupta et al., 2012). During cellulose hydrolysis, endoglucanases act first to split amorphous cellulose to create free chain ends. Exoglycanases then liberate cellobiose from crystalline regions, then, β-glucosidases convert cellobiose into glucose. (Wu et al., 2022).

Hemicellulases which include endo-β-1, 4-xylanase, β-xylosidase, α-arabinofuranosidase, β-mannanase, and α-galactosidase, disintegrates into branched xylans. Hemicellulose depolymerization begins when tightly linked to lignin and cellulose. Hemicellulose is degraded by a cocktail of hydrolytic enzymes that convert xylan into xylo-oligosaccharides, and then into xylose (Li et al., 2022). Accessory enzymes remove side-chain substitutes to facilitate complete hydrolysis. Microorganisms involved include filamentous fungi such as *Aspergillus* and *Trichoderma*, and some bacteria such as *B. pumilus*, and psychrotolerant *Paenibacillus* PXLY.

Lignin-Modifying Enzymes: Lignin poses the major barrier to hydrolytic attack of polysaccharides. Microorganisms (especially white-rot fungi, and some bacteria) secrete laccases (multi-copper oxidases), lignin peroxidases (LiP), manganese peroxidases (MnP), versatile peroxidases (VP), and other oxidoreductases. These enzymes oxidize phenolic and non-phenolic lignin subunits, generate radicals (or use mediators), cleave ether bonds (such as β-O-4 linkages), degrade cross-links, and reduce the physical and chemical shielding effect of lignin (Buraimoh et al. 2015a; Kumar and Chandra, 2020). Lignin Oxidation starts with the oxidative polymerization catalyzed by laccases and peroxidases that convert lignin into monomeric phenolics. The white –rot fungi are adjudged the most efficient lignin degraders due to secretion of extracellular enzymes and hyphal penetration. Asina et al. (2017) reviewed microbial strategies for degrading industrial lignin, particularly *Kraft lignin* and black liquor from paper mills. It highlights fungal and bacterial enzymatic systems (laccase, MnP, LiP) effective in lignin breakdown, challenges in process efficiency, and the need for optimized conditions and bioreactor systems for large-scale bioremediation and bio-product recovery. Bacterial growth and assessment of kraft-lignin degradation in submerged fermentation was carried out by Buraimoh et al. (2015a) using *Streptomyces* spp isolated from a tropical lagoon. The organisms utilized 23 to 99% kraft-lignin. Maximum values obtained for laccase and peroxidase activities were $9.5x10^{-2}$ and 400 μ mol mg^{-1} min^{-1} respectively.

Four sequential stages lead to microbial biocatalytic removal of lignocellulose (Figure 3). First, enzyme is secreted and adsorbed. This is when extracellular lignocellulolytic enzymes bind to the substrate surface. This is followed by depolymerization process where glycosidic and ether bonds are cleaved, releasing soluble sugars and aromatic intermediates. The soluble sugars are metabolically converted to organic acids, bioethanol, biohydrogen, and through glycolysis or fermentation. Then, the Synergistic cooperation between fungi and bacteria where the fungal delignification exposes cellulose fibers for bacterial cellulosomes or free enzymes to act (Chukwuma et al. 2021). Multienzyme cellulosomes, are surface-anchored complexes that couple substrate adhesion with hydrolysis. They are remarkably specific for crystalline cellulose, are employed by anaerobic bacteria such as *Clostridium thermocellum* and *Bacteroides* (Ding et al. 2020).

Buraimoh et al. (2015b) performed *in situ* and laboratory studies to determine the ability of bacterial to degrade lignin and carbohydrate components of sawdust, with a view to abating the impact of sawdust pollution on the lagoon ecosystem. In the *in-situ* study, 59.2% of sawdust was depleted whereas 82.5% of lignin component of the sawdust was depleted (Table 1). The maximum decrease in carbohydrate content was 85%. In a similar experiment under laboratory conditions, total weight losses ranging from 26 to 51% in the wood residues were reported.

Figure 3. Microbial degradation of lignocellulose

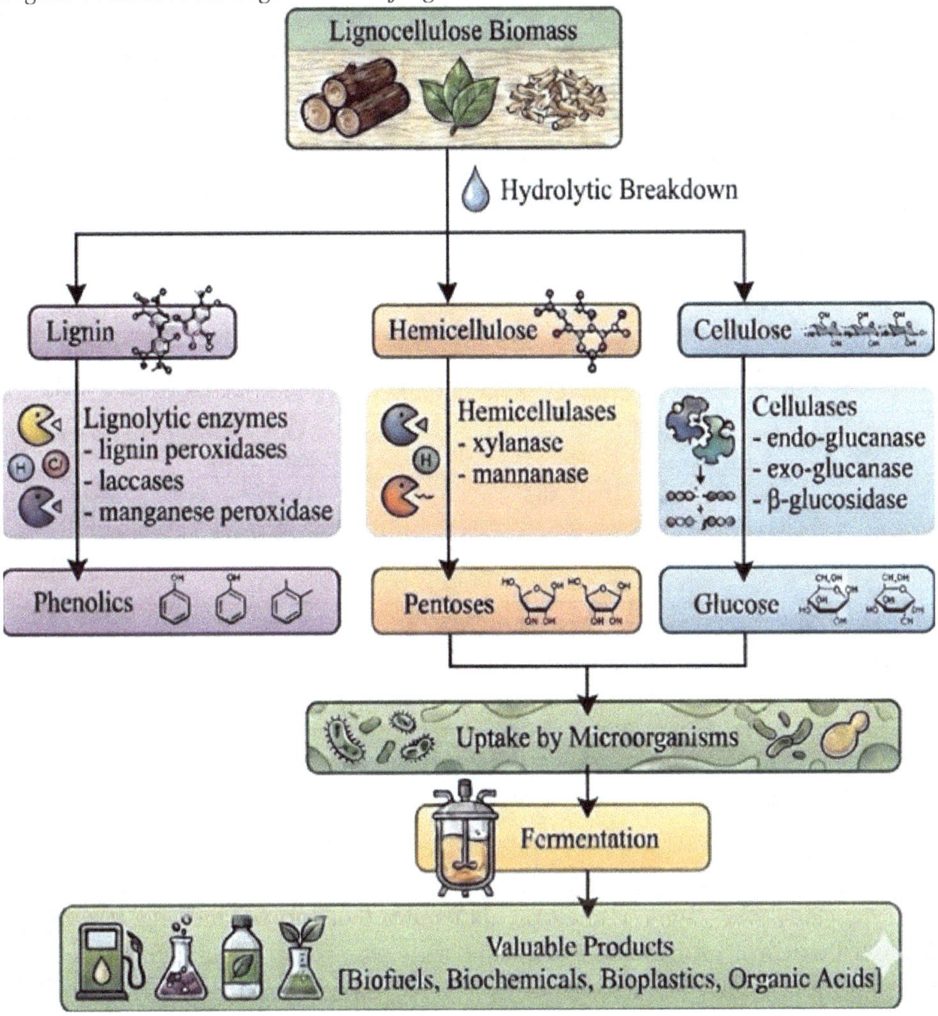

Table 1. Weight losses and bioremoval of lignin and carbohydrate components of wood residues under in-situ conditions

Organism	**Weight loss**		**Lignin degraded**		**Carbohydrate utilization**	
	%	**g d^{-1} cm^{-3}**	**%**	**g d^{-1} cm^{-3}**	**%**	**g d^{-1} cm^{-3}**
Streptomyces albogriseolus AOB KF977548	49.6	5.9×10^{-5}	47.51	1.037×10^{-6}	58	4.488×10^{-8}
Streptomyces aureus BOB KF977549	45.4	5.4×10^{-5}	44.98	9.817×10^{-7}	50	3.869×10^{-8}
Streptomyces coelicolor COB KF977550	46	5.48×10^{-5}	44.99	9.821×10^{-7}	48	3.714×10^{-8}
Streptomyces albus DOB KF977551	50.5	6.01×10^{-5}	44.99	9.821×10^{-7}	59	4.565×10^{-8}
Streptomyces pseudogriseolus EOB KF977552	49.7	5.92×10^{-5}	47.51	1.037×10^{-4}	50	3.869×10^{-8}
Bacillus bataviensis FOB KF977553	50.3	5.99×10^{-5}	50	1.091×10^{-6}	52	4.024×10^{-8}
Bacillus megaterium NOB KF977554	51.3	6.11×10^{-5}	52.49	1.146×10^{-6}	58	4.488×10^{-8}
Bacillus sp. OOB KF977555	25.95	5.19×10^{-3}	22.49	4.909×10^{-7}	28	2.167×10^{-8}
Paenibacillus sp. ROB KF977556	31.9	3.8×10^{-5}	22.49	4.909×10^{-7}	40	3.095×10^{-8}

Source: Buraimoh et al. (2015)

C. Cyanogens

Cyanogens are cyanide-containing molecules that can occur in organic forms as aliphatic or aromatic nitriles or as cyanogenic glucosides found in plants such as lotustralin and linamarin (Pandey and Yadav, 2015). They can also occur in inorganic forms, such as hydrogen cyanide (HCN) and potassium cyanide (KCN) (Rashid et al., 2016; Eze et al., 2020). Numerous industrial and natural environments commonly contain both organic and inorganic forms of cyanide (Berkinbayeva et al., 2025). The direct discharge of wastewaters containing nitriles and other cyanogenic compounds poses a severe health hazard, as most of them are highly genotoxic in character (Ogunyemi et al., 2019a). Cyanogens in wastewater and solid wastes have become a global problem, with their potential carcinogenicity mutagenicity and damage to organs in the human body. Consequently, removing nitriles and other sources of cyanide molecules from industrial effluents and the environment had become critical. Moreover, cyanide is a cytochrome c oxidase inhibitor that causes energy starvation and cell death by preventing respiration (Rashid et al., 2016). Long-term exposure from eating cassava or cyanide-laden environment is linked to goiter, neurological diseases, and konzo, a motor neuron disease that frequently occurs in places that utilize cassava as staple food (Amoako and Dzogbefia, 2011; Eze et al., 2020). Waste streams from cassava processing facilities contain both bound and free cyanide (Pandey and Yadav, 2015), while free cyanide is produced when hydrolytic enzymes act on the cyanogenic glucosides. Cyanide could accumulate in soil and groundwater arising from pollution from untreated cassava wastewater (Eze et al., 2020). Cassava peels are typically disposed of in open dumpsites along with municipal garbage while cassava steep-water and processing effluents go into unprotected drains, especially in developing countries.

Microbial Detoxification of Cyanogens

The primary processes of cyanogen degradation by microorganisms include hydrolytic, oxidative, sulfur-transfer, and fermentative pathways. Hydrolytic degradation is more common in *Fusarium* and *Bacillus* species isolated from cassava trash (Eze et al., 2020). Aerobic bacteria like *Pseudomonas* possess cyanide monooxygenase and nitrilase that convert cyanide to CO_2 and NH_3 (Rashid et al. 2016). Rhodanase catalyzes the transfer of sulfur from thiosulfate to cyanide, resulting in thiocyanate, a less hazardous product (Dash et al. 2009). Linamarin is broken down by lactic acid bacteria β-glucosidase during cassava fermentation to acetone cyanohydrin, which is then broken down to volatile HCN (Viwatpanich et al., 2013; Ogunyemi et al., 2019a). Ogunyemi et al. (2019b, 2019c) successively characterized two nitrile-metabolizing and linamarin-utilizing bacteria from solid wastes and cassava wastewater, respectively. The four test organisms were identified by cultural and biochemical characterization as well as the 16S RNA gene amplification method as *Bacillus* sp. (WOD8), *Corynebacterium* sp. (WOIS2), *Bacillus safensis* (WOB3), and *Bacillus safensis* (WOB7). When Glutanitrile was used as growth substrate, *Bacillus* sp. (WOD8) produced 4-cyanobutyric acid (4-CBA) and glutaric acid as metabolites in culture filtrates while *Corynebacterium* WOIS2 metabolized benzonitrile to produce benzoic acid as an intermediate. (Figure 4). These natural isolates may therefore show a brilliant prospect for use as seed organisms in the purification of aquatic and soil ecosystems polluted with cyanogenic compounds (Okafor et al., 2018).

Figure 4. Proposed pathway for the degradation of Glutaronitrile (A) and Benzonitrile (B) by Corynebacterium sp. strain WOIS2 and Bacillus sp. strain WOD8. Source: Ogunyemi et al. (2022)

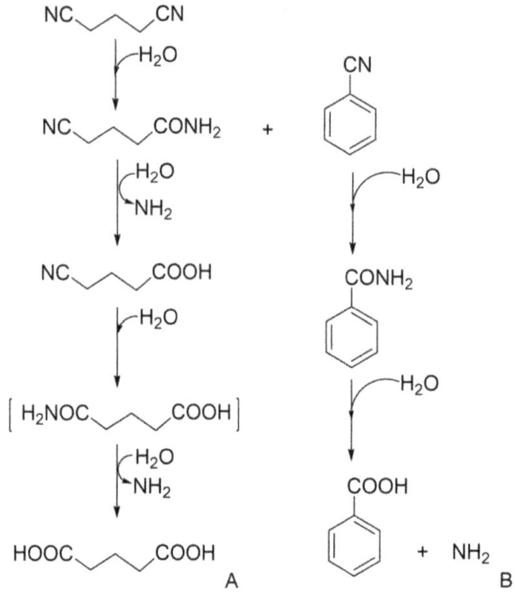

D. Oils and Petrochemicals

Natural Oils

Natural oils are chiefly lipids, which are composed of animal fats, plant-derived (vegetable) oils, greases and fatty acids. These are major constituents of industrial and domestic wastes particularly from food processing industries (Prasad and Manjunath, 2011) thereby contributing towards environmental contamination. The deposited vegetable oils can bring about severe anthropogenic effects such as blocking of pipes, oxygen depletion, oil film formation, odour problems (rancidity) and congested treatment filters, thus killing aquatic living organisms in their paths (Mudge, 1995). Biological treatment using lipolytic microorganisms through enzymatic processes had been the main approach in ridding the environment of lipid-rich food wastes (Doriet et al., 2013). Many aerobic and anaerobic microorganisms capable of degrading vegetable oils had been well studied comprising of bacterial and fungal species of primarily *Rhodococcus*, *Pseudomonas*, *Enterobacter aerogenes*, *Arthrobacter*, *Bacillus*, *Serratia*, *Staphylococcus*, *Burkhoderia*, *Acinetobacter*, *Lactobacillus*, *Aspergillus*, *Penicillium* and *Mucor* (Mongkolthanaruka and Dharmsthiti, 2002). The microorganisms secret extracellular lipase responsible for lipid biodegradation by hydrolyzing the triglycerides to release free fatty acids that are further oxidized to acetylcoA through β-oxidation (Tan and Gill, 1985). While the glycerol moiety generated is converted to glucose via series of chemical reactions for the energy-yielding cellular processes.

Synthetic Oils

Synthetic oils have found various commercial and industrial uses and most widespread are the polychlorinated biphenyls (PCBs) due to their chemical/physical stability and electrical insulating properties. PCBs are a family of compounds consisting of a biphenyl nucleus carrying 1-10 chlorine atoms. They are commonly used as hydraulic fluids, transformer fluids, plasticizers and fire retardants. They have been sold under trade names such as Aroclor, Phenoclor, Clophen and Askarel Oil. Typically, commercial PCB mixtures contain between 20 and 70 of the 209 theoretically possible congeners (chlorobiphenyls) that differ in the number and position of chlorine on the biphenyl structure and about 1.5 million tons of PCBs have been produced worldwide (Faroon et al., 2000). Significant quantity of PCBs had been released into the environment and accumulated in soils and sediments due to improper management of wastes and spills (Nogales et al., 1999). Despite reported toxicity, PCBs are biodegradable in the environment. Diverse aerobic PCB-degrading bacterial species including *Pseudomonas*, *Burkhoderia*, *Cupriavidus*, *Bacillus*, *Sphingomonas*, *Corynebacterium*, *Rhodococcus*, *Enterobacter* and *Ralstonia* (Adebusoye et al., 2008). Aerobic metabolism of PCB proceeds in two phases: the upper biphenyl pathway and the lower biphenyl pathway. In the upper pathway, the upper ring is degraded via a dioxygenase reaction to produce benzoic acid and 2-Hydroxypenta-2,4-dienoic acid. In the lower pathway, the second ring of biphenyl manifesting in benzoate, is further metabolized via catechol with ring cleavage, either utilizing the 'ortho' or 'meta' pathway to produce intermediates going into the TCA cycle (Pieper, 2005; Adebusoye, 2017).

Lubricating Oils

Lubricating oils have found diverse uses ranging from bearing lubrication and cooling in motor vehicles to hydraulic functions in industrial machines. Various viscosity grades of lubricating oils are produced from crude oil distillation and refining processes (Islam Sazzad, 2024), while the low-viscosity oils are degraded faster than high-viscosity oils (Amund and Adebiyi, 1991). However, synthetic lubricants have provided satisfactory lubricating functions in some applications. Ester-based synthetic lubricants used in outboard boat engines have offered some advantages over mineral oil-based ones with respect to their low pour point, high viscosity index combined with low volatility and high temperature resistance. The degradation of the ester-based synthetic lubricant Emkarate DE155 by strains of *Micrococcus roseus* and *Acinetobacter lwoffi* had been described (Wright et al., 1993; Amund, 1996). In an esterase-catalyzed reaction, Emkarate was metabolized into long chain fatty acids (octanoate and decanoate) and a non- degradable tri-hydroxyl compound identified as 1,1,1 Tris (hydroxymethyl) propane, which accumulates as a dead-end product (Amund, 1996).

Mineral Oils (Petroleum)

Petroleum-based products, especially crude oil, are released into the environment during exploration, production, refining, transport and storage. Crude oil seepage into the sea, inland waterways and soil surface is estimated at 600,000 metric tons annually (Kvenvolden and Cooper, 2003). Hydrocarbon release from anthropogenic activities is the mail cause of water and soil pollution, creating extensive damage to plant and animal tissues and may cause death. Hence, petroleum hydrocarbons are classified as priority pollutants (Costa et al., 2012). Commonly used technologies for soil and water remediation include mechanical clearing or mopping, burying, burning, evaporation, dispersion with chemical surfactants and washing. However, these technologies, apart from being expensive and laborious, do not achieve complete removal of the contaminants. Hence, the bioremediation technology, which employs the natural metabolic capabilities of microorganisms, which had been deployed to rid the environment of petroleum and other pollutants (Leahy and Colwell, 1990). The primary mechanisms by which petroleum is removed from the environment is microbial degradation employing the natural populations of microorganisms present in the soil or aquatic systems. These organisms include species of *Pseudomonas, Arthrobacter, Mycobacterium, Micrococcus, Alcaligenes, Flavobacterium, Bacillus, Sphingomonas, Acinetobacter, Corynebacterium, Aspergillus, Penicillium, Candida* and *Saccharomyces* (Chikere et al., 2011).

Three main approaches had been adopted in oil spill remediation as detailed in preceding chapters: (i), bioaugmentation in which oil-degrading organisms are seeded to supplement the existing microbial population of the area being treated or; (ii), biostimulation, whereby exogenous nutrients (fertilizers) are added to enhance the growth of indigenous oil-degrading strains in the polluted area, Bioremediation was successfully deployed in the cleanup of Prince William Sound shoreline in Alaska after the Exxon Valdez oil spill of 1989 (Prince, 1993; Mearns, 1997). The technology had similarly been used in the rehabilitation of polluted farmlands in Ogoniland of Rivers State in Nigeria (Chikere and Ekwuabu, 2014); (iii) Use of biosurfactants, which play the dual role of dispersing the oil, thus enhancing its bioavailability and also promote its degradation by micro-organisms.

Crude oil is a complex mixture made of hydrocarbons and non-hydrocarbon components classified as, alkanes, cycloalkanes, aromatics, polycyclic aromatics, asphaltenes and resins. Among the hydrocarbons, n-alkanes are the most biodegradable, while cycloalkanes and aromatics are less degradable. Prior to degradation, hydrocarbons are taken up into microbial cells as oil droplets with the aid of emulsifying agents (biosurfactants) produced by the cells (Hommel, 1990) and transported across the cell membrane to the site of enzymic oxidation. The initial enzymic attack on n-alkanes is catalyzed by Cytochrome P_{450} alkane hydroxylase or monooxygenase, which introduces oxygen molecule to produce the corresponding aliphatic alcohol, followed by an aldehyde and a carboxylic acid. Further degradation of the fatty acid via β-oxidation generates Acetyl Co-enzyme A, which is fed into the Tricarboxylic Acid (TCA) cycle for assimilation and energy transduction (Van Beilen and Funhoff, 2007). Microbial degradation of aromatics, on the other hand is catalyzed by dioxygenases, generating catechol or its derivatives as the first stable intermediate. Catechol is subsequently subjected to two types of aromatic ring cleavage pathways called the 'ortho' or 'meta' pathways to produce intermediates of the TCA cycle. This implies that diverse microorganisms are equipped with the enzymatic machinery to rid the polluted environments of hydrocarbons (Das and Chandran, 2011).

E. Heavy Metals

Heavy metals (e.g., cadmium, lead, mercury, chromium, arsenic, nickel) are toxic elements with high atomic weight and density. Unlike organic pollutants, they are non-biodegradable and persist in the environment, making them particularly hazardous (Briffa et al., 2020).

Types of Heavy Metals and Active Components

Heavy metals are dense elements that can be either essential or non-essential to biological systems. Essential heavy metals such as copper (Cu), zinc (Zn), iron (Fe), and manganese (Mn) are vital for enzyme function, metabolism, and antioxidant defense. However, when present in excess, they can generate reactive oxygen species (ROS), leading to oxidative stress and cellular damage. In contrast, non-essential heavy metals including lead (Pb), cadmium (Cd), arsenic (As), chromium (Cr), and mercury (Hg) have no known biological role and are toxic even at low concentrations. These metals persist in the environment and bioaccumulate in living organisms, posing serious ecological and health risks (Zhang et al., 2023).

The toxic effects of heavy metals arise mainly from their ionic forms, such as Pb^{2+}, Cd^{2+}, and Cr^{++}, which bind strongly to functional groups in proteins, enzymes, and nucleic acids. Ultimately, both essential and non-essential heavy metals can interfere with metabolic pathways and compromise organismal health when not properly regulated. Understanding these mechanisms is crucial for mitigating their environmental and biological impacts (Jomova et al., 2025).

Mechanisms of Action

Heavy metals exert their toxic effects through multiple biochemical pathways that disrupt normal cellular functions. They bind to sulfhydryl (-SH) groups in enzymes, thereby inhibiting key metabolic processes. Some metals displace essential cofactors in proteins for example, cadmium (Cd^{2+}) can replace zinc (Zn^{2+}), leading to enzyme inactivation. Additionally, heavy metals promote the generation of reactive oxygen species (ROS), which induce oxidative stress and damage cellular components. These reactive intermediates can also cause DNA mutations and compromise membrane integrity, ultimately impairing cellular viability and contributing to toxicity in living organisms (Mitra et al., 2022; Jomova et al., 2025).

Microbial Transformation (Biocatalysis for Heavy Metals)

Unlike organic pollutants, heavy metals cannot be degraded but can be transformed, immobilized, or removed from the environment by microbial processes. Microorganisms act as biocatalysts by altering the speciation, solubility, or toxicity of heavy metals. Key microbial groups include *Pseudomonas*, *Bacillus*, *Aspergillus*, and certain algae such as *Chlorella* and *Spirulina* (Zheng et al., 2024).

Mechanisms of Microbial Biocatalytic Removal

1. **Biosorption** – This is the passive binding of metal ions onto microbial cell walls via functional groups (carboxyl, hydroxyl, phosphate, amine).
2. **Bioaccumulation** – This involves the active uptake and intracellular sequestration of metals by living microbial cells.
3. **Bioprecipitation** – This is the conversion of soluble metal ions into insoluble forms (e.g., sulfide precipitation of cadmium by sulfate-reducing bacteria).
4. **Biotransformation (Redox Reactions):** This is the modification of a toxic metal into a less harmful one e.g *Chromium:* Cr^{++} (toxic) → Cr^{3+} (less toxic) by *Pseudomonas* spp. and *Mercury:* Hg^{2+} reduced to volatile Hg^0 by *merA* gene-encoded reductases.
5. **Complexation/Chelation** – This involves the secretion of microbial metabolites (siderophores, organic acids) that bind and immobilize metals.
6. **Biofilm-mediated Removal** – This is the use of microbial consortia in biofilms trap to detoxify metals through EPS (extracellular polymeric substances).

Case Studies

- **Chromium Reduction:** *Pseudomonas putida* reduces hexavalent chromium (Cr^{6+}) to Cr^{3+} in tannery wastewater, lowering toxicity significantly (Benites-Alfaro et al., 2025).
- **Lead and Cadmium Biosorption:** *Bacillus subtilis* and *Aspergillus niger* have demonstrated high biosorption capacities for Pb^{2+} and Cd^{2+}, respectively, in contaminated soils (El-Sharkawy et al., 2024).
- **Mercury Detoxification:** Engineered *Escherichia coli* strains expressing *merA* and *merB* genes convert Hg^{2+} and methylmercury into less toxic elemental Hg^0 (Shahpiri and Mohammadzadeh, 2018).

- **Algal Remediation:** Microalgae like *Chlorella* sp. accumulate Lead from textile wastewater, providing dual benefits of wastewater treatment and biomass production for bioenergy (Oyebamiji et al., 2021).

Challenges and Limitations

I. **Metal toxicity to microbes:** High concentrations inhibit microbial growth and enzyme activity.
II. **Bioavailability:** Metals bound in soil matrices are less accessible for microbial transformation.
III. **Environmental variability:** pH, salinity, and competing ions influence biosorption efficiency.
IV. **Scaling up:** Laboratory efficiencies often drop under field conditions due to mixed pollutants and ecological complexity.
V. **Disposal of biomass:** Metal-loaded microbial biomass must be safely handled to prevent secondary contamination.

3 OVERVIEW: MICROBES AS POWERHOUSE FOR ENVIRONMENTAL RESTORATION

3.1 Benefits of Microbial Interventions

3.1.1 Ecosystem Services: Enhancing Soil Formation, Nutrient Cycling, and Climate Change Mitigation

Soil microorganism including bacteria, fungi, and archaea function as the biological engines that regulate decomposition, mineralization, and organic matter stabilization within terrestrial ecosystems. Decomposition of plant residues by these microorganisms, release vital nutrients such as nitrogen, phosphorus, and sulfur, they facilitate humus formation, thereby improving soil structure, aggregation, and fertility (Das et al., 2022; Raupp et al., 2024). Microbial communities, soil biofilms, and complex aggregations of microorganisms embedded within extracellular polymeric substances (EPS) exert significant control over soil processes, contributing to nutrient availability, and uptake, aggregation, and resilience (Zhang et al., 2024). Extension of root systems by mycorrhizal fungi enhances nutrient, and mineral uptake and plant resilience. Microbial interventions also include nitrogen-fixing, carbon sequestration through organic matter stabilization, contributing to climate regulation, and sustainable agriculture. Soil microorganisms are vital regulators of carbon turnover, sequestration, and the production or consumption of greenhouse gases (GHGs) such as carbon dioxide (CO_2), methane (CH_4), and nitrous oxide (N_2O) (Don et al. 2023; Beattie et al., 2025). Microbes are central to global biogeochemical cycles, and their responses to environmental changes profoundly influence global GHG budgets. For instance, microbial metabolism in thawing permafrost accelerates the release of CO_2 and CH_4, amplifying atmospheric warming. Recognizing this, recent modeling approaches integrate microbial physiology and community dynamics into Earth system models to improve predictions of carbon–climate feedbacks (Lennon et al., 2024).

3.1.2 Microbial Interventions as Sustainable Alternatives to Chemical-Based Solutions

Traditional chemical-based methods for improving productivity and managing waste often leads to water and soil pollution, soil degradation, and greenhouse gas emissions. Whereas, microbial interventions provide a more sustainable alternative by transforming agricultural residues and wastes into valuable bioproducts (Xu et al., 2022). Agricultural residues can serve as rich carbohydrate feed stocks for microbial processes that generate bio-based products such as biochemical, biofuels, biofertilizers, biosurfactants, bioenzymes, biopesticides, and other value-added bioproducts (Kumar and Chandra, 2020; Álvarez-Castillo et al., 2021; Srivastava et al., 2023). These eco-friendly microbial bioprocesses reduce environmental pollution in addition to enhancing resource efficiency by promoting circular bioeconomy principles. Microbes such as *Pseudomonas*, *Bacillus*, *Aspergillus*, and *Trichoderma* produce powerful enzyme systems that catalyze the breakdown of complex organic and inorganic pollutants into harmless end products.

3.2 Challenges and Limitations of Sustainable Restoration through Microbial Interventions

Microbial interventions in Ecosystem degradation provide a sustainable pathway for restoring degraded urban and agricultural ecosystems by enabling the biocatalytic removal of pollutants, improvement of soil health, and enhancement of water quality. Many of the laboratory trials showed tremendous results; however, their application faces several challenges and limitations that must be addressed for effective large-scale deployment. These include:

3.2.1. Scalability Issues

While microbial approaches are highly efficient under controlled laboratory conditions, scaling them up to field applications presents significant hurdles. Large-scale implementation requires consistent microbial performance across heterogeneous environments, where pollutant concentrations, soil texture, and hydrological conditions vary widely. Mishra et al. (2024) gave an insight into why upscaling is a major challenge in Microbially Induced Carbonate Precipitation (MICP) techniques deployed in geotechnical engineering. Bioelectrochemical systems (BES), which integrate microbiology and electrochemistry to facilitate environmental remediation and resource recovery, highlighted upscaling as a major challenge. They, however, proposed innovations in reactor design and the incorporation of nanomaterials to enhance the efficiency and scalability of BES for field applications (Liang et al., 2025). Additionally, the mass production, stabilization, and delivery of microbial inoculants in cost-effective ways remain unresolved. A typical example is the Plant Growth Promotion Microorganisms (PGPMs) with promising capacity for sustainable agriculture, battling with scaling up production from laboratory to field conditions (Díaz-Rodríguez et al., 2025).

3.2.2. Environmental Sensitivity

The success of microbial processes is strongly influenced by environmental parameters such as pH, temperature, oxygen availability, nutrient content, and moisture levels. Mahalle et al. (2025) highlighted these conditions as the major challenges in the degradation of pesticides by either the microbial cells (fungi and bacteria) or the immobilized enzymes in the pesticide-contaminated environment. Even minor fluctuations can reduce microbial activity or enzyme efficiency, leading to inconsistent pollutant removal. Numerous studies demonstrate that even moderate deviations in pH, temperature, or moisture cause marked declines in microbial or enzyme-mediated pollutant removal. For instance, Lecca-Caballero et al. (2023) found that adjusting pH from 7 to 5 and temperature from 30 °C to 35 °C changed lead removal by *Serratia marcescens* dramatically. Furthermore, in mixed-contaminant environments, competition among pollutants or the presence of inhibitory substances may reduce microbial effectiveness. A typical example is the cytotoxicity effect exerted on microbial cells caused by the co-contamination of PAHs and heavy metals (Ali et al., 2021).

3.2.3. Persistence and Survival of Microbes

Introduced microbial strains often face competition with native microbial communities and may fail to establish long-term persistence in soil or aquatic ecosystems. Čaušević et al. (2024) demonstrated that bacterial inoculants (three *Pseudomonas* strains) failed to establish because resource niches were limited and introduced strains were overshadowed by the native microbes. Loss of viability, genetic drift, or predation by protozoa and other microorganisms may reduce the long-term effectiveness of bioremediation strategies. A study found that the addition of protozoa species significantly influenced the survival of introduced *Bacillus* strains in soil, demonstrating how predation (by protozoa) and microbiome interactions reduce long-term persistence of inoculants (Mawarda et al., 2023).

3.2.4. Regulatory and Policy Barriers

Despite advances in microbial biotechnology, regulatory frameworks for environmental release of microorganisms remain limited in many regions. Concerns over biosafety, ecological risks, and public acceptance slow the translation of laboratory findings into field-scale applications. Clear guidelines for approval, monitoring, and safe disposal of metal-loaded or pollutant-associated biomass are often lacking (Lensch et al., 2024).

3.2.5. Economic and Technical Constraints

The cost of developing microbial consortia, immobilized enzyme systems, or bioreactors can be high compared to conventional physicochemical methods. Additionally, specialized infrastructure and technical expertise are required, which may not be readily available in resource-limited settings, especially in developing regions (Kuppan et al., 2024).

3.2.6. Long-Term Effectiveness and Monitoring

Even when successful, microbial interventions may not fully mineralize pollutants or may transform them into intermediate products that still pose environmental risks. Long-term monitoring and post-treatment assessment are essential to ensure ecosystem recovery, but these are often neglected due to financial or logistical constraints (Maqsood et al., 2024).

In summary, while microbial interventions hold great promise for the sustainable restoration of urban and agricultural ecosystems, challenges highlighted above must be addressed. Overcoming these limitations will require advances in microbial engineering, supportive policy frameworks, interdisciplinary collaboration, and field-based validation to ensure reliable and sustainable outcomes.

4 FUTURE PERSPECTIVES

Successful microbiological restoration of polluted urban and agricultural ecosystems requires creation of databases for environmental risk assessment by regulatory agencies to ameliorate greater environmental impact arising from deployment or activation of native microorganisms. It is incumbent on governmental agencies to formulate policies that would encourage the recycling of agricultural residues and organic biomass into value-added products in order to mitigate environmental pollution and public health. Governmental interventions and policies would encourage sustainable waste management practices in urban and industrial settings, while the application of genetically-engineered microbes to stimulate recovery of polluted sites would come under appropriate scrutiny.

5 CONCLUSIONS

Microbial interventions represent a powerful and sustainable strategy for restoring degraded urban and agricultural ecosystems. The extraordinary metabolic variety of microorganisms allows for the conversion, detoxification, or immobilization of numerous pollutants, such as pesticides, hydrocarbons, cyanogenic substances, lignocellulosic residues, and heavy metals. Beyond pollutant removal, microbes play essential roles in soil formation, nutrient cycling, carbon sequestration, and overall ecosystem resilience, making them central to long-term environmental sustainability. Advances in microbiome engineering further expand the potential of microbial applications. Both bottom-up (synthetic consortia) and top-down (community selection and enrichment) approaches offer complementary pathways for designing efficient and functionally stable microbiomes tailored to specific environmental challenges. The development of microbial inoculants for agriculture and bioremediation, combined with modern monitoring tools such as metagenomics and functional biomarkers, strengthens the feasibility of deploying microbial technologies in real-world settings. However, several limitations must be addressed for widespread adoption. Environmental variability, competition with native microorganisms, scalability challenges, regulatory constraints, and long-term monitoring requirements can limit field performance. Additionally, ensuring ecological safety and preventing unintended consequences of engineered systems remain critical considerations. Overall, integrating ecological knowledge, engineering innovation, and supportive policy frameworks will be essential for translating laboratory successes into reliable field applications. With continued interdisciplinary research and responsible implementation, microbial-based technologies hold immense promise for achieving resilient ecosystems, sustainable agriculture, and enhanced environmental quality.

REFERENCES

Adebusoye, S. A. (2017). Biological degradation of 4-chlorobenzoic acid by a PCB-metabolizing bacterium through a pathway not involving chlorocatechol. *Biodegradation, 28*(10), 37–51. DOI: 10.1007/s10532-016-9776-3 PMID: 27766437

Adebusoye, S. A., Ilori, M. O., Picardal, F. W., & Amund, O. O. (2008). Characterization of multiple novel aerobic Polychlorinated Biphenyl (PCB)-utilizing bacterial strains ndigenous to Tropical African Soils. *Biodegradation, 19*(1), 145–159. DOI: 10.1007/s10532-007-9122-x PMID: 17534725

Ahsan, Z., Kalsoom, U., Bhatti, H. N., Aftab, K., Khalid, N., & Bilal, M. (2021). Enzyme-assisted bioremediation approach for synthetic dyes and polycyclic aromatic hydrocarbons degradation. *Journal of Basic Microbiology, 61*(11), 960–981. DOI: 10.1002/jobm.202100218 PMID: 34608659

Aimar, F. (2023). Limiting soil sealing and depaving: Local actions for regenerating public spaces to build green infrastructures. In: Giudice, B., Novarina, G., Voghera, A. (eds), *Green infrastructure*. The urban book series. Springer, Cham. DOI: 10.1007/978-3-031-28772-5_11

Ali, M., Song, X., Ding, D., Wang, Q., Zhang, Z., & Tang, Z. (2022). Bioremediation of PAHs and heavy metals co-contaminated soils: Challenges and enhancement strategies. *Environmental Pollution, 295*, 118686. DOI: 10.1016/j.envpol.2021.118686 PMID: 34920044

Alpaidze, L., & Pace, R. (2021). Ecosystem services provided by urban forests in the southern Caucasus region: A modeling study in Tbilisi, Georgia. *Climate (Basel), 9*(11), 157. DOI: 10.3390/cli9110157

Álvarez-Castillo, E., Felix, M., Bengoechea, C., & Guerrero, A. (2021). Proteins from agri-food industrial biowastes or co-products and their applications as green materials. *Foods, 10*(5), 981. DOI: 10.3390/foods10050981 PMID: 33947093

Amoako, T. N., & Dzogbefia, V. P. (2011). Effect of fermentation on the quality and detoxification of cassava mash. *International Food Research Journal, 18*(2), 695–699.

Amund, O. O. (1996). Utilization and degradation of an ester-based synthetic lubricant by *Acinetobacter lwoffi*. *Biodegradation, 7*(2), 91–95. DOI: 10.1007/BF00114621 PMID: 8882802

Amund, O. O., & Adebiyi, A. G. (1991). Effect of viscosity on the biodegradability of automotive lubricating oils. *Tribology International, 24*(4), 235–237. DOI: 10.1016/0301-679X(91)90049-F

Asina, F., Brzonova, I., Kozliak, E., Kubátová, A., & Ji, Y. (2017). Microbial treatment of industrial lignin: Successes, problems and challenges. *Renewable & Sustainable Energy Reviews, 77*, 1179–1205. DOI: 10.1016/j.rser.2017.03.098

Bardgett, R. D., & van der Putten, W. H. (2014). Belowground biodiversity and ecosystem functioning. *Nature, 515*(7528), 505–511. DOI: 10.1038/nature13855 PMID: 25428498

Bautista-Cruz, A., Aquino-Bolaños, T., Hernández-Canseco, J., & Quiñones-Aguilar, E. E. (2024). Cellulolytic aerobic bacteria isolated from agricultural and forest soils: An overview. *Biology (Basel), 13*(2), 102. DOI: 10.3390/biology13020102 PMID: 38392320

Beattie, G., Edlund, A., Esiobu, N., Gilbert, J., Nicolaisen, M., Jansson, J., Jensen, P., Keiluweit, M., Lennon, J. T., Martiny, J., Minnis, V., Newman, D., Peixoto, R., Schadt, C., & van-der Meer, J. (2025). Soil microbiome interventions for carbon sequestration and climate mitigation. *mSystems, 10*(1), e01129–e24. DOI: 10.1128/msystems.01129-24 PMID: 39692482

Benites-Alfaro, E., Lizarzaburu Aguinaga, D., Cabrera Carranza, C., Jave Nakayo, J., Suca-Apaza, G. R., Atuncar Yrribari, J. F., Lopez Bulnes, J., & Munoz-Ccuro, F. (2025). Biological method with *Pseudomonas Putida* for Chromium VI reduction in chromium plating process wastewater. *Chemical Engineering Transactions, 117*, 427–432. DOI: 10.3303/CET25117072

Berkinbayeva, A., Saulebekkyzy, S., Kenzhaliyev, B., Sailed, K., Yessengaziyev, A., Nurtazina, N., Karim, D., & Birlikzhan, Y. (2025). Sodium percarbonate for eco-efficient cyanide detoxification in gold mining tailings. *Metals, 15*(10), 1162. DOI: 10.3390/met15101162

Bochenek, A. D., Klemm, K. and Woźna, M. (2022). Assessment of effectiveness of selected adaptation actions to climate change: The example of the new centre of Lodz. *Budownictwo i Architektura 21*(4), 025-042. .DOI: 10.35784/bud-arch.3150

Bonoli, A., Conte, A., Maglionico, M., & Stojkov, I. (2013). Green roofs for sustainable water management in urban areas. *Environmental Engineering and Management Journal, 12*(S11), 153–156.

Boro, M., Sannyasi, S., Chettri, D., & Verma, A. K. (2022). Microorganisms in biological control strategies to manage microbial plant pathogens: A review. *Archives of Microbiology, 204*(11), 666. DOI: 10.1007/s00203-022-03279-w PMID: 36214917

Bouzouidja, R., Leconte, F., Kiss, M., Pierret, M., Pruvot, C., Détriché, S., Berthelot, J. C., & Rousseau, D. (2021). Experimental comparative study between conventional and green parking lots: Analysis of subsurface thermal behavior under warm and dry summer conditions. *Atmosphere (Basel)*, *12*(8), 994. DOI: 10.3390/atmos12080994

Bretzel, F., Vannucchi, F., Pezzarossa, B., Paraskevopoulou, A., & Romano, D. (2024). Establishing wildflower meadows in anthropogenic landscapes. *Frontiers in Horticulture*, *2*, 1248785. DOI: 10.3389/fhort.2023.1248785

Briffa, J., Sinagra, E., & Blundell, R. (2020). Heavy metal pollution in the environment and their toxicological effects on humans. *Heliyon*, *6*(9), e04691. DOI: 10.1016/j.heliyon.2020.e04691 PMID: 32964150

Buraimoh, O. M., Amund, O. O., & Ilori, M. O. (2015a). Kraft lignin degradation by autochthonous *Streptomyces* strains isolated from a tropical lagoon ecosystem. *Journal of Microbiology, Biotechnology and Food Sciences*, *5*(3), 248253. DOI: 10.15414/jmbfs.2015/16.5.3.248-253

Buraimoh, O. M., Ilori, M. O., Amund, O. O., Michel, F. C.Jr, & Grewal, S. K. (2015b). Assessment of bacterial degradation of lignocellulosic residues (sawdust) in a tropical estuarine microcosm using improvised floating raft equipment. *International Biodeterioration & Biodegradation*, *104*, 186–193. DOI: 10.1016/j.ibiod.2015.06.010

Calfapietra, C. (2020). Nature-based solutions for microclimate regulation and air quality. In: *Nature-based solutions: State of the art in EU-funded projects*. DOI: 10.2777/383904

Čaušević, S., Dubey, M., Morales, M., Salazar, G., Sentchilo, V., Carraro, N., Ruscheweyh, H. J., Sunagawa, S., & van der Meer, J. R. (2024). Niche availability and competitive loss by facilitation control proliferation of bacterial strains intended for soil microbiome interventions. *Nature Communications*, *15*(1), 2557. DOI: 10.1038/s41467-024-46933-1 PMID: 38519488

CBD (Convention on Biological Diversity) (2010). Aichi Biodiversity Targets of the Strategic Plan 2011–2020.

Chen, M., Li, Q., Liu, C., Meng, E., & Zhang, B. (2025). Microbial degradation of lignocellulose for sustainable biomass utilization and future research perspectives. *Sustainability (Basel)*, *17*(9), 4223. DOI: 10.3390/su17094223

Chia, X. K., Hadibarata, T., Kristanti, R. A., Jusoh, M. N. H., Tan, I. S., & Foo, H. C. Y. (2024). The function of microbial enzymes in breaking down soil contaminated with pesticides: A review. *Bioprocess and Biosystems Engineering*, *47*(5), 597–620. DOI: 10.1007/s00449-024-02978-6 PMID: 38456898

Chikere, C., & Ekwuabu, C. (2014). Culture-dependent characterization of hydrocarbon-utilizing bacteria in selected crude oil impacted sites in Bodo, Ogoniland, Nigeria. *African Journal of Environmental Science and Technology*, *8*(6), 401–406. DOI: 10.5897/AJEST2014.1707

Chikere, C. B., Okpokwasili, G. C., & Chikere, B. O. (2011). Monitoring of microbial hydrocarbon remediation in the soil. 3. *Biotechnology*, *1*(3), 117–138. DOI: 10.1007/s13205-011-0014-8 PMID: 22611524

Chukwuma, O. B., Rafatullah, M., Tajarudin, H. A., & Ismail, N. (2021). Review on bacterial contribution to lignocellulose breakdown into useful bio-products. *International Journal of Environmental Research and Public Health*, *18*(11), 6001. DOI: 10.3390/ijerph18116001 PMID: 34204975

Cornu, J. Y., Huguenot, D., Jézéquel, K., Lollier, M., & Lebeau, T. (2017). Bioremediation of copper-contaminated soils by bacteria. *World Journal of Microbiology & Biotechnology*, *33*(2), 26. DOI: 10.1007/s11274-016-2191-4 PMID: 28044274

Costa, A. S., Romao, L. P. C., Aranjo, B. R., Lucas, S. C. O., Maciel, S. T. A., Wisniewski, A.Jr, & Alexandre, M. R. (2012). Environmental strategies to remove volatile fractions (BTEX) from petroleum industry wastewater using biomass. *Bioresource Technology*, *105*, 31–39. DOI: 10.1016/j.biortech.2011.11.096 PMID: 22178487

Cristiano, E., Urru, S., Farris, S., Ruggiu, D., Deidda, R., & Viola, F. (2020). Analysis of potential benefits on flood mitigation of a CAM green roof in Mediterranean urban areas. *Building and Environment*, *183*, 107179. DOI: 10.1016/j.buildenv.2020.107179

Das, N., & Chandran, P. (2011). Microbial degradation of petroleum contaminants: An overview. *Biotechnology Research International*, *941810*, 1–3. DOI: 10.4061/2011/941810 PMID: 21350672

Das, P. P., Singh, K. R., Nagpure, G., Mansoori, A., Singh, R. P., Ghazi, I. A., Kumar, A., & Singh, J. (2022). Plant-soil-microbes: A tripartite interaction for nutrient acquisition and better plant growth for sustainable agricultural practices. *Environmental Research*, *214*, 113821. DOI: 10.1016/j.envres.2022.113821 PMID: 35810815

Dash, R. R., Gaur, A., & Balomajumder, C. (2009). Cyanide in industrial wastewaters and its removal: A review on biotreatment. *Journal of Hazardous Materials, 163*(1), 1–11. DOI: 10.1016/j.jhazmat.2008.06.051 PMID: 18657360

De Souza, M. L., Seffernick, J., Martinez, B., Sadowsky, M. J., & Wackett, L. P. (1998). The atrazine catabolism genes atzABC are widespread and highly conserved. *Journal of Bacteriology, 180*(7), 1951–1954. DOI: 10.1128/JB.180.7.1951-1954.1998 PMID: 9537398

Delgado-Baquerizo, M., Oliverio, A. M., Brewer, T. E., Benavent-González, A., Eldridge, D. J., Bardgett, R. D., Maestre, F. T., Singh, B. K., & Fierer, N. (2018). A global atlas of the dominant bacteria found in soil. *Science, 359*(6373), 320–325. DOI: 10.1126/science.aap9516 PMID: 29348236

Derakhshan Nejad, Z., Jung, M. C., & Kim, K. H. (2018). Remediation of soils contaminated with heavy metals with an emphasis on immobilization technology. *Environmental Geochemistry and Health, 40*(3), 927–953. DOI: 10.1007/s10653-017-9964-z PMID: 28447234

Díaz-Rodríguez, A. M., Parra Cota, F. I., Cira Chávez, L. A., García Ortega, L. F., Estrada Alvarado, M. I., Santoyo, G., & de los Santos-Villalobos, S. (2025). Microbial inoculants in sustainable agriculture: Advancements, challenges, and future directions. *Plants, 14*(2), 191. DOI: 10.3390/plants14020191 PMID: 39861545

Don, A., Seidel, F., Leifeld, J., Kätterer, T., Martin, M., Pellerin, S., Emde, D., Seitz, D., & Chenu, C. (2023). Carbon sequestration in soils and climate change mitigation-Definitions and pitfalls. *Global Change Biology, 29*(12), e16983. DOI: 10.1111/gcb.16983 PMID: 37905459

Donati, G. F. A., Bolliger, J., Psomas, A., Maurer, M., & Bach, P. M. (2022). Reconciling cities with nature: Identifying local blue-green infrastructure interventions for regional biodiversity enhancement. *Journal of Environmental Management, 316*, 115254. DOI: 10.1016/j.jenvman.2022.115254 PMID: 35576714

Dong, C., Tiwari, A., Anisha, G. S., Chen, C., Singh, A., Haldar, D., Patel, A. K., & Singhania, R. R. (2023). Laccase: A potential biocatalyst for pollutant degradation. *Environmental Pollution, 319*, 120999. DOI: 10.1016/j.envpol.2023.120999 PMID: 36608728

Dong, X., Chu, Y., Tong, Z., Sun, M., Meng, D., Yi, X., Feng, Y., Xiao, R., & Chen, H. (2024). Mechanisms of adsorption and functionalization of biochar for pesticides: A review. *Ecotoxicology and Environmental Safety, 272*, 116019. DOI: 10.1016/j.ecoenv.2024.116019 PMID: 38295734

Dorst, H., van der Jagt, A., Raven, R., & Runhaar, H. (2019). Urban greening through nature-based solutions: Key characteristics of an emerging concept. *Sustainable Cities and Society, 49*, 101620. DOI: 10.1016/j.scs.2019.101620

Douglass, J. F., Radosevich, M., & Tuovinen, O. H. (2017). Microbial attenuation of atrazine in agricultural soils: Biometer assays, bacterial taxonomic diversity, and catabolic genes. *Chemosphere, 176*, 352–360. DOI: 10.1016/j.chemosphere.2017.02.102 PMID: 28273542

El-Sharkawy, R. M., Khairy, M., Abbas, M. H. H., Zaki, M. E. A., & El-Hadary, A. E. (2024). Innovative optimization for enhancing Pb^{2+} biosorption from aqueous solutions using *Bacillus subtilis. Frontiers in Microbiology, 15*, 1384639. DOI: 10.3389/fmicb.2024.1384639 PMID: 39176280

European Commission. (2013). Green infrastructure (GI): Enhancing Europe's natural capital. In: *Communication from the Commission to the European Parliament, the European Economic Council, the European Economic and Social Committee and the Committee of the Regions*. DOI: 10.1017/CBO9781107415324.004

Eze, C. N., Anyanwu, B. N., & Onwurah, I. N. E. (2020). Biodegradation of cyanide in cassava mill effluent using indigenous bacterial isolates. *Environmental Monitoring and Assessment, 192*(6), 389. PMID: 32447581

Falkowski, P. G. (2015). Biogeochemical controls and feedbacks on ocean primary production. *Science, 350*(6262), 763–767. PMID: 9660741

Faroon, O., Jones, D., & De Rosa, C. (2000). Effects of polychlorinated biphenyls on the nervous system. *Toxicology and Industrial Health, 16*(7-8), 305–333. DOI: 10.1177/074823370001600708 PMID: 11693948

Fernández-Braña, A., Salgado, L., Gallego, J. L. R., Afif, E., Boente, C., & Forján, R. (2023). Phytoremediation potential depends on the degree of soil pollution: A case study in an urban brownfield. *Environmental Science and Pollution Research International, 30*(25), 67708–67719. DOI: 10.1007/s11356-023-26968-5 PMID: 37118389

Gawai, D. (2018). Role of fungi as biocontrol agents for the control of plant diseases in sustainable agriculture. In *Fungi and their role in sustainable development: current perspectives* (pp. 283–291). Springer. DOI: 10.1007/978-981-13-0393-7_16

Guidi Nissim, W., Castiglione, S., Guarino, F., Pastore, M. C., & Labra, M. (2023). Beyond cleansing: Ecosystem services related to phytoremediation. *Plants, 12*(5), 1031. DOI: 10.3390/plants12051031 PMID: 36903892

Gupta, P., Samant, K., & Sahu, A. (2012). Isolation of cellulose-degrading bacteria and determination of their cellulolytic potential. *International Journal of Microbiology*, *2012*, 1–5. DOI: 10.1155/2012/578925 PMID: 22315612

Hansen, R., & Pauleit, S. (2014). From multifunctionality to multiple ecosystem services? A conceptual framework for multifunctionality in green infrastructure planning for urban areas. *Ambio*, *43*(4), 516–529. DOI: 10.1007/s13280-014-0510-2 PMID: 24740622

Hommel, R. K. (1990). Formation and phylogenetic role of biosurfactants. *Journal of Applied Microbiology*, *89*(1), 158–219.

Hong, J., Boussetta, N., Enderlin, G., Merlier, F., & Grimi, N. (2022). Degradation of residual herbicide atrazine in agri-food and washing water. *Food*, *11*(16), 2416. DOI: 10.3390/foods11162416 PMID: 36010414

Hou, D., Al-Tabbaa, A., O'Connor, D., Hu, Q., Zhu, Y. G., Wang, L., Kirkwood, N., Ok, Y. S., Tsang, D. C. W., Bolan, N. S., & Rinklebe, J. (2023). Sustainable remediation and redevelopment of brownfield sites. *Nature Reviews. Earth & Environment*, *4*(4), 271–286. DOI: 10.1038/s43017-023-00404-1

Jacobs, D. F., Oliet, J. A., Aronson, J., Bolte, A., Bullock, J. M., Donoso, P. J., Landhausser, S. M., Madsen, P., Peng, S., Rey-Benayas, J. M., & Weber, J. C. (2015). Restoring forests: What constitutes success in the twenty-first century? *New Forests*, *46*(5–6), 601–614. DOI: 10.1007/s11056-015-9513-5

Janeeshma, E., Habeeb, H., Sinha, S., Arora, P., Chattaraj, S., Das Mohapatra, P., Panneerselvam, P., & Mitra, D. (2024). Enzymes-mediated solid waste management: A sustainable practice for recycling. *Waste Management Bulletin*, *1*(4), 104–113. DOI: 10.1016/j.wmb.2023.10.007

Jomova, K., Alomar, S. Y., Nepovimova, E., Kuca, K., & Valko, M. (2025). Heavy metals: Toxicity and human health effects. *Archives of Toxicology*, *99*(1), 153–209. DOI: 10.1007/s00204-024-03903-2 PMID: 39567405

Kato, H., Miura, D., Kato, M., & Shimizu, M. (2024). Metabolic mechanism of lignin-derived aromatics in white-rot fungi. *Applied Microbiology and Biotechnology*, *108*(1), 532. DOI: 10.1007/s00253-024-13371-4 PMID: 39661194

Kloepper, J. W., Ryu, C. M., & Zhang, S. (2004). Induced systemic resistance and promotion of plant growth by *Bacillus* spp. *Phytopathology*, *94*(11), 1259–1266. DOI: 10.1094/PHYTO.2004.94.11.1259 PMID: 18944464

Köhl, J., Kolnaar, R., & Willem, R. (2019). Mode of action of microbial biological control agents against plant diseases: Relevance beyond efficacy. *Frontiers in Plant Science*, *10*, 845. DOI: 10.3389/fpls.2019.00845 PMID: 31379891

Koonin, E. V., Dolja, V. V., & Krupovic, M. (2015). Origins and evolution of viruses of eukaryotes: The ultimate modularity. *Virology*, *479-480*, 2–25. DOI: 10.1016/j.virol.2015.02.039 PMID: 25771806

Kuiper, I., Lagendijk, E. L., Bloemberg, G. V., & Lugtenberg, B. J. J. (2004). Rhizoremediation: A beneficial plant-microbe interaction. *Molecular Plant-Microbe Interactions*, *17*(1), 6–15. DOI: 10.1094/MPMI.2004.17.1.6 PMID: 14714863

Kumar, A., & Chandra, R. (2020). Ligninolytic enzymes and its mechanisms for degradation of lignocellulosic waste in environment. *Heliyon*, *6*(2), e03170. DOI: 10.1016/j.heliyon.2020.e03170 PMID: 32095645

Kuppan, N., Padman, M., Mahadeva, M., Srinivasan, S., & Devarajan, R. (2024). A comprehensive review of sustainable bioremediation techniques: Eco friendly solutions for waste and pollution management. *Waste Management Bulletin*, *2*(3), 154–171. DOI: 10.1016/j.wmb.2024.07.005

Kvenvolden, K. A., & Cooper, C. K. (2003). Natural Seepage of Oils into the Marine Environment. *Geo-Marine Letters*, *23*(3-4), 140–146. DOI: 10.1007/s00367-003-0135-0

la Cecilia, D., Malvolti, G., Despini, F., Teggi, S., Bigi, A., & Bach, P. (2020). Urban heat island mitigation scenarios by blue-green infrastructures modelling. *Proceedings of International Conference on Environmental Science and Technology 18*. DOI: 10.30955/gnc2023.00287

Landa-Acuña, D., Acosta, R. A. S., Hualpa Cutipa, E., Vargas de la Cruz, C., & Luis Alaya, B. (2020). Bioremediation: A low-cost and clean-green technology for environmental management. In Shah, M. (Ed.), *Microbial bioremediation & biodegradation*. Springer., DOI: 10.1007/978-981-15-1812-6_7

Leahy, J. G., & Colwell, R. R. (1990). microbial degradation of hydrocarbons in the environment. *Microbiological Reviews*, *54*(3), 305–315. DOI: 10.1128/mr.54.3.305-315.1990 PMID: 2215423

Lecca-Caballero, D., Vega-Moreno, E., Cabanillas-Chirinos, L., Diaz Del Aguila, K., Rojas-Villacorta, W., Salvatierra-Espinola, W., Naveda, R. N., Rojas-Flores, S., & De La Cruz-Noriega, M. (2023). An i*n vitro* study of the effects of temperature and pH on lead bioremoval using *Serratia marcescens*. *Sustainability (Basel)*, *15*(19), 14048. DOI: 10.3390/su151914048

Lennon, J. T., Abramoff, R. Z., Allison, S. D., Burckhardt, R. M., DeAngelis, K. M., Dunne, J. P., Frey, S. D., Friedlingstein, P., Hawkes, C. V., Hungate, B. A., Khurana, S., Kivlin, S. N., Levine, N. M., Manzoni, S., Martiny, A. C., Martiny, J. B. H., Nguyen, N. K., Rawat, M., Talmy, D., & Zakem, E. J. (2024). Priorities, opportunities, and challenges for integrating microorganisms into Earth system models for climate change prediction. *mBio*, *15*(5), e00455–e24. DOI: 10.1128/mbio.00455-24 PMID: 38526088

Lensch, A., Lindfors, H. A., Duwenig, E., Fleischmann, T., Hjort, C., Kärenlampi, S. O., McMurtry, L., Melton, E., Andersen, M. R., Skinner, R., Wyss, M., & van Kranenburg, R. (2024). Safety aspects of microorganisms deliberately released into the environment. *EFB Bioeconomy Journal*, *4*, 100061. DOI: 10.1016/j.bioeco.2023.100061

Li, M.-R., Zhan, F.-D., Chen, J.-J., Zu, Y.-Q., & Li, Y. (2020). Atrazine degradation pathway and genes of Arthrobacter sp. FM326. *Polish Journal of Environmental Studies*, *29*(5), 3683–3689. DOI: 10.15244/pjoes/115326

Li, Y. Y., Song, W. Y., Han, X. Y., Wang, Y. C., Rao, S. Q., Zhang, Q., Zhou, J. W., Li, J. H., Liu, S., & Du, G. C. (2022). Recent progress in key lignocellulosic enzymes: Enzyme discovery, molecular modifications, production, and enzymatic biomass saccharification. *Bioresource Technology*, *363*, 127986. DOI: 10.1016/j.biortech.2022.127986 PMID: 36126851

Liang, M., Luo, G., & Lei, P. (2025). Recent advances in bioelectrochemical approaches for sustainable environmental remediation. *International Journal of Electrochemical Science*, *20*(8), 101077. DOI: 10.1016/j.ijoes.2025.101077

Madhuri, R. J., Saraswathi, M., Gowthami, K., Bhargavi, M., Divya, Y., & Deepika, V. (2019). Recent approaches in the production of novel enzymes from environmental samples by enrichment culture and metagenomic approach. In Buddolla, V. (Ed.), *Recent developments in Applied Microbiology and Biochemistry* (pp. 251–262). Academic Press., DOI: 10.1016/B978-0-12-816328-3.00019-2

Madigan, M. T., Bender, K. S., Buckley, D. H., Sattley, W. M., & Stahl, D. A. (2018). *Brock Biology of Microorganisms* (15th ed.). Pearson Education.

Mahalle, S., Bhende, R. S., Bokade, P., Bajaj, A., & Dafale, N. A. (2025). Emerging microbial remediation methods for rejuvenation of pesticide-contaminated sites. *Total Environment Microbiology*, *1*(3), 100026. DOI: 10.1016/j.temicr.2025.100026

Maqsood, Q., Hussain, N., Sumrin, A., Ali, S. W., Tariq, M. R., & Mahnoor, M. (2024). Monitoring and abatement of synthetic pollutants using engineered microbial systems. *Discover Life*, *54*(1), 9. DOI: 10.1007/s11084-024-09652-7

Martinez, B., Tomkins, J., Wackett, L. P., Wing, R., & Sadowsky, M. J. (2001). Complete nucleotide sequence and organization of the atrazine catabolic plasmid pADP-1 from *Pseudomonas* sp. ADP. *Journal of Bacteriology*, *183*(19), 5684–5697. DOI: 10.1128/JB.183.19.5684-5697.2001 PMID: 11544232

Marycz, M., Brillowska-Dąbrowska, A., Muñoz, R. A., & Gębicki, J. (2022). State of the art review on the use of fungi in biofiltration to remove volatile hydrophobic pollutants. *Reviews in Environmental Science and Biotechnology*, *21*(1), 225–246. DOI: 10.1007/s11157-021-09608-7

Mawarda, P. C., Le Roux, X., Acosta, M. U., van Elsas, J. D., & Salles, J. F. (2023). The impact of protozoa addition on the survivability of *Bacillus* inoculants and soil microbiome dynamics. *ISME Communications*, *2*(1), 82. DOI: 10.1038/s43705-022-00166-9 PMID: 37938668

Mearns, A. J. (1997). Cleaning oiled shores: Putting bioremediation to test. *Spill Science & Technology Bulletin*, *4*(4), 209–217. DOI: 10.1016/S1353-2561(98)00026-7

Megharaj, M., & Naidu, R. (2017). Soil and brownfield bioremediation. *Microbial Biotechnology*, *10*(5), 1244–1249. DOI: 10.1111/1751-7915.12840 PMID: 28834380

Mendes, R., Garbeva, P., & Raaijmakers, J. M. (2013). The rhizosphere microbiome: Significance of plant beneficial, plant pathogenic, and human pathogenic microorganisms. *FEMS Microbiology Reviews*, *37*(5), 634–663. DOI: 10.1111/1574-6976.12028 PMID: 23790204

Mihalakakou, G., Souliotis, M., Papadaki, M., Menounou, P., Dimopoulos, P., Kolokotsa, D., Gobakis, K., Alexandri, E., Koronaki, I., Paravantis, J., Giannakopoulos, E., Kalaitzoglou, A., Karagiozis, D., & Santamouris, M. (2023). Green roofs as a nature-based solution for improving urban sustainability: Progress and perspectives. *Renewable & Sustainable Energy Reviews*, *180*, 113306. DOI: 10.1016/j.rser.2023.113306

Mishra, P. N., Jain, S., Bore, T., Chang, I., Kwon, Y., Wang, Y., Dash, H. R., Kumar, A., Tiwari, S., Jiang, N., Das, S. K., & Scheuermann, A. (2024). Biological perspectives in geotechnics: Application and monitoring. *Journal of Rock Mechanics and Geotechnical Engineering*, *16*(7), 2854–2878. DOI: 10.1016/j.jrmge.2023.10.007

Mishra, T., Tiwari, P. B., Kanchan, S., & Kesheri, M. (2025). Advances in microbial bioremediation for effective wastewater treatment. *Water (Basel)*, *17*(22), 3196. DOI: 10.3390/w17223196

Mitra, S., Chakraborty, A. J., Tareq, A. M., Emran, T. B., Nainu, F., Khusro, A., Idris, A. M., Khandaker, M. U., Osman, H., Alhumaydhi, F. A., & Simal-Gandara, J. (2022). Impact of heavy metals on the environment and human health: Novel therapeutic insights to counter toxicity. *Journal of King Saud University. Science, 34*(3), 101865. DOI: 10.1016/j.jksus.2022.101865

Mongkolthanaruka, W., & Dharmsthiti, S. (2002). Biodegradation of lipid-rich wastewater by a mixed bacterial consortium. *International Biodeterioration & Biodegradation, 50*(2), 101–103. DOI: 10.1016/S0964-8305(02)00057-4

Morán, A. C., Müller, A., Manzano, M., & González, B. (2010). Simazine treatment history determines a significant herbicide degradation potential in soils that is not improved by bioaugmentation with *Pseudomonas* sp. ADP. *Journal of Applied Microbiology, 101*(1), 26–35. DOI: 10.1111/j.1365-2672.2006.02990.x PMID: 16834588

Mudge, S. M. (1995). Deleterious effects from accidental spillages of vegetable oils. *Spill Science & Technology Bulletin, 2*(2-3), 187–191. DOI: 10.1016/S1353-2561(96)00019-9

Nogales, B., Moore, E. R. B., Abraham, W. R., & Timmis, K. N. (1999). Identification of the metabolically-active members of a bacterial community in a polychlorinated biphenyl polluted Moorland soil. *Environmental Microbiology, 1*(3), 199–212. DOI: 10.1046/j.1462-2920.1999.00024.x PMID: 11207739

Ogunyemi, A. K., Buraimoh, O. M., Ogunyemi, B. C., Samuel, T. A., Ilori, M. O., & Amund, O. O. (2022). Nitrilase gene detection and nitrile metabolism in two bacterial strains associated with waste streams in Lagos, Nigeria. *Bulletin of the National Research Center, 46*(1), 151. DOI: 10.1186/s42269-022-00804-1

Ogunyemi, A. K., Samuel, T. A., Amund, O. O., & Ilori, M. O. (2019a). Cassava wastewater and solid waste leachate as cyanogenic substrates for the growth of nitrile and linamarin-utilizing bacteria. *Journal of Tropical Life Science, 9*(1), 79–87. DOI: 10.11594/jtls.09.01.11

Ogunyemi, A. K., Samuel, T. A., Amund, O. O., & Ilori, M. O. (2019b). Nitrile-metabolizing bacterial strains associated with municipal waste tips in the Lagos metropolis, Nigeria. Polish Journal of Natural Sciences. *Polish Journal of Natural Sciences, 34*(2), 257–272. https://ir.unilag.edu.ng/handle/123456789/6750

Ogunyemi, A. K., Samuel, T. A., Amund, O. O., & Ilori, M. O. (2019c). Characterization of linamarin-utilizing bacterial strains associated with detoxification of cyanogens in waste effluents. *Malaysian Applied Biology, 48*(3), 157–165. https://jms.mabjournal.com/index.php/mab/article/view/1851

Okafor, N., Ijioma, B., & Ijeoma, C. (2018). Microbial diversity and detoxification potential of cassava waste dumpsite soils. *Environmental Research Journal, 12*(3), 122–130.

Okoh, A. I., Sibanda, T., & Gusha, S. S. (2010). Inadequately treated wastewater as a source of human enteric viruses in the environment. *International Journal of Environmental Research and Public Health, 7*(6), 2620–2637. DOI: 10.3390/ijerph7062620 PMID: 20644692

Omotayo, A. E., Ilori, M. O., Amund, O. O., Ghosh, D., Roy, K., & Radosevich, M. (2011). Establishment and characterization of atrazine degrading cultures from Nigerian agricultural soil using traditional and Bio-Sep bead enrichment techniques. *Applied Soil Ecology, 48*(1), 63–70. DOI: 10.1016/j.apsoil.2011.01.006

Omotayo, A. E., Ilori, M. O., Obayori, O. S., & Amund, O. O.Omotayo, A. E.Ilori, M. O.Obayori, O. S.Amund, O. O. (2016). Influence of pH, temperature and nutrient addition on the degradation of atrazine by *Nocardioides* spp. isolated from agricultural soil in Nigeria. *Malaysian Journal of Microbiology, 12*(4), 270–278. DOI: 10.21161/mjm.81315

Omotayo, A. E., Ilori, M. O., Radosevich, M., & Amund, O. O. (2013). Metabolism of atrazine in liquid cultures and soil microcosms by *Nocardioides* strains isolated from a contaminated Nigerian agricultural soil. *Soil & Sediment Contamination, 22*(4), 365–375. DOI: 10.1080/15320383.2013.733444

Omotayo, A. E., Jebe-Abdullahi, F. T., & Kukoyi, G. A. (2025). Circulation of materials in urban agroecosystem. Canva.com https://www.canva.com

Omotayo, A. E., & Kukoyi, G. A. (2025). Biocatalytic degradation pathway of atrazine. Canva.com https://www.canva.com

Omotayo, A. E., Shonubi, O. O., Towuru, E. G., Babalola, S. E., Ilori, M. O., & Amund, O. O. (2014). Rhizoremediation of hydrocarbon-contaminated soil by *Paspalum vaginatum* (Sw.) and its associated bacteria. *International Research Journal of Microbiology, 5*(1), 1–7. DOI: 10.14303/irjm.2014.0102

Oro, C. E. D., Saorin Puton, B. M., Venquiaruto, L. D., Dallago, R. M., & Tres, M. V. (2024). Effective microbial strategies to remediate contaminated agricultural soils and conserve functions. *Agronomy (Basel), 14*(11), 2637. DOI: 10.3390/agronomy14112637

Orta-Ortiz, M. S., & Geneletti, D. (2023). Prioritizing urban nature-based solutions to support scaling-out strategies: A case study in Las Palmas de Gran Canaria. *Environmental Impact Assessment Review, 102*, 107158. DOI: 10.1016/j.eiar.2023.107158

Oyebamiji, O. O., Boeing, W. J., Holguin, F. O., Ilori, O. M., & Amund, O. O. (2019). Green microalgae cultured in textile wastewater for biomass generation and biodetoxification of heavy metals and chromogenic substances. *Bioresource Technology Reports*, *7*, 100247. DOI: 10.1016/j.biteb.2019.100247

Oyebamiji, O. O., Corcoran, A. A., Pérez, E. N., Ilori, M. O., Amund, O. O., Holguin, F. O., & Boeing, W. J. (2021). Lead tolerance and bioremoval by four strains of green algae from Nigerian fish ponds. *Algal Research*, *58*, 102403. DOI: 10.1016/j.algal.2021.102403

Pandey, S. B., & Yadav, R. S. (2015). Microbial degradation of cyanide compounds: A review. *International Journal of Advanced Research*, *3*(5), 1223–1231.

Parte, S. G., Mohekar, A. D., & Kharat, A. S. (2017). Microbial degradation of pesticide: A review. *African Journal of Microbiological Research*, *11*(24), 992–1012. DOI: 10.5897/AJMR2016.8402

Paul, M., White, J. R., Duance, C., Wolf, J., Gregory, T., & Charles, R. M. (2006). Hydrocarbons in a crude oil-contaminated soil. *Water, Air, and Soil Pollution*, *169*(1-4), 207–220. DOI: 10.1007/s11270-006-2194-0

Petrova, S., Velcheva, I., Nikolov, B., Angelov, N., Hristozova, G., Zaprjanova, P., Mitova, T., & Petrov, P. (2022). Nature-based solutions for the sustainable management of urban soils and quality of life improvements. *Land (Basel)*, *11*(4), 569. DOI: 10.3390/land11040569

Pieper, D. A. (2005). Aerobic degradation of polychlorinated biphenyl. *Applied Microbiology and Biotechnology*, *67*(2), 170–191. DOI: 10.1007/s00253-004-1810-4 PMID: 15614564

Prasad, M. P., & Manjunath, K. (2011). Comparative study on biodegradation of lipid-rich wastewater using lipase-producing bacterial species. *Indian Journal of Biotechnology*, *10*(1), 121–124.

Prince, R. C. (1993). Petroleum spill bioremediation in marine environments. *Critical Reviews in Microbiology*, *19*(4), 217–242. DOI: 10.3109/10408419309113530 PMID: 8305136

Qattan, S. Y. A. (2025). Harnessing bacterial consortia for effective bioremediation: Targeted removal of heavy metals, hydrocarbons and persistent pollutants. *Environmental Sciences Europe*, *37*(1), 85. DOI: 10.1186/s12302-025-01103-y

Rafeeq, H., Afsheen, N., Rafique, S., Arshad, A., Intisar, M., Hussain, A., Bilal, M., & Iqbal, H. M. N. (2023). Genetically engineered microorganisms for environmental remediation. *Chemosphere*, *310*, 136751. DOI: 10.1016/j.chemosphere.2022.136751 PMID: 36209847

Ramanathan, S. S., Gannon, T. W., Everman, W. J., & Locke, A. M. (2022). Atrazine, mesosulfuron-methyl, and topramezone persistence in North Carolina soils. *Agronomy Journal*, *114*(2), 1068–1079. DOI: 10.1002/agj2.21041

Rashid, N., Ur Rehman, M. S., Han, J. I., & Park, J. Y. (2016). Current status, issues, and developments in cyanide biodegradation. *Biotechnology Advances*, *34*(5), 874–881. PMID: 27188926

Rastogi, R. P., & Madamwar, D. (2021). Biotechnological potential of microbial consortia for sustainable environmental management. *Bioresource Technology*, *337*, 125411.

Raupp, P., Carrillo, Y., & Nielsen, U. N. (2024). Soil health to enhance ecological restoration and conservation. *Journal of Sustainable Agriculture and Environment*, *3*(4), e70022. DOI: 10.1002/sae2.70022

Russo, A., Escobedo, F. J., Cirella, G. T., & Zerbe, S. (2017). Edible green infrastructure: An approach and review of provisioning ecosystem services and disservices in urban environments. *Agriculture, Ecosystems & Environment*, *242*, 53–66. DOI: 10.1016/j.agee.2017.03.026

Saini, S., Tewari, S., Dwivedi, J., & Sharma, V. (2023). Biofilm-mediated wastewater treatment: A comprehensive review. *Materials Advances*, *4*(6), 1415–1443. DOI: 10.1039/D2MA00945E

Sala, O. E., Chapin, F. S., Armesto, J. J., Berlow, E., Bloomfield, J., Dirzo, R., Huber-Sanwald, E., Huenneke, L. F., Jackson, R. B., Kinzig, A., Leemans, R., Lodge, D. M., Mooney, H. A., Oesterheld, M., Poff, N. L., Sykes, M. T., Walker, B. H., Walker, M., & Wall, D. H. (2000). Global biodiversity scenarios for the year 2100. *Science*, *287*(5459), 1770–1774. DOI: 10.1126/science.287.5459.1770 PMID: 10710299

Sazzad, I. (2024). Advancing sustainable lubricating oil management: Re-refining techniques, market insights, innovative enhancements, and conversion to fuel. *Heliyon*, *10*(20), e39248. DOI: 10.1016/j.heliyon.2024.e39248 PMID: 39640706

Schulz, S., Brankatschk, R., Dumig, A., Kogel-Knabner, I., Schloter, M., & Zeyer, J. (2013). The role of microorganisms at different stages of ecosystem development for soil formation. *Biogeosciences*, *10*(6), 3983–3996. DOI: 10.5194/bg-10-3983-2013

Sesay, F., Sesay, R. E. V., Kamara, M., Li, X., & Niu, C. (2025). Biodegradation of pharmaceutical contaminants in wastewater using microbial consortia: Mechanisms, applications, and challenges. *Journal of Environmental Management, 384*, 125564. DOI: 10.1016/j.jenvman.2025.125564 PMID: 40306218

Shahpiri, A., & Mohammadzadeh, A. (2018). Mercury removal by engineered *Escherichia coli* cells expressing different rice metallothionein isoforms. *Annals of Microbiology, 68*(3), 145–152. DOI: 10.1007/s13213-018-1326-2

Sharafi, K., Moradi, M., Azari, A., Sharafi, H., & Pirsaheb, M. (2016). Comparative evaluation of parasitic removal in municipal wastewater using constructed wetland and extended aeration–activated sludge system in Kermanshah province, Iran. *International Journal of Health and Life Sciences, 2*(1), 16–21.

Sharma, P., Bano, A., Singh, S. P., Dubey, N. K., Chandra, R., & Iqbal, H. M. N. (2022). Recent advancements in microbial-assisted remediation strategies for toxic contaminants. *Cleaner Chemical Engineering, 2*, 100020. DOI: 10.1016/j.clce.2022.100020

Sharma, P. R., Chopra, A., Cameotra, S. S., & Suri, C. R. (2010). Efficient biotransformation of herbicide diuron by bacterial strain *Micrococcus* sp. PS-1. *Biodegradation, 21*(6), 979–987. https://api.semanticscholar.org/CorpusID:29519247. DOI: 10.1007/s10532-010-9357-9 PMID: 20401520

Silvestrini, F., Fratini, F., & Magini, R. (2021). Planning with nature: Sustainable urban prototypes for Portuense district in Rome, Italy. *WIT Transactions on Ecology and the Environment, 253*, 15–26. Advance online publication. DOI: 10.2495/SC210021

Singh, B. K., & Walker, A. (2006). Microbial degradation of organophosphorus compounds. *FEMS Microbiology Reviews, 30*(3), 428–471. DOI: 10.1111/j.1574-6976.2006.00018.x PMID: 16594965

Singh, J. S., & Gupta, V. K. (2018). Soil microbial biomass: A key soil driver in management of ecosystem functioning. *The Science of the Total Environment, 634*, 497–500. DOI: 10.1016/j.scitotenv.2018.03.373 PMID: 29635193

Singh, M., Singh, D., Gupta, A., Pandey, K. D., Singh, P. K., & Kumar, A. (2019). Plant growth promoting rhizobacteria: application in biofertilizers and biocontrol of phytopathogens. In *PGPR Amelioration in Sustainable Agriculture* (pp. 41–66). Woodhead Publishing. DOI: 10.1016/B978-0-12-815879-1.00003-3

Singh, P., Suri, C. R., & Cameotra, S. S. (2004). Isolation of a member of *Acinetobacter* species involved in atrazine degradation. *Biochemical and Biophysical Research Communications, 317*(3), 697–702. DOI: 10.1016/j.bbrc.2004.03.112 PMID: 15081396

Singh, S., Kumar, V., Chauhan, A., Datta, S., Wani, A. B., Singh, N., & Singh, J. (2018). Toxicity, degradation and analysis of the herbicide atrazine. *Environmental Chemistry Letters, 16*(1), 211–237. DOI: 10.1007/s10311-017-0665-8

Siripattanakul, S., Wirojanagud, W., McEvoy, J., Limpiyakorn, T., & Khan, E. (2010). Atrazine degradation by stable mixed cultures enriched from agricultural soil and their characterization. *Journal of Applied Microbiology, 106*(3), 986–992. DOI: 10.1111/j.1365-2672.2008.04075.x PMID: 19191954

Smith, S. E., & Read, D. J. (2008). *Mycorrhizal Symbiosis* (3rd ed.). Academic Press.

Srivastava, N., Singh, R., Sing, P., Ahmad, I., Singh, R. P., Rai, A. K., Asiri, M., & Gupta, V. K. (2023). Recent advances on lignocellulosic bioresources and their valorization in biofuels production: Challenges and viability assessment. *Environmental Technology & Innovation, 29*, 103037. DOI: 10.1016/j.eti.2023.103037

Strong, L. C., Rosendahl, C., Johnson, G., Sadowsky, M. J., & Wackett, L. P. (2002). *Arthrobacter aurescens* TC1 metabolizes diverse s-triazine ring compounds. *Applied and Environmental Microbiology, 68*(12), 5973–5980. DOI: 10.1128/AEM.68.12.5973-5980.2002 PMID: 12450818

Suding, K., Higgs, E., Palmer, M., Callicott, J. B., Anderson, C. B., Baker, M., Gutrich, J. J., Hondula, K. L., LaFevor, M. C., Larson, B. M. H., Randall, A., Ruhl, J. B., & Schwartz, K. Z. S. (2015). Committing to ecological restoration. *Science, 348*(6235), 638–640. DOI: 10.1126/science.aaa4216 PMID: 25953995

Suyal, D. C., Soni, R., Sai, S., & Goel, R. (2016). Microbial inoculants as biofertilizer. In *Microbial inoculants in sustainable agricultural productivity* (pp. 311–318). Springer. DOI: 10.1007/978-81-322-2647-5_18

Syuhadah, B. H., & Raikhan, N. H. N. (2022). *Pseudomonas aeruginosa* NR.22 - local hydrocarbon degrader species originated from crude oil non-exploration site: Isolation, survivability and hydrocarbon degradation study. *Desalination and Water Treatment, 257*, 260–269. DOI: 10.5004/dwt.2022.28473

Tan, K. H., & Gill, C. O. (1985). Batch Growth of *Saccaromycopsis lipolytica* on animal fats. *Applied Microbiology and Biotechnology, 21*(5), 292–298. DOI: 10.1007/BF00252707

Thauer, R. K., Kaster, A. K., Seedorf, H., Buckel, W., & Hedderich, R. (2008). Methanogenic archaea: Ecologically relevant differences in energy conservation. *Nature Reviews. Microbiology*, *6*(8), 579–591. DOI: 10.1038/nrmicro1931 PMID: 18587410

Topp, E., Zhu, H., Nour, S. M., Houot, S., Lewis, M., & Cuppels, D. (2000). Characterization of an atrazine-degrading *Pseudaminobacter* sp. isolated from Canadian and French agricultural soils. *Applied and Environmental Microbiology*, *66*(7), 2773–2782. DOI: 10.1128/AEM.66.7.2773-2782.2000 PMID: 10877767

Valois, L., Brachet, A., Schiopu, N., & Barot, S. (2023). Performance assessment of the ecosystem services provided by urban nature-based solutions: Focus on rainwater management. *IOP Conference Series. Earth and Environmental Science*, *1196*(1), 012028. DOI: 10.1088/1755-1315/1196/1/012028

Van Beilen, J. B., & Funhoff, E. G. (2007). Alkane hydroxylases involved in microbial alkane degradation. *Applied Microbiology and Biotechnology*, *74*(1), 13–21. DOI: 10.1007/s00253-006-0748-0 PMID: 17216462

van den Hoogen, J., Geisen, S., Routh, D., Ferris, H., Traunspurger, W., Wardle, D. A., de Goede, R. G. M., Adams, B. J., Ahmad, W., Andriuzzi, W. S., Bardgett, R. D., Bonkowski, M., Campos-Herrera, R., Cares, J. E., Caruso, T., de Brito Caixeta, L., Chen, X., Costa, S. R., Creamer, R., & Crowther, T. W. (2019). Soil nematode abundance and functional group composition at a global scale. *Nature*, *572*(7768), 194–198. DOI: 10.1038/s41586-019-1418-6 PMID: 31341281

Vannucchi, F., Bretzel, F., Pini, R., & Rumble, H. (2021). Less is more: Soil and substrate quality as an opportunity for urban greening and biodiversity conservation. In Catalano, C., Andreucci, M. B., Guarino, R., Bretzel, F., Leone, M., & Pasta, S. (Eds.), *Urban services to ecosystems. Future city* (Vol. 17, pp. 207–224). Springer., DOI: 10.1007/978-3-030-75929-2_11

Viwatpanich, K., Siritanon, T., & Tongsook, S. (2013). Detoxification of cassava wastewater using microbial fermentation. *Journal of Environmental Science and Health. Part B, Pesticides, Food Contaminants, and Agricultural Wastes*, *48*(6), 460–466.

Walters, S. A., & Midden, K. S. (2018). Sustainability of urban agriculture: Vegetable production on green roofs. *Agriculture*, *8*(11), 168. DOI: 10.3390/agriculture8110168

Wright, M. A., Taylor, F., Randles, S. J., Brown, D. E., & Higgins, I. J. (1993). Biodegradation of a Synthetic Lubricant by *Micrococcus roseus*. *Applied and Environmental Microbiology*, *59*(4), 1072–1076. DOI: 10.1128/aem.59.4.1072-1076.1993 PMID: 8476283

Wu, D., Wei, Z. M., Mohamed, T. A., Zheng, G. G., Qu, F. T., Wang, F., Zhao, Y., & Song, C. H. (2022). Lignocellulose biomass bioconversion during composting: Mechanism of action of lignocellulase, pretreatment methods and future perspectives. *Chemosphere, 286*, 131635. DOI: 10.1016/j.chemosphere.2021.131635 PMID: 34346339

Wu, Q., Huang, Y., Irga, P., Kumar, P., Li, W., Wei, W., Feng, H., Jiang, Y., Schnelle-Kreis, J., Zimmermann, R., Tang, M., & Xie, X. (2024). Synergistic control of urban heat island and urban pollution island effects using green infrastructure. *Journal of Environmental Management, 370*, 122985. DOI: 10.1016/j.jenvman.2024.122985 PMID: 39461153

Xiang, L., Harindintwali, J. D., Wang, F., Redmile-Gordon, M., Chang, S. X., Fu, Y., Feng, Y., Zeng, X. C., Su, S. M., Tian, C., Lammel, D. R., Basiliko, N., Zhang, W., Ok, Y. S., Borken, W., Ludley, K., Xie, S., Sun, H., Wei, Z., & Jiang, X. (2022). Integrating biochar, bacteria, and plants for sustainable remediation of soils contaminated with organic pollutants. *Environmental Science & Technology, 56*(23), 16546–16566. DOI: 10.1021/acs.est.2c02976 PMID: 36301703

Xie, Y., Yu, X., Ng, N. C., Li, K., & Fang, L. (2018). Exploring the dynamic correlation of landscape composition and habitat fragmentation with surface water quality in the Shenzhen River and deep bay cross-border watershed, China. *Ecological Indicators, 90*, 231–246. DOI: 10.1016/j.ecolind.2017.11.051

Xu, M., Yang, M., Sun, H., Gao, M., Wang, Q., & Wu, C. (2022). Bioconversion of biowaste into renewable energy and resources: A sustainable strategy. *Environmental Research, 214*, 113929. DOI: 10.1016/j.envres.2022.113929 PMID: 35868577

Yaashikaa, P. R., Senthil Kumar, P., Jeevanantham, S., & Saravanan, R. (2022). A review on bioremediation approach for heavy metal detoxification and accumulation in plants. *Environmental Pollution, 301*, 119035. DOI: 10.1016/j.envpol.2022.119035 PMID: 35196562

Yale, R. L., Sapp, M., Sinclair, C. J., & Moir, J. W. (2017). Microbial changes linked to the accelerated degradation of the herbicide atrazine in a range of temperate soils. *Environmental Science and Pollution Research International, 24*(8), 7359–7374. DOI: 10.1007/s11356-017-8377-y PMID: 28108915

Zareba, A., Krzeminska, A., & Kozik, R. (2021). Urban vertical farming as an example of nature-based solutions supporting a healthy society living in the urban environment. *Resources, 10*(11), 109. DOI: 10.3390/resources10110109

Zdarta, J., Jankowska, K., Bachosz, K., Degórska, O., Kaźmierczak, K., Nguyen, L. N., Nghiem, L. D., & Jesionowski, T. (2021). Enhanced Wastewater Treatment by Immobilized Enzymes. *Current Pollution Reports*, *7*(2), 167–179. DOI: 10.1007/s40726-021-00183-7

Zhang, B., Hu, X., Zhao, D., Wang, Y., Qu, J., Tao, Y., Kang, Z., Yu, H., Zhang, J., & Zhang, Y. (2024). Harnessing microbial biofilms in soil ecosystems: Enhancing nutrient cycling, stress resilience, and sustainable agriculture. *Journal of Environmental Management*, *370*, 122973. DOI: 10.1016/j.jenvman.2024.122973 PMID: 39437688

Zhang, B., Xu, J., Sun, M., Yu, P., Ma, Y., Xie, L., & Chen, L. (2023). Comparative secretomic and proteomic analysis reveal multiple defensive strategies developed by *Vibrio cholerae* against the heavy metal (Cd^{2+}, Ni^{2+}, Pb^{2+}, and Zn^{2+}) stresses. *Frontiers in Microbiology*, *14*, 1294177. DOI: 10.3389/fmicb.2023.1294177 PMID: 37954246

Zheng, X., Lin, H., Du, D., Li, G., Alam, O., Cheng, Z., Liu, X., Jiang, S., & Li, J. (2024). Remediation of heavy metals polluted soil environment: A critical review on biological approaches. *Ecotoxicology and Environmental Safety*, *284*, 116883. DOI: 10.1016/j.ecoenv.2024.116883 PMID: 39173222

Chapter 8
Microbial Resilience in Response to Climate Change and Pollution:
Implications for Ecosystem Stability

Oladele Ogunseitan
http://orcid.org/0000-0003-1317-6219
University of California, Irvine, USA

ABSTRACT

Since living organisms emerged on Earth, prokaryotes have contributed to major shifts in global climate, including the production and the balance of other atmospheric gases that facilitated the evolution of eukaryotes. The history of environmental evolution is inextricable from the history of microbial evolution. Extremely diverse microbial communities inhabit all ecosystems that have been explored on Earth, and these communities continue to serve as the foundational architects of ecosystem stability. Yet it remains unclear how anthropogenic climate change is affecting microbial diversity and community structure as they face unprecedented challenges from the dual pressure of anthropogenic toxic environmental pollution. This chapter examines the complex interplay between environmental stressors and microbial resilience, exploring how prokaryotes adapt, survive, create, and maintain ecosystem functions. The chapter addresses key questions about planetary boundaries and the role of microorganisms in sustainability of safe operating zones for novel entities and climate change.

DOI: 10.4018/979-8-2600-2100-2.ch008

1 INTRODUCTION

In the epic natural history of life on Earth, microorganisms, mostly invisible to the unaided human eyes, stand as both the progenitors and protagonists of the most consequential phases. As we delve deeper into understanding the intricate relationships between microbial communities and ecosystem stability, we uncover a story that spans billions of years and continues to unfold in real-time across every conceivable habitat on our planet. This chapter builds upon the foundational concepts explored throughout this book, weaving together the threads of microbial diversity, environmental adaptation, and ecosystem resilience into a comprehensive understanding of how microbial communities respond to the unprecedented challenges of the Anthropocene and threats to safe operating zones of planetary boundaries (Ogunseitan, 2025; Steffen et al., 2015).

This chapter situates the putative resilience of microorganisms and their communities centrally in the mission of this book to illuminate the remarkable capabilities of microorganisms and their potential applications in addressing some of humanity's most pressing environmental challenges. From bioremediation to biotechnology, from carbon cycling to climate regulation, *observed* microorganisms have demonstrated remarkable capacity to transform, adapt, and innovate in ways that continue to surprise and inspire researchers worldwide. The emphasis on "observed" in the last sentence points to the limitations of the conclusion that microbial communities are resilient because human capacity to observe the structures and functions of such communities are in fact severely limited. We have no direct evidence of microbial extinctions on the planet largely because we are incapable of knowing with the aid of methods that have otherwise served the scientific method very well. The goal here is to expand our limited exploration by examining how these same remarkable capabilities manifest as resilience mechanisms when microbial communities face the dual pressures of climate change and environmental pollution.

If we accept the view that microorganisms are the foundation of ecosystem structure and function, it follows that microbial diversity serves as the foundation upon which ecosystem resilience is built. The staggering variety of metabolic pathways, stress tolerance mechanisms, and adaptive strategies found within microbial communities provides ecosystems with an unparalleled capacity to maintain function in the face of environmental perturbations. This diversity is not merely a catalog of different species; it represents a vast library of evolutionary solutions to environmental challenges, accumulated over billions of years of trial, adaptation, and environmental selection (Ogunseitan, 2005; Ogunseitan, 2026).

The concept of functional redundancy within microbial communities exemplifies this diversity-resilience relationship. When environmental stressors eliminate certain microbial populations, functionally similar organisms can often step in to maintain critical ecosystem processes. This biological insurance policy operates at multiple scales, from the molecular level where different enzymes can catalyze similar reactions, to the community level where different species can perform equivalent ecological roles.

Building upon documented knowledge of microbial capabilities, we now turn our attention to how these capabilities translate into resilience under stress, including toxic chemical pollution and climate change based on the human perspective.

1.1 Defining Microbial Resilience in Environmental Context

One way to think about microbial resilience is to consider the resurgence of antibiotic-resistant bacterial pathogens nearly a century after the discovery of penicillin, followed by seemingly permanent defeat of some communicable diseases in human populations. From the human perspective, bacterial resistance to antibiotics is the antithesis of resilience of global health systems effectiveness. However, from the evolutionary perspective, the emergence and dissemination of antibiotic resistance in bacterial populations and communities is evidence of resilience (Ogunseitan, 2022; Ogunseitan et al. 2022; Petrillo & Ogunseitan, 2018).

Microbial resilience encompasses both resistance and recovery mechanisms, but the distinction between these concepts becomes particularly nuanced when dealing with microbial communities. Resistance refers to the ability of a community to maintain its structure and function despite environmental perturbations. In microbial systems, this might manifest as the maintenance of key metabolic processes even when community composition shifts, or the preservation of critical species despite environmental stress.

Recovery, on the other hand, describes the capacity of a community to return to its original state following a disturbance. For microbial communities, recovery can be remarkably rapid due to their short generation times and high reproductive rates. However, the definition of "original state" becomes complex when dealing with communities that naturally exhibit high temporal variability and whose composition may be influenced by stochastic factors. Therefore, the dimensions of resilience include temporal, spatial, functional, and componential (Allison & Martiny, 2008; Shade et al., 2014; Shade et al., 2018).

The understanding of temporal scales of microbial responses to environmental stimuli is limited by the technical capabilities of the methods of observation. Microbial communities can respond to environmental changes on timescales ranging from sub-seconds to decades. Immediate responses might include changes in gene expression or metabolic activity, while longer-term responses could involve population shifts, evolutionary adaptations, or community restructuring (Griffiths &

Philippot, 2013). Understanding resilience requires consideration of these multiple temporal scales and their interactions.

Spatial scales of microbial resilience may be understood from the perspective of epidemiology, defined as the patterns of specific communicable diseases in a human population; or from the perspective of microbial geography whereby the prevailing environmental conditions strongly influence the presence or absence of certain microorganisms. For example, influential environmental parameters may be due to aridity, temperature, acidity, or incident light spectrum. A community of microorganisms might appear resilient at the ecosystem scale while experiencing significant disruption at the microhabitat scale, or vice versa. The hierarchical organization of microbial communities, from individual cells to biofilms to ecosystem-wide networks, means that resilience mechanisms can operate at multiple spatial scales simultaneously.

Functional redundancy represents one of the most important aspects of microbial resilience. This concept refers to the presence of functional or metabolism groups that can perform similar ecological roles. When environmental stressors, including toxic chemical exposures block or inactivates a particular metabolic process, other biochemical pathways may be induced from dormancy or accelerated (Lennon & Jones, 2011). The high diversity of microbial communities typically provides substantial functional redundancy, contributing to their overall resilience.

However, resilience due to functional redundancy may be compromised by deficiencies in the componential dimension of resilience. Certain ecosystem functions may be performed by only a few specialized organisms. These keystone species are essential components of ecosystems because they form unique functional groups which may represent potential vulnerabilities in otherwise resilient systems. Understanding the distribution of functional and componential redundancy across different ecosystem processes is crucial for predicting community responses to environmental stress.

The maintenance of ecosystem services represents the ultimate measure of microbial resilience from a human perspective. Ecosystem services provided by microbial communities include nutrient cycling, pollutant degradation, soil formation, climate regulation, and disease suppression. Resilient microbial communities maintain these services even under environmental stress, though the specific organisms providing the services may change.

1.2 The Anthropocene Complicates the Microbiocene Legacy

The history of life on our planet to date is dominated by microbial life forms. The Microbiocene period will probably never end, whereby metabolic activities of microorganisms continue to transform the environmental parameters that create opportunities for other life forms including humans. The transformation includes the presence of oxygen in the atmosphere, and the balance of atmospheric gases including carbon dioxide. The degradation of leaf litter in soils, and the transformation of various types of waste products from living organisms and emissions from geologic processes. However, the Anthropocene epoch now overlaps the Microbiocene, and presents microbial communities with an unprecedented combination of environmental stressors that challenge their resilience in ways that have no historical precedent. Unlike the seemingly gradual environmental changes that have characterized most of Earth's history, anthropogenic changes are occurring at rates that may exceed the adaptive capacity of even the most resilient microbial communities. These changes have been captured by the Planetary boundaries concept, whereby human activities have reportedly led to the transgression of seven out of nine safe operating zones, including climate change and novel entities defined as anthropogenic chemicals (Ogunseitan, 2025). How microbial communities respond to these transgressions of the Anthropocene's influence on planetary health is the subject of extensive multidisciplinary research.

First, consider global climate change which represents one of the most pervasive stressors affecting microbial communities globally. Rising temperatures, altered precipitation patterns, increased frequency of extreme weather events, and changing atmospheric composition all directly impact microbial habitats. These changes occur across multiple spatial and temporal scales, from local microhabitat modifications to global atmospheric alterations, and from seasonal variations to multi-decadal trends (Intergovernmental Panel on Climate Change, 2022).

The complexity of climate change impacts on microbial communities stems from the interconnected nature of environmental variables. Temperature changes affect not only microbial physiology directly but also influence water availability, nutrient cycling rates, and the physical structure of habitats. Similarly, altered precipitation patterns affect soil moisture, pH, and the transport of nutrients and pollutants, all of which influence microbial community structure and function (Bradford et al., 2008; Cavicchioli et al., 2019).

Ocean acidification, driven by increased concentration of CO_2 in the atmosphere, represents a particularly significant challenge for marine microbial communities (Doney et al., 2009). The changing pH of seawater affects the physiology of many marine microorganisms, particularly those involved in calcium carbonate formation and dissolution (Meron et al., 2011). These changes can cascade through marine food webs and biogeochemical cycles, with implications for global carbon and nutrient cycling (Pollack et al., 2009; Webster et al., 2016).

Secondly, novel entities such as toxic chemical pollutants of the environment adds another layer of stress to climate-challenged microbial communities. Unlike climate change, which affects environmental conditions broadly, toxic pollution often creates localized hotspots of extreme stress. However, the global distribution of pollutants through atmospheric, soil, and oceanic transport means that even remote environments are not immune to contamination (Ogunseitan, 2023; Ogunseitan, 2024).

The diversity of pollutants presents microbial communities with a bewildering array of chemical challenges. Heavy metals, organic pollutants, pharmaceuticals, plastics, and engineered nanomaterials each present unique toxicological challenges and require different adaptive responses. The simultaneous presence of multiple pollutants can create complex interactions that are difficult to predict and may overwhelm natural detoxification mechanisms.

Persistent organic pollutants (POPs) represent a particularly challenging class of contaminants due to their resistance to degradation and tendency to bioaccumulate. These compounds can persist in the environment for decades or longer, providing sustained stress to microbial communities. The global distribution of POPs means that even pristine environments may harbor these contaminants at levels that could affect sensitive microbial processes.

Emerging contaminants, including pharmaceuticals, personal care products, and engineered nanomaterials, present novel challenges for which microbial communities may lack pre-existing adaptive mechanisms. These compounds often have no natural analogs, meaning that microorganisms must evolve entirely new detoxification pathways or tolerance mechanisms to cope with their presence.

The synergistic effects of climate change and pollution represent perhaps the greatest challenge to microbial resilience. Environmental stressors rarely act in isolation, and their combined effects can be greater than the sum of their individual impacts. For example, elevated temperatures may increase the toxicity of certain pollutants, while pollution-stressed communities may be less able to cope with climate-related stressors (Ogunseitan, 1996; Khalili et al., 2016; Ogunseitan, 2019).

Cumulative impacts represent another dimension of the dual stressor challenge. Microbial communities may be able to cope with individual stressors or short-term perturbations, but the cumulative effect of multiple stressors over extended periods may gradually erode their resilience. This erosion may not be immediately apparent, as communities may maintain function while losing diversity or adaptive capacity.

The unprecedented nature of current environmental changes means that historical analogies may provide limited guidance for predicting microbial responses. While microbial communities have survived previous mass extinctions and climate changes, the current combination of rapid climate change and novel pollutants represents a unique challenge in Earth's history.

2 THEORETICAL FRAMEWORKS: RESILIENCE AND MICROBIAL ECOLOGY

The theoretical foundation for understanding microbial resilience draws from ecological resilience theory (ERT), which has evolved considerably since its introduction by Holling (1973). Notably, the 6th Assessment Report of the Intergovernmental Panel on Climate Change (IPCC) defines resilience as "not just the ability to maintain essential function, identity and structure, but also the capacity for transformation" (Intergovernmental Panel on Climate Change, 2022). This theoretical framework provides essential conceptual tools for understanding how microbial communities respond to environmental perturbations and maintain ecosystem function under stress. The inclusion of capacity for transformation as part of general understanding of resilience resonates with biotechnological applications of microbial communities for environmental restoration, including bioremediation of polluted ecosystems.

It is helpful to differentiate "engineering resilience", the earliest conceptualization of resilience, which focuses on the speed of return to equilibrium following a disturbance, from "ecological resilience". This perspective assumes that ecosystems have a single stable state and that resilience can be measured by how quickly the system returns to this state after perturbation. While this concept has limited applicability to complex ecological systems, it remains relevant for understanding certain aspects of microbial community dynamics, particularly in highly controlled or simplified systems. In microbial communities, engineering resilience might manifest as the rapid recovery of specific metabolic functions following a disturbance. For example, soil microbial communities often show rapid recovery of respiration rates following drought stress, even if community composition remains altered. This type of resilience is particularly important for maintaining ecosystem services that depend on specific microbial processes (Schimel, 2018).

In contrast, "ecological resilience" recognizes that ecosystems may have multiple stable states and focuses on the amount of disturbance a system can absorb before shifting to an alternative state. This perspective is more applicable to complex microbial communities, which often exhibit multiple stable configurations and can undergo regime shifts in response to environmental changes. The concept of stability domains is central to ecological resilience theory. These domains represent the range of conditions within which a system maintains its current state. For microbial communities, stability domains might be defined by environmental variables such as temperature, pH, nutrient availability, or concentrations of antibiotics or other toxic pollutants. Understanding these domains is crucial for predicting when communities might undergo regime shifts that are important, for example in soils where repeated and prolonged application of fertilizers and toxic pesticides may stimulate shifts with decreasing capacity to support agricultural services.

The concept of regime shifts is particularly relevant to understanding microbial resilience. Regime shifts represent transitions between alternative stable states, often triggered by environmental changes that push the system beyond critical thresholds (Scheffer et al., 2001; Scheffer et al., 2003). In microbial communities, regime shifts might involve transitions between different dominant functional groups, changes in community structure, or shifts in ecosystem function. Hysteresis effects are often associated with regime shifts, meaning that the conditions required to reverse a shift may be different from those that caused it. This has important implications for restoration efforts, as simply removing the stressor that caused a regime shift may not be sufficient to restore the original community state.

Adaptive cycles also represent another key concept in resilience theory, describing the cyclical nature of ecosystem development and reorganization. The adaptive cycle consists of four phases: growth, conservation, release, and renewal. In microbial communities, these cycles might occur over various timescales, from the daily cycles of activity in response to resource availability to longer-term cycles of community development and disturbance. During the growth phase, microbial communities accumulate biomass and diversity, often following colonization of new habitats or recovery from disturbance. The conservation phase is characterized by stable community structure and efficient resource utilization, but also by reduced flexibility and increased vulnerability to disturbance. The release phase involves the breakdown of established structures, often triggered by environmental stress or resource depletion. Finally, the renewal phase represents a period of reorganization and innovation, during which new community structures may emerge.

Clearly, the concept of ecological resilience has rich theoretical underpinnings, and research continues on how to assess and apply the understanding of the concept at the intersections characterizing social-ecological systems. However, most of the published literature on the topic address macroorganisms and their communities. Microbial communities exhibit distinctive characteristics that warrant caution in making direct analogies and comparisons with the preponderance of examples and case studies on ecological resilience (Coyte et al., 2015; Faust & Raes, 2012). In particular, understanding microbial community dynamics requires consideration of the unique characteristics that distinguish microbial (prokaryotes) communities from communities defined by macroorganisms (eukaryotes). These characteristics, including rapid generation times, horizontal gene transfer, dormancy capabilities, and the potential for massive population sizes within small spatial scales have implications for how microorganisms respond to overlapping stressor such as climate change and toxic chemical pollution.

3 MECHANISMS OF MICROBIAL RESILIENCE

Observations of microbial resilience in natural ecosystems and under artificial conditions are helpful in refining theoretical underpinnings of resilience and microbial ecology. Laboratory experiments have also been helpful in unraveling the molecular mechanisms underlying microbial resilience. Primary mechanisms of response to stressful environmental conditions have been studied in detail for a limited number of microorganisms. The mechanisms operate at the cellular level through physiological processes; at the levels of genetic and evolutionary adaptation; at population-level dynamics; and at aggregate community levels (Cardinale et al., 2012; Loreau et al., 2001; Louca et al., 2018; Wittebolle et al., 2009). The operation of these mechanisms is not mutually exclusive, and often act in concert to produce a coordinated response. However, for the sake of clarity, it is important to separately consider the various categories of mechanisms of microbial resilience (Figure 16.1).

Resilience at the Cellular Level: Microbial resilience at the cellular level depends on sophisticated physiological mechanisms that allow individual organisms to survive and function under adverse conditions. These mechanisms represent millions of years of evolutionary refinement and provide the foundation for community-level resilience. For example, phylogenetically-conserved heat shock proteins and chaperones represent one of the most universal and important cellular stress responses. These proteins help maintain proper protein folding under stressful conditions and can refold proteins that are damaged to restore their function. The rapid induction of heat shock proteins allows cells to survive temperature stress, chemical exposure, and other harsh conditions (Pollack et al., 2009). The heat shock response is highly conserved across all domains of life, indicating its fundamental importance for cellular survival. Different heat shock proteins have specialized functions, with some acting as molecular chaperones that assist in protein folding, while others help transport damaged proteins to degradation pathways.

Antioxidant systems and reactive oxygen species (ROS) management are crucial for protecting cells from oxidative damage. Environmental stressors often lead to increased ROS production, which can damage DNA, proteins, and lipids. Microbial cells possess sophisticated antioxidant systems, including enzymes like catalase and superoxide dismutase, as well as small molecule antioxidants like glutathione (Ogunseitan & LeBlanc, 2005; Ogunseitan et al., 2001; Prosser, 2015; Renella et al., 2014). The regulation of antioxidant systems is tightly controlled and responds rapidly to changes in oxidative stress levels. Cells can increase antioxidant production within minutes of stress exposure, providing immediate protection against oxidative damage.

Figure 1. Diagram illustrating the multi-scale integration of cellular, genetic, population, community and network-level mechanisms that enable microorganisms to survive, adapt and thrive under dynamic environmental conditions and stressors including toxic pollution and climate change.

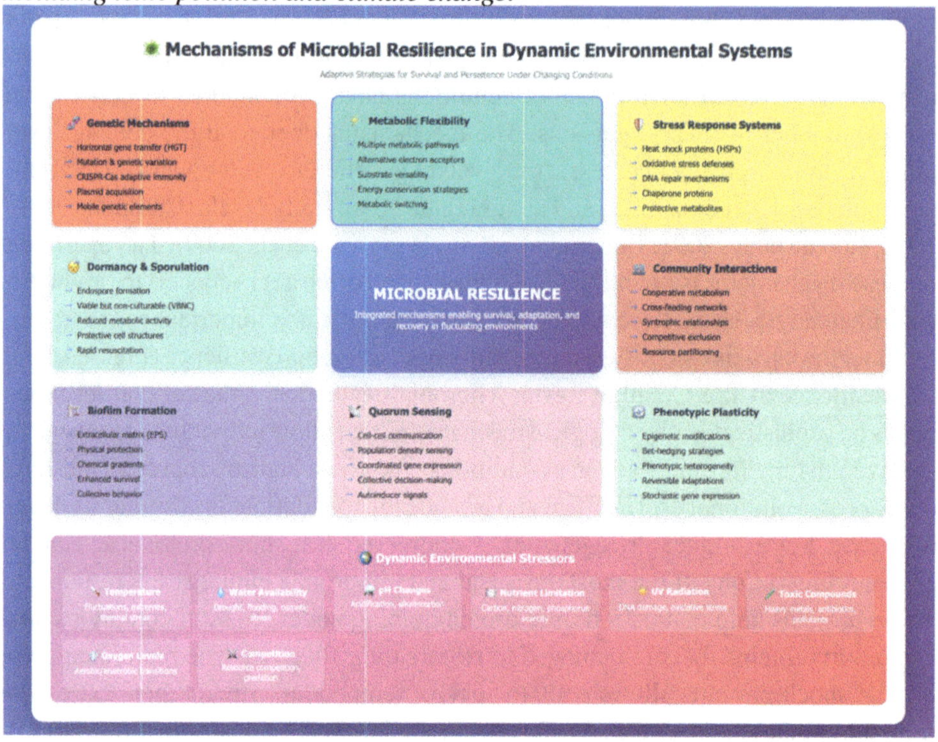

Membrane composition adjustments allow microorganisms to maintain membrane integrity and function under changing environmental conditions. Temperature stress, for example, can be countered by altering the fatty acid composition of membranes to maintain optimal fluidity. Similarly, osmotic stress can be managed through changes in membrane lipid composition. The ability to rapidly modify membrane composition gives microorganisms remarkable flexibility in responding to environmental changes. Some organisms can complete membrane remodeling within hours of stress exposure, allowing them to maintain function under conditions that would be lethal to less adaptive organisms (Storz & Hengge-Aronis, 2000).

Metabolic pathway flexibility enables microorganisms to switch between different energy sources and biosynthetic pathways depending on environmental conditions. This flexibility is particularly important under stress conditions when preferred substrates may be unavailable or when energy conservation becomes critical. Many microorganisms possess multiple metabolic pathways for key processes, allowing them to maintain function even when some pathways are inhibited by environmental stress. This redundancy provides a buffer against environmental perturbations and enhances overall cellular resilience (Ogunseitan, 1998; Ogunseitan, 2000; Ogunseitan et al., 1999).

Spore formation and germination represent one of the most extreme examples of microbial stress tolerance. Bacterial endospores can survive conditions that would kill vegetative cells, including extreme temperatures, radiation, desiccation, and chemical exposure. The ability to form spores allows populations to persist through extended periods of adverse conditions. The spore formation process involves the complete reorganization of cellular contents and the synthesis of specialized protective structures. This process requires significant energy investment but provides unparalleled protection against environmental stress. Spore germination is triggered by specific environmental cues that indicate favorable conditions for growth. This allows spore-forming organisms to remain dormant during stress periods and rapidly resume activity when conditions improve (Matlock et al., 2019).

Viable but non-culturable (VBNC) states represent a less extreme form of dormancy in which cells remain alive but cannot be cultured using standard laboratory techniques. Cells in VBNC states have reduced metabolic activity but retain the ability to resume normal function when conditions become favorable. The VBNC state may be more common than previously recognized and could represent an important survival strategy for many microorganisms. Understanding the factors that trigger entry into and exit from VBNC states is important for predicting microbial survival and recovery under stress conditions (Evans and Wallenstein, 2014).

Cryptobiosis and anhydrobiosis refer to states of suspended animation in which metabolic activity is virtually undetectable. These states are typically induced by extreme desiccation and allow organisms to survive complete water loss for extended periods. The mechanisms underlying cryptobiosis involve the replacement of cellular water with protective molecules that maintain cellular structure in the absence of water. When water becomes available, these organisms can rapidly resume normal metabolic activity. Environmental disturbances may eliminate active community members while leaving dormant stages intact. These dormant stages can serve as a reservoir for community recovery when conditions improve. The diversity and abundance of dormant propagules in microbial communities can significantly influence recovery trajectories following disturbance (de Vries et al., 2012).

Genetic and Evolutionary Adaptations: The genetic flexibility of microorganisms provides them with unparalleled capacity for rapid adaptation to environmental changes. This flexibility operates through multiple mechanisms, from point mutations to horizontal gene transfer, and can produce adaptive responses on timescales ranging from hours to years (Ochman et al., 2000; Ogunseitan, 2008). Mutation rates under stress often increase significantly, providing more opportunities for natural selection. Many microorganisms possess stress-induced mutagenesis systems that increase mutation rates when cells are under stress, potentially accelerating the evolution of stress tolerance mechanisms. The SOS response in bacteria represents one well-studied example of stress-induced mutagenesis. When cells experience DNA damage or other severe stress, the SOS response activates error-prone DNA repair mechanisms that increase mutation rates. While this increases the risk of harmful mutations, it also increases the probability of beneficial mutations that could enhance survival.

Horizontal gene transfer acceleration under stress conditions can rapidly spread beneficial traits through microbial communities (Aminov, 2011). Stress conditions often increase the frequency of horizontal gene transfer, possibly because stressed cells are more likely to take up DNA from the environment or because transfer mechanisms are activated as survival strategies. The rapid spread of antibiotic resistance genes through bacterial communities provides a dramatic example of how horizontal gene transfer can accelerate adaptation. Resistance genes can spread through entire communities within days or weeks, much faster than would be possible through vertical inheritance alone (Davies & Davies, 2010; Martinez, 2008).

Plasmid-mediated resistance represents a particularly important mechanism for rapid adaptation to chemical stress. Plasmids carrying resistance genes can be transferred between organisms, allowing sensitive cells to rapidly acquire resistance mechanisms. The modular nature of many resistance plasmids allows them to accumulate multiple resistance genes, providing protection against multiple stressors (Thomas & Nielsen, 2005).

Epigenetic modifications represent a relatively recently recognized mechanism of rapid adaptation that can bridge short-term physiological responses and long-term evolutionary changes. These modifications can alter gene expression without changing DNA sequence and can sometimes be inherited across generations. DNA methylation and other epigenetic modifications can providc a form of cellular memory that allows organisms to respond more rapidly to repeated stress exposure. This priming effect can enhance survival during subsequent stress events and may contribute to population-level adaptation.

Population Genetic Responses: Genetic diversity maintenance is crucial for population resilience, as it provides the raw material for adaptation to changing environmental conditions. Populations with higher genetic diversity are generally more resilient to environmental stress because they are more likely to contain individuals with beneficial traits. However, environmental stress can also reduce genetic diversity through population bottlenecks and selective sweeps. Balancing the maintenance of genetic diversity with the need to adapt to specific stressors represents a fundamental challenge for population survival (Ogunseitan, 2026).

Bottleneck effects and founder events can dramatically reduce genetic diversity in stressed populations. When populations are reduced to small sizes by environmental stress, they may lose much of their genetic diversity through random genetic drift. The loss of genetic diversity during bottleneck events can reduce the adaptive potential of populations and make them more vulnerable to future stressors. However, if the surviving individuals are well-adapted to the stressful conditions, the population may be better able to cope with similar stressors in the future.

Gene flow and migration patterns can help maintain genetic diversity in stressed populations by introducing new genetic variants from other populations. However, gene flow can also introduce maladaptive genes that reduce local adaptation. The balance between gene flow and local adaptation represents a key factor in population resilience. Moderate levels of gene flow may be optimal, providing genetic diversity while allowing for local adaptation to specific environmental conditions. Local adaptation versus gene flow represents a fundamental tension in population genetics. Local adaptation can enhance population fitness in specific environments, but it may reduce fitness in other environments and limit the ability to respond to environmental changes. The spatial scale of environmental variation and the dispersal ability of organisms influence the relative importance of local adaptation versus gene flow. In highly variable environments, gene flow may be more important for maintaining population resilience (Shade et al., 2014).

Community-Level Resilience Mechanisms: Community-level resilience mechanisms operate through the interactions among species and the emergent properties of community structure. These mechanisms can provide resilience that exceeds the sum of individual species' resilience capabilities. Species substitution patterns describe how the decline of one species can be compensated by the increase of another species that performs similar ecological functions. This mechanism allows communities to maintain ecosystem function even when species composition changes significantly. The effectiveness of species substitution depends on the degree of functional similarity between species and the availability of suitable replacement species in the community. Communities with high functional redundancy are more likely to maintain function through species substitution (Cases et al., 2005).

Functional group stability may be more important for ecosystem resilience than species-level stability. If multiple species within a functional group can perform similar roles, the loss of individual species may have minimal impact on ecosystem function. The identification and protection of functional groups rather than individual species may be a more effective approach for maintaining ecosystem resilience. This approach recognizes that ecosystem function depends more on the presence of key functional groups than on specific species.

Niche complementarity effects occur when species use different resources or occupy different niches, reducing competition and allowing for higher total community productivity and stability. This mechanism can enhance community resilience by reducing the likelihood that multiple species will be affected by the same stressor. Communities with high niche complementarity may be more resilient to environmental stress because they make more efficient use of available resources and are less likely to experience competitive exclusion under stress conditions.

Portfolio effects in community stability arise from the statistical averaging of population fluctuations across multiple species. When species respond differently to environmental variation, their combined response may be more stable than the response of any individual species. The portfolio effect is strongest when species responses to environmental variation are negatively correlated, meaning that when some species decline, others increase. This negative correlation can arise from differences in environmental tolerances or resource requirements.

Network Stability and Modularity: Interaction network robustness describes the ability of ecological networks to maintain their structure and function despite the loss of species or interactions. Robust networks are those that can tolerate the removal of nodes (species) or edges (interactions) without losing connectivity or function. The robustness of ecological networks depends on their structural properties, including connectivity patterns, degree distributions, and the presence of highly connected hub species. Networks with scale-free or small-world properties may be particularly robust to random species loss but vulnerable to targeted removal of hub species.

Keystone species protection is crucial for maintaining network stability, as these species play disproportionately important roles in maintaining community structure and function. The loss of keystone species can trigger cascading effects that lead to community collapse. Identifying and protecting keystone species requires understanding their roles in community networks and the consequences of their loss. This may involve experimental removal studies or network analysis to identify species with high connectivity or unique functional roles.

Modular community organization can enhance stability by limiting the spread of disturbances through the network. In modular networks, species are organized into distinct clusters with strong internal connections but weaker connections between clusters.

When a disturbance affects one module, it may not necessarily spread to other modules, allowing the community to maintain partial function even under stress. This compartmentalization can prevent system-wide collapse and facilitate recovery.

Cascade failure prevention mechanisms help limit the spread of disturbances through ecological networks. These mechanisms may include redundant pathways, buffering species, or network structures that limit the propagation of effects. Understanding and enhancing cascade failure prevention mechanisms may be important for maintaining ecosystem stability under increasing environmental stress. This may involve managing network structure or protecting key species that serve as buffers against disturbance propagation.

Cross-protection mechanisms allow organisms that have been exposed to one type of stress to better tolerate other types of stress. This phenomenon suggests that stress response systems may have overlapping functions that provide broad protection against multiple stressors. Multiple stress tolerance can arise through the evolution of general stress response systems that provide protection against various types of environmental challenges. These systems may be more energetically efficient than maintaining separate response systems for each type of stress.

The evolution of multiple stress tolerance may be particularly important in environments where organisms face multiple simultaneous stressors. The ability to respond to multiple stressors with a single response system may provide a significant adaptive advantage.

Stress memory and priming describe the ability of organisms to "remember" past stress events and respond more quickly or effectively to subsequent stress. This memory can be maintained through epigenetic modifications, changes in gene expression, or the accumulation of stress response proteins (Ogunseitan, 1993).

Stress priming can allow organisms to respond more rapidly to recurring stressors, potentially reducing the damage caused by stress events. This mechanism may be particularly important in variable environments where similar stressors occur repeatedly.

Cross-resistance mechanisms allow organisms that have evolved resistance to one stressor to also resist other, chemically unrelated stressors. This phenomenon may arise through the evolution of general detoxification systems or stress tolerance mechanisms. Cross-resistance can provide unexpected benefits when organisms encounter novel stressors that are chemically similar to stressors they have previously encountered. However, cross-resistance may also impose costs if resistance mechanisms interfere with normal cellular processes.

Adaptive stress response networks describe the coordinated regulation of multiple stress response systems that allows organisms to mount appropriate responses to different types and intensities of stress. These networks integrate information about environmental conditions and coordinate cellular responses accordingly.

The evolution of sophisticated stress response networks allows organisms to fine-tune their responses to environmental challenges, potentially reducing the costs of stress responses while maintaining protection against harmful conditions. Many of the mechanisms described in this section operate simultaneously to respond to individual or coincident stressors. It is likely that under highly general and dynamic conditions represented by climate change and exposure to a heterogenous mixture of toxic pollutants, a wide range of pre-existing mechanisms will be deployed to respond in microbial communities. However, climate change and pollution also represent opportunities for innovation through mutation and natural selection in microbial communities. This is the context in which the following sections are presented.

4 CLIMATE CHANGE IMPACTS ON MICROBIAL COMMUNITIES

Abrupt climate change is not a new phenomenon on Earth, and microbial communities have survived such episodes, and they will survive the current anthropogenic climate change due to the combustion of fossil fuels and the accumulation of greenhouse gases in the atmosphere. The question is whether microbial communities upon which human societies depend will be sufficiently resilient to reduce the consequences of climate change. Temperature represents one of the most fundamental environmental variables affecting microbial life, influencing everything from enzyme kinetics and membrane fluidity to protein stability and metabolic rates. As global temperatures continue to rise due to anthropogenic climate change, microbial communities worldwide are expected to respond to thermal stress that challenges their physiological capacities and forces rapid adaptation or community reorganization within ranges that may not be reversible in the context of resilience.

The relationship between temperature and microbial activity follows predictable patterns based on fundamental biochemical principles. Enzyme activity typically increases with temperature up to an optimal point, beyond which thermal denaturation causes rapid decline in function. However, this simple relationship becomes complex when considered at the community level, where different species have different thermal optima and where temperature affects multiple physiological processes simultaneously (Canizares et al., 2011; Khalili et al., 2016).

Thermal adaptation in microorganisms involves multiple mechanisms operating at different organizational levels. At the cellular level, heat shock responses provide immediate protection against thermal stress through the production of chaperone proteins that help maintain protein structure and function. These responses can be triggered within minutes of temperature elevation and provide short-term protection that allows cells to survive acute thermal stress (Pollack et al., 2009).

Membrane adaptation represents another crucial mechanism of thermal tolerance. As temperatures rise, cell membranes become more fluid, potentially compromising their integrity and function. Microorganisms can respond by altering the fatty acid composition of their membranes, increasing the proportion of saturated fatty acids or branched-chain fatty acids that maintain membrane stability at higher temperatures.

Protein thermostability can be enhanced through various mechanisms, including changes in amino acid composition, increased disulfide bonding, and the production of compatible solutes that stabilize protein structure. These adaptations may require longer timescales to implement, as they often involve changes in gene expression or even evolutionary changes in protein sequences. To illustrate microbial resilience in the context of climate change, four case studies are examined, including drought responses, arctic permafrost, and coral reefs in the marine ecosystem.

Drought Responses in Soil Microbiomes: Soil microbiomes represent some of the most diverse and functionally important microbial communities on Earth, playing crucial roles in nutrient cycling, plant health, and ecosystem productivity. These communities are particularly vulnerable to drought stress, which can disrupt soil structure, reduce nutrient availability, and alter the physical and chemical environment in ways that profoundly affect microbial survival and function (Barnard et al., 2013).

The immediate responses of soil microbiomes to drought stress involve a combination of physiological adjustments and behavioral changes that help organisms cope with reduced water availability. These responses can be observed within hours to days of the onset of drought conditions and represent the first line of defense against water stress.

Osmotic stress tolerance mechanisms in soil microorganisms include the rapid accumulation of compatible solutes, changes in membrane composition to reduce water loss, and the production of extracellular polymeric substances (EPS) that help retain moisture in the immediate vicinity of cells. These mechanisms allow many soil microorganisms to maintain metabolic activity even under moderately dry conditions.

The production of EPS represents a particularly important drought response mechanism. These polymers can absorb and retain water, creating localized zones of higher moisture content around microbial cells and communities. EPS production often increases under drought stress, suggesting that this represents an active adaptation to water limitation.

Spore formation and dormancy strategies represent more extreme responses to severe drought stress. Many soil bacteria and fungi can produce resistant spores or enter dormant states that allow them to survive extended periods without water. These dormancy strategies effectively pause biological activity until conditions become favorable for growth and reproduction (Matlock & Ogunseitan, 2022).

Bacterial endospores represent one of the most extreme examples of drought tolerance in soil microorganisms. These highly resistant structures can survive complete desiccation for years or even decades, allowing populations to persist through extended drought periods. The ability to form spores provides a significant survival advantage in environments with unpredictable water availability. Fungal spores and sclerotia serve similar functions, allowing fungal populations to survive drought stress and rapidly recolonize soils when moisture returns. The production of these resistant structures often increases under drought stress, suggesting that fungi actively prepare for extended dry periods.

Recovery dynamics following rewetting represent a critical phase in drought response cycles. When water returns to drought-stressed soils, microbial communities must rapidly reactivate metabolic processes, repair cellular damage, and reestablish community structure and function. The speed and completeness of this recovery can determine the long-term impacts of drought on soil ecosystem function. The rewetting process often triggers a pulse of microbial activity as dormant organisms become active and begin metabolizing accumulated organic matter. This "Birch effect" can lead to rapid increases in soil respiration and nutrient mineralization, temporarily increasing the availability of nutrients for plant uptake. However, recovery is not always complete or immediate. Severe or prolonged drought can cause permanent changes in soil microbial community composition, with some species failing to recover and others becoming dominant in the post-drought community. These changes can have lasting effects on soil function and ecosystem productivity.

The resilience of soil microbiomes to drought stress varies considerably depending on soil type, climate history, and management practices. Soils with high organic matter content often show greater drought resilience, as organic matter helps retain moisture and provides energy sources for microbial communities during stress periods. Agricultural management practices can significantly influence drought resilience in soil microbiomes. Practices that increase soil organic matter, such as cover cropping and reduced tillage, can enhance the drought tolerance of soil microbial communities. Conversely, practices that reduce soil organic matter or disrupt soil structure can increase vulnerability to drought stress.

Arctic Permafrost Microbial Communities: The Arctic permafrost represents one of the most dramatic examples of climate change impacts on microbial communities, where rising temperatures are causing fundamental shifts in community structure and function with global implications for carbon cycling and climate feedback (Monteux et al., 2018).

Permafrost soils contain vast amounts of organic carbon that have been frozen for thousands to millions of years. As global temperatures rise, permafrost thaw exposes this organic matter to microbial decomposition, potentially releasing significant amounts of carbon dioxide and methane to the atmosphere. The microbial communities responsible for this decomposition are themselves undergoing rapid changes as their frozen habitat transforms into an active soil ecosystem. Thaw-induced community shifts in permafrost soils involve the activation of dormant microorganisms, the immigration of new species from adjacent habitats, and the establishment of new biogeochemical processes. The transition from frozen to thawed conditions represents a fundamental change in habitat characteristics, including temperature, moisture, oxygen availability, and substrate accessibility (Schurr et al., 2015). The activation of dormant microorganisms represents one of the most immediate responses to permafrost thaw. Many microorganisms can survive freezing in dormant states and become active when conditions become favorable. This activation can lead to rapid increases in microbial biomass and activity as previously inaccessible organic matter becomes available for decomposition (Mackelprang et al., 2011).

Immigration of new species from adjacent habitats can introduce novel metabolic capabilities and interaction networks to thawing permafrost communities. These immigrant species may be better adapted to the new environmental conditions and may outcompete resident species, leading to significant changes in community composition. Methane production represents one of the most significant biogeochemical consequences of permafrost thaw. Methanogenic archaea, which produce methane under anaerobic conditions, can become dominant in waterlogged thawed permafrost soils. The methane they produce is a potent greenhouse gas, creating a positive feedback loop that accelerates climate change. The implications for global climate feedback are profound. Permafrost soils contain an estimated 1,700 billion tons of carbon, roughly twice the amount currently in the atmosphere. Even a small fraction of this carbon released as greenhouse gases could significantly accelerate climate change, making permafrost thaw one of the most important climate tipping points (Lenton et al., 2008).

Research in permafrost systems has revealed the complexity of microbial responses to thaw. Different types of permafrost (continuous vs. discontinuous, mineral vs. organic soils) show different patterns of microbial community change. The depth of thaw, drainage conditions, and vegetation cover all influence how microbial communities respond to warming. Temperature sensitivity of different microbial processes varies considerably, with some processes showing exponential increases with temperature while others plateau or even decline at higher temperatures. This differential sensitivity can lead to changes in the relative importance of different biogeochemical processes as temperatures rise. The role of plant-microbe interactions in permafrost systems adds another layer of complexity. As permafrost thaws, plant communities may change, altering the quantity and quality of organic inputs to soil microbial communities. These changes can cascade through the microbial food webs, affecting decomposition rates and greenhouse gas production.

Coral Reef Microbial Resilience: Coral reefs represent some of the most biodiverse and economically important marine ecosystems on Earth, supporting an estimated 25% of all marine species despite occupying less than 1% of ocean area. These ecosystems are built on complex symbiotic relationships between corals, algae, and diverse microbial communities, all of which are threatened by ocean acidification and other climate change impacts (Pollack et al., 2009).

The coral holobiont, consisting of the coral animal, symbiotic algae (zooxanthellae), and associated microbial communities, represents a highly integrated ecosystem that depends on precise biochemical and physiological coordination among its components. Ocean acidification disrupts this coordination by affecting the physiology of each component and the interactions between them. pH stress on symbiotic relationships within the coral holobiont can manifest in several ways. The coral animal's ability to calcify and build its skeleton is directly impacted by reduced pH, as lower carbonate ion concentrations make it more difficult to precipitate calcium carbonate. This can weaken coral skeletons and make them more vulnerable to physical damage and bioerosion.

The symbiotic algae that provide much of the coral's energy through photosynthesis can also be affected by pH changes. While some studies suggest that increased CO_2 availability might enhance photosynthesis, the overall stress of acidification often disrupts the delicate balance of the symbiotic relationship, potentially leading to bleaching events where corals expel their algal partners.

Microbial community restructuring represents one of the most significant responses to ocean acidification in coral reef systems. The diverse microbial communities associated with corals play crucial roles in nutrient cycling, disease resistance, and overall coral health. Changes in pH can alter the composition and function of these communities, with cascading effects on coral physiology and survival.

Research has shown that ocean acidification can shift coral-associated microbial communities toward more pathogenic compositions, potentially increasing coral susceptibility to disease. The mechanisms underlying these shifts are complex and may involve direct effects of pH on microbial physiology, indirect effects through changes in coral physiology, and altered competitive relationships among microbial species (Pollack et al., 2009). The buffering capacity of coral reef environments adds another layer of complexity to acidification impacts. Coral reefs naturally experience significant daily and seasonal pH fluctuations due to photosynthesis and respiration cycles. This natural variability may provide some resilience to acidification, as reef organisms are already adapted to pH variation.

However, the magnitude and persistence of anthropogenic acidification may exceed the natural adaptive capacity of reef systems. The combination of acidification with other stressors, particularly warming temperatures, creates synergistic effects that can overwhelm natural resilience mechanisms. Implications for coral health and recovery are profound and far-reaching. Acidification can reduce coral growth rates, weaken skeletal structures, and increase susceptibility to disease and physical damage. These effects can reduce the ability of coral reefs to recover from other disturbances, such as bleaching events, storms, or pollution.

The loss of coral reef ecosystems would have cascading effects on marine biodiversity and human communities that depend on reef resources. Coral reefs provide coastal protection, support fisheries, and generate billions of dollars in tourism revenue annually. The microbial communities that support these ecosystems represent a critical but often overlooked component of reef resilience.

Research into coral reef microbial resilience is revealing potential strategies for enhancing reef survival under acidification. These include identifying naturally resilient coral genotypes and their associated microbial communities, developing probiotics that could enhance coral health, and implementing management strategies that reduce other stressors and allow reefs to better cope with acidification.

5 TOXIC POLLUTION STRESS AND MICROBIAL RESILIENCE

The industrial revolution led to major leaps in the quality of life for many sectors in human societies, including food production, transportation, knowledge generation, and innovation in manufacturing consumer goods. However, the adverse consequences of the industrial revolution include point- and non-point emissions of toxic chemicals into air, water, and soil systems that ultimately produce disease in human populations and in other life forms in a shared global ecosystem. The recognition of these consequences has led to research paradigms such as "One Health" which focus on the interconnection of life forms and the need to explore the intersection to prevent and respond to emergencies. The inclusion of environmental systems and microbial communities have not been adequately explored in the One Health paradigm, and neither is the relevance of toxic environmental pollution, in comparison with zoonosis (Ogunseitan, 2023; Zinsstag et al., 2011). This section considers notorious case studies to illustrate current state and gaps in knowledge regarding the impacts of widespread pollution on microbial resilience. The case studies include toxic metals, mine drainage, organic chemical pollutants, plastics, and emerging pollutants such as nanomaterials.

Toxic Metal Contamination Responses: Heavy metal contamination represents one of the most persistent and widespread forms of environmental pollution, affecting microbial communities across diverse ecosystems from mining sites to urban environments. Unlike organic pollutants that can be degraded, heavy metals are elements that cannot be destroyed and persist indefinitely in the environment, creating long-term selective pressures that drive microbial adaptation and community restructuring (Bruins et al., 2000; Hibbert & Ogunseitan, 2014). The toxicity of heavy metals to microorganisms stems from their ability to disrupt essential cellular processes through multiple mechanisms. Heavy metals can bind to sulfur-containing amino acids in proteins, disrupting protein structure and function. They can generate reactive oxygen species that damage cellular components, interfere with enzyme activity by displacing essential metal cofactors, and disrupt membrane integrity by binding to phospholipids (Ogunseitan, 2016; Ogunseitan, 2017).

Despite their toxicity, many microorganisms have evolved sophisticated mechanisms to tolerate or even thrive in metal-contaminated environments. These mechanisms include active efflux systems that pump metals out of cells, sequestration systems that bind metals to prevent their interaction with sensitive cellular components, and biochemical modifications that reduce metal uptake or toxicity (Chern et al., 2007).

Metal efflux systems represent one of the most important mechanisms of metal tolerance in microorganisms. These systems use energy to actively transport metals out of cells, maintaining intracellular metal concentrations below toxic levels. Different efflux systems are specific for different metals, and organisms may possess multiple systems to handle various metal stressors.

Metal sequestration mechanisms involve the binding of metals to specific molecules that prevent their interaction with sensitive cellular targets. Metallothioneins are small, cysteine-rich proteins that can bind multiple metal ions and are commonly produced by metal-tolerant organisms. Other sequestration mechanisms include the binding of metals to cell wall components or their precipitation as insoluble compounds. Biofilm formation often increases under metal stress, as the extracellular polymeric substances that make up biofilms can bind metals and reduce their bioavailability to cells within the biofilm. This community-level response can protect sensitive organisms and allow communities to persist in metal-contaminated environments Giller et al., 1998).

Mine Tailings Microbial Communities: Mine tailings represent some of the most extreme metal-contaminated environments on Earth, often containing concentrations of heavy metals that are orders of magnitude higher than background levels. These environments provide natural laboratories for studying microbial adaptation to metal stress and offer insights into the limits of microbial resilience.

The environmental conditions in mine tailings are characterized not only by extreme metal concentrations but also by low pH, high salinity, and limited organic matter availability. This combination of stressors creates a harsh environment that only the most tolerant microorganisms can survive, leading to the development of highly specialized microbial communities.

Metal tolerance mechanisms in mine tailings communities often involve multiple resistance systems operating simultaneously. Organisms may possess efflux pumps for several different metals, produce multiple types of metal-binding proteins, and modify their cellular physiology to minimize metal uptake and toxicity. The evolution of metal tolerance in mine tailings communities can occur remarkably rapidly, with significant increases in tolerance observed within years or decades of contamination. This rapid evolution is facilitated by the strong selective pressure imposed by metal toxicity and the large population sizes and rapid generation times of microorganisms (Ogunseitan, 1998).

Horizontal gene transfer plays a crucial role in the spread of mechanisms of metal tolerance or resistance in microbial communities. Resistance genes are often located on mobile genetic elements such as plasmids or transposons that can be transferred between organisms, allowing the rapid dissemination of beneficial traits. Dominance of microbial strains in mine tailings involves both the selection of pre-existing metal-tolerant organisms and the evolution of new tolerance mechanisms. The relative importance of these processes depends on the history of contamination, the severity of metal stress, and the diversity of the initial microbial community. The structure of mine tailings microbial communities often differs dramatically from those in uncontaminated environments, with reduced diversity and dominance by a few highly metal-tolerant species. However, despite their low diversity, these communities can maintain important ecosystem functions such as nutrient cycling and organic matter decomposition.

Biogeochemical cycling alterations in mine tailings environments reflect both the direct effects of metals on microbial physiology and the indirect effects of altered community structure. Metal contamination can inhibit key biogeochemical processes such as nitrogen fixation and methane production, while also promoting others such as sulfur oxidation.

The oxidation of sulfide minerals in mine tailings by chemolithotrophic bacteria can generate acid mine drainage, creating extremely acidic conditions that further stress microbial communities. This process represents a positive feedback loop where microbial activity creates conditions that are increasingly hostile to most forms of life. Despite the harsh conditions, mine tailings communities can provide important ecosystem services, including the stabilization of tailings materials, the prevention of metal leaching, and the preparation of substrates for eventual revegetation. Understanding and harnessing these natural processes is important for mine site remediation and restoration.

Research in mine tailings environments has revealed remarkable examples of microbial adaptation and has contributed to the development of biotechnologies for metal extraction and environmental remediation. The extreme metal tolerance of tailings microorganisms makes them valuable for applications in bioleaching, bioremediation, and the recovery of valuable from waste materials (Hemme et al., 2010).

Organic Chemical Pollutants: Organic pollutants represent a diverse class of contaminants that includes everything from naturally occurring compounds released in excessive quantities to entirely synthetic molecules with no natural analogs. The ability of microbial communities to degrade these compounds represents one of the most remarkable examples of biological adaptation and provides the foundation for many bioremediation technologies (Megharaj et al., 2011). The degradation of organic pollutants by microorganisms involves complex biochemical pathways that break down large, often toxic molecules into smaller, less harmful compounds. These pathways may be constitutive, meaning they are always present in the organism, or inducible, meaning they are produced only in response to the presence of the target compound.

Enzyme evolution plays a crucial role in the development of new degradation capabilities. When microorganisms are exposed to novel organic compounds, existing enzymes may have some activity toward these compounds, even if they evolved to act on different substrates (Canizares-Gonzalez et al., 2011). Through mutation and selection, these enzymes can evolve enhanced activity toward the new substrates, eventually leading to efficient degradation pathways.

The evolution of new degradation pathways often involves the recruitment and modification of existing metabolic pathways. Microorganisms may adapt pathways that originally evolved for the metabolism of natural compounds to handle synthetic pollutants with similar chemical structures. This process can occur relatively rapidly when the pollutant is chemically similar to natural substrates. Horizontal gene transfer accelerates the spread of degradation capabilities through microbial communities. Genes encoding degradative enzymes are often located on mobile genetic elements that can be transferred between organisms, allowing the rapid dissemination of new capabilities throughout a community.

Plastic-Degrading Marine Bacteria: The accumulation of plastic waste in marine environments represents one of the most visible and persistent pollution problems of the modern era. An estimated 8 million tons of plastic enter the oceans annually, where it accumulates in gyres, washes up on beaches, and breaks down into microplastics that pervade marine food webs. The discovery of marine bacteria capable of degrading plastic polymers represents a potential breakthrough in addressing this global environmental challenge (Schwarzenbach et al., 2006). The degradation of plastic polymers by microorganisms was long thought to be impossible due to the synthetic nature of these materials and their resistance to natural degradation processes. However, recent research has identified several bacterial species capable of breaking down various types of plastic, including polyethylene terephthalate (PET), polyurethane, and even some forms of polyethylene (Danso et al., 2018).

Enzyme evolution and horizontal gene transfer have played crucial roles in the development of plastic degradation capabilities. The enzymes responsible for plastic degradation often show similarity to enzymes involved in the degradation of natural polymers, suggesting that they evolved from pre-existing metabolic pathways. The discovery of PETase, an enzyme that can break down PET plastic, in the bacterium *Ideonella sakaiensis* represents a landmark in understanding plastic biodegradation. This enzyme appears to have evolved relatively recently, possibly in response to the increasing prevalence of PET in the environment. The enzyme shows structural similarity to cutinases, which are involved in the degradation of plant cuticles, suggesting an evolutionary origin in natural polymer degradation.

Community-level degradation networks in marine environments may involve multiple bacterial species working together to completely mineralize plastic polymers. Some bacteria may specialize in the initial breakdown of polymer chains, while others metabolize the resulting oligomers and monomers. This division of labor can enhance the overall efficiency of plastic degradation. The marine environment presents unique challenges for plastic degradation, including low temperatures, high salinity, and limited nutrient availability. Marine plastic-degrading bacteria must be adapted to these conditions while maintaining the ability to produce and secrete the enzymes necessary for polymer breakdown. Biofilm formation on plastic surfaces may play an important role in marine plastic degradation. Bacteria growing in biofilms on plastic debris can create localized environments with enhanced enzyme concentrations and may be protected from the harsh conditions of the open ocean. The extracellular polymeric substances in biofilms may also help solubilize plastic degradation products. The implications for ocean plastic pollution are potentially enormous, though significant challenges remain. The rates of natural plastic degradation by marine bacteria are currently very slow compared to the rate of plastic input to the oceans. Additionally, the complete mineralization of plastic polymers requires specific environmental conditions that may not be present throughout the marine environment.

Research into enhancing natural plastic degradation capabilities is ongoing, with approaches including the engineering of more efficient degradative enzymes, the development of bacterial consortia optimized for plastic degradation, and the identification of environmental conditions that promote natural degradation processes. The discovery of plastic-degrading bacteria has also inspired biotechnological applications beyond marine cleanup. These organisms and their enzymes are being investigated for use in plastic recycling, waste treatment, and the development of biodegradable plastic alternatives. However, the ecological implications of widespread plastic degradation in marine environments require careful consideration. The breakdown products of plastic degradation may themselves have environmental impacts, and the large-scale modification of marine microbial communities to enhance plastic degradation could have unintended consequences for marine ecosystems.

Emerging Contaminants and Novel Stressors: Emerging contaminants represent a rapidly growing category of environmental pollutants that includes pharmaceuticals, personal care products, endocrine-disrupting compounds, and engineered nanomaterials. These substances present unique challenges for microbial communities because they often have no natural analogs and may persist in the environment at concentrations that can affect sensitive biological processes (Janssen et al., 2005).

Pharmaceutical compound impacts on microbial communities are of particular concern because many of these substances are specifically designed to be biologically active. Antibiotics can directly kill or inhibit sensitive bacteria, potentially altering community composition and promoting antibiotic resistance. Hormones and other bioactive compounds can disrupt microbial physiology even at very low concentrations. The presence of antibiotics in the environment creates selective pressure for the development and spread of antibiotic resistance genes. These genes can be transferred between bacteria through horizontal gene transfer, potentially creating reservoirs of resistance that could affect human and animal health (Berendonk et al., 2015; Pollack et al., 2009).

Microplastic interactions with microbial communities represent another emerging area of concern. Microplastics can serve as surfaces for biofilm formation, potentially transporting microorganisms to new environments. They can also adsorb other pollutants, creating concentrated sources of contamination that affect attached microbial communities. The ingestion of microplastics by microorganisms can cause physical damage and may interfere with normal cellular processes. The breakdown products of plastic degradation may also have toxic effects on microbial communities, though research in this area is still in its early stages (Yoshida et al., 2016; Zettler et al., 2013).

Nanomaterial effects on microbial communities depend on the specific properties of the materials, including their size, surface chemistry, and reactivity. Some nanomaterials can penetrate cell membranes and disrupt cellular processes directly, while others may have more subtle effects on community structure and function. The antimicrobial properties of some nanomaterials, such as silver nanoparticles, can significantly alter microbial community composition (Colman et al., 2013). While these effects may be beneficial in some applications, they can also disrupt natural microbial communities and ecosystem processes.

6 CONCLUSIONS

Efforts to understand critical thresholds in microbial community responses to environmental stress is crucial for predicting when communities might undergo irreversible changes or collapse. The concept of planetary boundaries and safe operating zones emerged out of this types of efforts (Ogunseitan, 2025). Thresholds across planetary boundaries represent points beyond which small additional stresses can trigger large, often irreversible changes in community structure and function. Stress intensity and duration effects interact in complex ways to determine microbial community responses. Low-intensity stress applied over long periods may have different effects than high-intensity stress applied briefly. Understanding these interactions is important for predicting threshold responses.

Chronic low-level stress may gradually erode community resilience without triggering obvious changes in community structure or function. This erosion may not be apparent until additional stress pushes the community beyond a critical threshold.

Acute high-intensity stress may trigger immediate changes in community structure but may not necessarily lead to long-term changes if the community can recover once the stress is removed. However, if the stress exceeds critical thresholds, recovery may not be possible.

Community composition tipping points occur when environmental stress causes fundamental shifts in the dominant species or functional groups within a community. These shifts may be difficult to reverse even after the original stressor is removed.

The identification of community composition tipping points requires long-term monitoring of community structure and function under different stress conditions. Early warning indicators may include changes in diversity, evenness, or the abundance of key species.

Functional collapse indicators represent measurable changes in ecosystem processes that signal the loss of critical microbial functions. These indicators may be more relevant for ecosystem management than changes in community composition because they directly relate to ecosystem services. Examples of functional collapse indicators might include reduced rates of nutrient cycling, decreased decomposition rates, or loss of pollutant degradation capacity. These indicators can provide early warning of ecosystem dysfunction. Early warning systems for microbial community collapse could be developed based on understanding of critical thresholds and collapse indicators. These systems could help managers take preventive action before irreversible changes occur. Effective early warning systems require real-time monitoring of key indicators, understanding of threshold relationships, and rapid response capabilities. The development of such systems represents an important frontier in ecosystem management.

Climate change and persistent pollutants may lead to cascading effects which occur when changes in one component of an ecosystem trigger changes in other components, potentially leading to system-wide collapse. In microbial systems, cascading effects can occur when the loss of key functional groups or species triggers changes throughout the community. Microbial failure propagation describes how the loss of function in one part of a microbial community can spread to other parts of the community through interaction networks. This propagation can lead to system-wide collapse if critical nodes or connections are disrupted.

The structure of microbial interaction networks influences how failures propagate through communities. Networks with scale-free properties may be particularly vulnerable to targeted attacks on highly connected hub species but resistant to random species loss.

Ecosystem service disruptions can occur when cascading effects in microbial communities lead to the loss of critical ecosystem functions. These disruptions can have significant economic and social consequences, particularly for services that are difficult to replace technologically. Examples of ecosystem service disruptions might include the loss of water purification capacity in wetlands, reduced soil fertility in agricultural systems, or increased greenhouse gas emissions from soils. The economic costs of these disruptions can be substantial (Karr et al., 2000). Socio-economic consequences of microbial system collapse can be severe, particularly in systems that provide essential services for human well-being (Costanza et al., 1997). The collapse of soil microbial communities in agricultural systems, for example, could threaten food security (Wall et al., 2012). Understanding and quantifying the socioeconomic consequences of microbial system collapse is important for justifying investments in ecosystem protection and restoration. However, these consequences are often difficult to predict and quantify.

Recovery potential assessment involves evaluating the likelihood and timescale of ecosystem recovery following collapse. Some systems may have high recovery potential if key species or functions can be restored, while others may be permanently altered. Factors that influence recovery potential include the severity and duration of the disturbance, the availability of colonizing species, the degree of habitat degradation, and the presence of alternative stable states. Understanding these factors is crucial for restoration planning.

There is an opportunity now to intensify development of effective indicators of microbial community resilience is crucial for ecosystem management and conservation. These indicators should provide early warning of potential problems and help guide management interventions. Microbial diversity metrics are commonly used as indicators of ecosystem health and resilience. Higher diversity is generally associated with greater resilience, but the relationship is not always straightforward and may depend on the type of diversity measured and the specific ecosystem context.

Species richness, evenness, and functional diversity all provide different perspectives on community diversity and may respond differently to environmental stress. A comprehensive assessment of diversity should consider multiple metrics and their relationships to ecosystem function. Functional stability measures focus on the maintenance of ecosystem processes rather than community composition. These measures may be more relevant for ecosystem management because they directly relate to ecosystem services and human well-being. Examples of functional stability measures might include the stability of nutrient cycling rates, decomposition processes, or pollutant degradation capacity. These measures can provide insights into ecosystem resilience that are not captured by diversity metrics alone.

Network analysis approaches can provide insights into community structure and stability that are not available from traditional diversity measures. Network metrics such as connectivity, modularity, and centrality can help identify key species and predict community responses to disturbance. The application of network analysis to microbial communities is still developing, but it holds promise for understanding community dynamics and predicting resilience. However, network analysis requires detailed information about species interactions that may be difficult to obtain. Real-time monitoring systems for microbial communities are becoming increasingly feasible with advances in molecular techniques and sensor technology. These systems could provide early warning of community changes and help guide adaptive management strategies.

The development of real-time monitoring systems requires integration of multiple technologies including environmental sensors, molecular analysis techniques, and data processing systems. The cost and complexity of these systems may limit their application to high-priority ecosystems.

Extreme environments provide unique opportunities to study microbial resilience under conditions that approach the limits of life. These environments can provide insights into fundamental resilience mechanisms and potential applications for biotechnology and astrobiology. Extremophile microorganisms have evolved remarkable adaptations to survive in environments characterized by extreme temperature, pH, salinity, pressure, or radiation. These adaptations provide insights into the limits of microbial resilience and potential mechanisms for enhancing resilience in other systems. Stress tolerance mechanisms in extremophiles often involve multiple, redundant systems that provide protection against various types of stress. These mechanisms may include specialized proteins, modified cellular structures, and unique metabolic pathways.

The study of extremophile stress tolerance mechanisms has led to important biotechnological applications, including the development of enzymes that function under extreme conditions and the design of stress-tolerant organisms for industrial applications. Community stability principles derived from extreme environment studies may be applicable to other ecosystems facing environmental stress. Extreme environments often support low-diversity communities dominated by highly specialized organisms, yet these communities can be remarkably stable.

Model system development using extreme environments can provide controlled conditions for studying resilience mechanisms and testing theoretical predictions. These systems can help bridge the gap between laboratory studies and complex natural ecosystems. Reference community selection involves identifying target microbial communities that restoration efforts should aim to establish. This selection may be based on historical conditions, nearby undisturbed sites, or desired ecosystem functions. The selection of reference communities is challenging because microbial communities are highly variable in space and time, and historical baselines may not be appropriate under changing environmental conditions. Functional rather than compositional targets may be more appropriate.

DECLARATION OF CONFLICT-OF-INTEREST

The author has no conflict of interest in this work.

ACKNOWLEDGEMENTS

I am deeply grateful for all my mentors and mentees on the quest to understand microbial diversity and the contributions of the "unseen world" to our visible world. I am grateful for the resources available for my current academic work through the University of California Presidential Chair funds; and the World Institute for Sustainable Development of Materials (WISDOM).

REFERENCES

Allison, S. D., & Martiny, J. B. H. (2008). Resistance, resilience, and redundancy in microbial communities. *Proceedings of the National Academy of Sciences of the United States of America, 105*(Supplement 1), 11512–11519. DOI: 10.1073/pnas.0801925105 PMID: 18695234

Aminov, R. I. (2011). Horizontal gene exchange in environmental microbiota. *Frontiers in Microbiology, 2*, 158. DOI: 10.3389/fmicb.2011.00158 PMID: 21845185

Barnard, R. L., Osborne, C. A., & Firestone, M. K. (2013). Responses of soil bacterial and fungal communities to extreme desiccation and rewetting. *The ISME Journal, 7*(11), 2229–2241. DOI: 10.1038/ismej.2013.104 PMID: 23823489

Berendonk, T. U., Manaia, C. M., Merlin, C., Fatta-Kassinos, D., Cytryn, E., Walsh, F., Bürgmann, H., Sørum, H., Norström, M., Pons, M. N., Kreuzinger, N., Huovinen, P., Stefani, S., Schwartz, T., Kisand, V., Baquero, F., & Martinez, J. L. (2015). Tackling antibiotic resistance: The environmental framework. *Nature Reviews. Microbiology, 13*(5), 310–317. DOI: 10.1038/nrmicro3439 PMID: 25817583

Bradford, M. A., Davies, C. A., Frey, S. D., Maddox, T. R., Melillo, J. M., Mohan, J. E., Reynolds, J. F., Treseder, K. K., & Wallenstein, M. D. (2008). Thermal adaptation of soil microbial respiration to elevated temperature. *Ecology Letters, 11*(12), 1316–1327. DOI: 10.1111/j.1461-0248.2008.01251.x PMID: 19046360

Bruins, M. R., Kapil, S., & Oehme, F. W. (2000). Microbial resistance to metals in the environment. *Ecotoxicology and Environmental Safety, 45*(3), 198–207. DOI: 10.1006/eesa.1999.1860 PMID: 10702338

Canizares-Gonzalez, R., Benitez, E., & Ogunseitan, O. A. (2011). Molecular analyses of beta-glucosidase diversity and function in soil. *European Journal of Soil Biology, 47*(1), 1–8. DOI: 10.1016/j.ejsobi.2010.11.002

Cardinale, B. J., Duffy, J. E., Gonzalez, A., Hooper, D. U., Perrings, C., Venail, P., Narwani, A., Mace, G. M., Tilman, D., Wardle, D. A., Kinzig, A. P., Daily, G. C., Loreau, M., Grace, J. B., Larigauderie, A., Srivastava, D. S., & Naeem, S. (2012). Biodiversity loss and its impact on humanity. *Nature*, *486*(7401), 59–67. DOI: 10.1038/nature11148 PMID: 22678280

Cases, I., & de Lorenzo, V. (2005). Genetically modified organisms for the environment: Stories of success and failure and what we have learned from them. *International Microbiology*, *8*(3), 213–222. https://digital.csic.es/bitstream/10261/1986/1/Cases.pdf PMID: 16200500

Cavicchioli, R., Ripple, W. J., Timmis, K. N., Azam, F., Bakken, L. R., Baylis, M., Behrenfeld, M. J., Boetius, A., Boyd, P. W., Classen, A. T., Crowther, T. W., Danovaro, R., Foreman, C. M., Huisman, J., Hutchins, D. A., Jansson, J. K., Karl, D. M., Koskella, B., Mark Welch, D. B., & Webster, N. S. (2019). Scientists' warning to humanity: Microorganisms and climate change. *Nature Reviews. Microbiology*, *17*(9), 569–586. DOI: 10.1038/s41579-019-0222-5 PMID: 31213707

Chern, E. C., Tsai, D. W., & Ogunseitan, O. A. (2007). Deposition of glomalin-related soil proteins and sequestered toxic metals into watersheds. *Environmental Science & Technology*, *41*(10), 3566–3572. DOI: 10.1021/es0628598 PMID: 17547179

Colman, B. P., Arnaout, C. L., Anciaux, S., Gunsch, C. K., Hochella, M. F.Jr, Kim, B., Lowry, G. V., McGill, B. M., Reinsch, B. C., Richardson, C. J., Unrine, J. M., Wright, J. P., Yin, L., & Bernhardt, E. S. (2013). Low concentrations of silver nanoparticles in biosolids cause adverse ecosystem responses under realistic field scenario. *PLoS One*, *8*(2), e57189. DOI: 10.1371/journal.pone.0057189 PMID: 23468930

Costanza, R., d'Arge, R., de Groot, R., Farber, S., Grasso, M., Hannon, B., Limburg, K., Naeem, S., O'Neill, R. V., Paruelo, J., Raskin, R. G., Sutton, P., & van den Belt, M. (1997). The value of the world's ecosystem services and natural capital. *Nature*, *387*(6630), 253–260. DOI: 10.1038/387253a0

Coyte, K. Z., Schluter, J., & Foster, K. R. (2015). The ecology of the microbiome: Networks, competition, and stability. *Science*, *350*(6261), 663–666. DOI: 10.1126/science.aad2602 PMID: 26542567

Danso, D., Schmeisser, C., Chow, J., Zimmermann, W., Wei, R., Leggewie, C., Li, X., Hazen, T., & Streit, W. R. (2018). New insights into the function and global distribution of polyethylene terephthalate (PET)-degrading bacteria and enzymes in marine and terrestrial metagenomes. *Applied and Environmental Microbiology*, *84*(8), e02773–e17. DOI: 10.1128/AEM.02773-17 PMID: 29427431

Davies, J., & Davies, D. (2010). Origins and evolution of antibiotic resistance. *Microbiology and Molecular Biology Reviews : MMBR*, *74*(3), 417–433. DOI: 10.1128/MMBR.00016-10 PMID: 20805405

de Vries, F. T., Liiri, M. E., Bjørnlund, L., Bowker, M. A., Christensen, S., Setälä, H. M., & Bardgett, R. D. (2012). Soil bacterial networks are less stable under drought than fungal networks. *Nature Communications*, *3*, 2409. DOI: 10.1038/s41467-018-05516-7 PMID: 30072764

Doney, S. C., Fabry, V. J., Feely, R. A., & Kleypas, J. A. (2009). Ocean acidification: The other CO_2 problem. *Annual Review of Marine Science*, *1*(1), 169–192. DOI: 10.1146/annurev.marine.010908.163834 PMID: 21141034

Evans, S. E., & Wallenstein, M. D. (2014). Climate change alters ecological strategies of soil bacteria. *Ecology Letters*, *17*(2), 155–164. DOI: 10.1111/ele.12206 PMID: 24261594

Faust, K., & Raes, J. (2012). Microbial interactions: From networks to models. *Nature Reviews. Microbiology*, *10*(8), 538–550. DOI: 10.1038/nrmicro2832 PMID: 22796884

Giller, K. E., Witter, E., & McGrath, S. P. (1998). Toxicity of heavy metals to microorganisms and microbial processes in agricultural soils: A review. *Soil Biology & Biochemistry*, *30*(10-11), 1389–1414. DOI: 10.1016/S0038-0717(97)00270-8

Griffiths, B. S., & Philippot, L. (2013). Insights into the resistance and resilience of the soil microbial community. *FEMS Microbiology Reviews*, *37*(2), 112 129. DOI: 10.1111/j.1574-6976.2012.00343.x PMID: 22568555

Hemme, C. L., Deng, Y., Gentry, T. J., Fields, M. W., Wu, L., Barua, S., Barry, K., Tringe, S. G., Watson, D. B., He, Z., Hazen, T. C., Tiedje, J. M., Rubin, E. M., & Zhou, J. (2010). Metagenomic insights into evolution of a heavy metal-contaminated groundwater microbial community. *The ISME Journal*, *4*(5), 660–672. DOI: 10.1038/ismej.2009.154 PMID: 20182523

Hibbert, K., & Ogunseitan, O. A. (2014). Risks of toxic ash from artisanal mining of discarded cellphones. *Journal of Hazardous Materials*, *278*, 1–7. DOI: 10.1016/j.jhazmat.2014.05.089 PMID: 24937657

Holling, C. S. (1973). Resilience and stability of ecological systems. *Annual Review of Ecology and Systematics*, *4*(1), 1–23. DOI: 10.1146/annurev.es.04.110173.000245

Intergovernmental Panel on Climate Change. (2022). *Climate change 2022: Impacts, adaptation, and vulnerability. Contribution of Working Group II to the Sixth Assessment Report of the Intergovernmental Panel on Climate Change*. Cambridge University Press., DOI: 10.1017/9781009325844

Janssen, D. B., Dinkla, I. J. T., Poelarends, G. J., & Terpstra, P. (2005). Bacterial degradation of xenobiotic compounds: Evolution and distribution of novel enzyme activities. *Environmental Microbiology*, *7*(12), 1868–1882. DOI: 10.1111/j.1462-2920.2005.00966.x PMID: 16309386

Karr, J. R., Fausch, K. D., Angermeier, P. L., Yant, P. R., & Schlosser, I. J. (2000). Assessing ecological integrity in running waters: A method and its rationale. *Special Publication of the Illinois Natural History Survey, 5*, 1-19. https://www.researchgate.net/publication/245507787

Khalili, B., Ogunseitan, O. A., Goulden, M. L., & Allison, S. D. (2016). Interactive effects of precipitation manipulation and nitrogen addition on soil properties in California grassland and shrubland. *Applied Soil Ecology*, *107*, 144–153. DOI: 10.1016/j.apsoil.2016.05.018

Lennon, J. T., & Jones, S. E. (2011). Microbial seed banks: The ecological and evolutionary implications of dormancy. *Nature Reviews. Microbiology*, *9*(2), 119–130. DOI: 10.1038/nrmicro2504 PMID: 21233850

Lenton, T. M., Held, H., Kriegler, E., Hall, J. W., Lucht, W., Rahmstorf, S., & Schellnhuber, H. J. (2008). Tipping elements in the Earth's climate system. *Proceedings of the National Academy of Sciences of the United States of America, 105*(6), 1786–1793. DOI: 10.1073/pnas.0705414105 PMID: 18258748

Loreau, M., Naeem, S., Inchausti, P., Bengtsson, J., Grime, J. P., Hector, A., Hooper, D. U., Huston, M. A., Raffaelli, D., Schmid, B., Tilman, D., & Wardle, D. A. (2001). Biodiversity and ecosystem functioning: Current knowledge and future challenges. *Science*, *294*(5543), 804–808. DOI: 10.1126/science.1064088 PMID: 11679658

Louca, S., Polz, M. F., Mazel, F., Albright, M. B. N., Huber, J. A., O'Connor, M. I., Ackermann, M., Hahn, A. S., Srivastava, D. S., Crowe, S. A., Doebeli, M., & Parfrey, L. W. (2018). Function and functional redundancy in microbial systems. *Nature Ecology & Evolution*, *2*(6), 936–943. DOI: 10.1038/s41559-018-0519-1 PMID: 29662222

Mackelprang, R., Waldrop, M. P., DeAngelis, K. M., David, M. M., Chavarria, K. L., Blazewicz, S. J., Rubin, E. M., & Jansson, J. K. (2011). Metagenomic analysis of a permafrost microbial community reveals a rapid response to thaw. *Nature*, *480*(7377), 368–371. DOI: 10.1038/nature10576 PMID: 22056985

Martinez, J. L. (2008). Antibiotics and antibiotic resistance genes in natural environments. *Science, 321*(5887), 365–367. DOI: 10.1126/science.1159483 PMID: 18635792

Matlock, M., Hopfer, S., & Ogunseitan, O. A. (2019). Communicating risk for a climate-sensitive disease: A case study of Valley Fever in Central California. *International Journal of Environmental Research and Public Health, 16*(17), 3254. DOI: 10.3390/ijerph16183254 PMID: 31491843

Megharaj, M., Ramakrishnan, B., Venkateswarlu, K., Sethunathan, N., & Naidu, R. (2011). Bioremediation approaches for organic pollutants: A critical perspective. *Environment International, 37*(8), 1362–1375. DOI: 10.1016/j.envint.2011.06.003 PMID: 21722961

Meron, D., Atias, E., Iasur Kruh, L., Elifantz, H., Minz, D., Fine, M., & Banin, E. (2011). The impact of reduced pH on the microbial community of the coral *Acropora eurystoma. The ISME Journal, 5*(1), 51–60. DOI: 10.1038/ismej.2010.102 PMID: 20668489

Monteux, S., Weedon, J. T., Blume-Werry, G., Gavazov, K., Jassey, V. E. J., Johansson, M., Keuper, F., Olid, C., & Dorrepaal, E. (2018). Long-term in situ permafrost thaw effects on bacterial communities and potential aerobic respiration. *The ISME Journal, 12*(9), 2129–2141. DOI: 10.1038/s41396-018-0176-z PMID: 29875436

Ochman, H., Lawrence, J. G., & Groisman, E. A. (2000). Lateral gene transfer and the nature of bacterial innovation. *Nature, 405*(6784), 299–304. DOI: 10.1038/35012500 PMID: 10830951

Ogunseitan, O. A. (1993). Direct extraction of proteins from environmental samples. *Journal of Microbiological Methods, 17*(4), 273–281. DOI: 10.1016/0167-7012(93)90056-N

Ogunseitan, O. A. (1996). Protein profile variation in cultivated and native freshwater microorganisms exposed to chemical environmental pollutants. *Microbial Ecology, 31*(3), 291–304. DOI: 10.1007/BF00171573 PMID: 8661530

Ogunseitan, O. A. (1998). Protein method for investigating mercuric reductase gene expression in aquatic environments. *Applied and Environmental Microbiology, 64*(2), 695–702. DOI: 10.1128/AEM.64.2.695-702.1998 PMID: 9464410

Ogunseitan, O. A. (2000). Microbial proteins as biomarkers of ecosystem health. In Scow, K., Fogg, G., Hinton, D., & Johnson, M. L. (Eds.), *Integrated assessment of ecosystem health* (pp. 207–222). CRC Press., DOI: 10.1201/9780429104404

Ogunseitan, O. A. (2005). *Microbial diversity* (Foreword by L. Margulis). Wiley-Blackwell Publishing. Print ISBN:9780632047086, Online ISBN:9780470750490

Ogunseitan, O. A. (2008). Genetic transduction in freshwater ecosystems. *Freshwater Biology*, *53*(6), 1228–1239. DOI: 10.1111/j.1365-2427.2007.01857.x

Ogunseitan, O. A. (2016). Power failure: The battered legacy of leaded batteries. *Environmental Science & Technology*, *50*(17), 8997–8998. DOI: 10.1021/acs.est.6b03174 PMID: 27467531

Ogunseitan, O. A. (2017). Mercury safety reform in the 21st century: Advancing the new framework for toxic substances control. *Environment*, *59*(4), 4–13. DOI: 10.1080/00139157.2017.1325299

Ogunseitan, O. A. (2019). Embracing global warmth and climate resilience through green chemistry legislation. *Hastings Environmental Law Journal, 25*(2), 301–318. https://repository.uclawsf.edu/hastings_environmental_law_journal/vol25/iss2/5

Ogunseitan, O. A. (2022). One Health and the environment: From conceptual framework to implementation science. *Environment*, *64*(2), 4–24. DOI: 10.1080/00139157.2022.2021792

Ogunseitan, O. A. (2023). Chemicals management approach to sustainable development of materials. *MRS Bulletin*, *48*(5), 491–497. DOI: 10.1557/s43577-023-00518-3

Ogunseitan, O. A. (2024). Situating One Health at the center of the climate agenda. *Environment*, *66*(4), 4–16. DOI: 10.1080/00139157.2017.1325299

Ogunseitan, O. A. (2025). Perspective—Planetary boundaries for recalcitrant materials and toxic chemical pollutants: Specifications for sustainable safe operating zones. *Frontiers in Environmental Science*, *13*, 1593947. DOI: 10.3389/fenvs.2025.1593947

Ogunseitan, O. A. (2026). Bacterial diversity, introduction to. In J. B. Wolf & C. A. de M. Russo (Eds.), *Encyclopedia of evolutionary biology* (2nd ed., Vol. 5, pp. 42–47). Elsevier, Academic Press. https://doi.org/DOI: 10.1016/B978-0-443-15750-9.00073-2

Ogunseitan, O. A., Gussin, G., Jimah, T., Wang, S., Fenny, A., Bada-Alambedji, R., & Andre, B. (2022). *The ten commandments of antibiotic stewards*. CABI One Health Cases., DOI: 10.1079/onehealthcases.2022.0003

Ogunseitan, O. A., & LeBlanc, J. (2005). Environmental proteomics: Methods and applications for aquatic ecosystems. *Molecular Microbial Ecology Manual*, *4*(10), 1027–1046.

Ogunseitan, O. A., LeBlanc, J., & Dalmasso, E. (2001). Microbial community proteomics. In Rochelle, P. A. (Ed.), *Environmental molecular microbiology* (pp. 125–140). Horizon Scientific Press.

Ogunseitan, O. A., & Matlock, M. (2022). *Mouldy planet: Fungi and One Health.* CABI One Health Case Studies., DOI: 10.1079/onehealthcases.2022.0004

Ogunseitan, O. A., Yang, S. L., & Scheinbach, E. (1999). The delta-aminolevulinate dehydratase of marine *Vibrio alginolyticus* is resistant to lead (Pb). *The Biological Bulletin, 197*(2), 283–284. DOI: 10.2307/1542653 PMID: 28281822

Petrillo, J. E., & Ogunseitan, O. A. (2018). Emerging issues in the environmental context of antibiotic-resistance. *Environment International, 116*, 39–42. DOI: 10.1016/j.envint.2018.03.049 PMID: 29649775

Pollack, K., Balasz, K., & Ogunseitan, O. A. (2009). Proteomic assessment of caffeine effects on coral symbionts. *Environmental Science & Technology, 43*(6), 2085–2091. DOI: 10.1021/es802617f PMID: 19368218

Prosser, J. I. (2015). Dispersing misconceptions and identifying opportunities for the use of 'omics' in soil microbial ecology. *Nature Reviews. Microbiology, 13*(7), 439–446. DOI: 10.1038/nrmicro3468 PMID: 26052662

Renella, G., Ogunseitan, O. A., Giagnoni, L., & Arenella, M. (2014). Environmental proteomics: A long march in the pedosphere. *Soil Biology & Biochemistry, 69*, 34–37. DOI: 10.1016/j.soilbio.2013.10.035

Scheffer, M., Carpenter, S., Foley, J. A., Folke, C., & Walker, B. (2001). Catastrophic shifts in ecosystems. *Nature, 413*(6856), 591–596. DOI: 10.1038/35098000 PMID: 11595939

Scheffer, M., & Carpenter, S. R. (2003). Catastrophic regime shifts in ecosystems: Linking theory to observation. *Trends in Ecology & Evolution, 18*(12), 648–656. DOI: 10.1016/j.tree.2003.09.002

Schimel, J. P. (2018). Life in dry soils: Effects of drought on soil microbial communities and processes. *Annual Review of Ecology, Evolution, and Systematics, 49*(1), 409–432. DOI: 10.1146/annurev-ecolsys-110617-062614

Schuur, E. A. G., McGuire, A. D., Schädel, C., Grosse, G., Harden, J. W., Hayes, D. J., Hugelius, G., Koven, C. D., Kuhry, P., Lawrence, D. M., Natali, S. M., Olefeldt, D., Romanovsky, V. E., Schaefer, K., Turetsky, M. R., Treat, C. C., & Vonk, J. E. (2015). Climate change and the permafrost carbon feedback. *Nature, 520*(7546), 171–179. DOI: 10.1038/nature14338 PMID: 25855454

Schwarzenbach, R. P., Escher, B. I., Fenner, K., Hofstetter, T. B., Johnson, C. A., von Gunten, U., & Wehrli, B. (2006). The challenge of micropollutants in aquatic systems. *Science, 313*(5790), 1072–1077. DOI: 10.1126/science.1127291 PMID: 16931750

Shade, A., Dunn, R. R., Blowes, S. A., Keil, P., Bohannan, B. J. M., Herrmann, M., Küsel, K., Lennon, J. T., Sanders, N. J., Storch, D., & Chase, J. (2018). Macroecology to unite all life, large and small. *Trends in Ecology & Evolution, 33*(10), 731–744. DOI: 10.1016/j.tree.2018.08.005 PMID: 30209011

Shade, A., Jones, S. E., Caporaso, J. G., Handelsman, J., Knight, R., Fierer, N., & Gilbert, J. A. (2014). Conditionally rare taxa disproportionately contribute to temporal changes in microbial diversity. *mBio, 5*(4), e01371–e14. DOI: 10.1128/mBio.01371-14 PMID: 25028427

Shade, A., Peter, H., Allison, S. D., Baho, D. L., Berga, M., Bürgmann, H., Huber, D. H., Langenheder, S., Lennon, J. T., Martiny, J. B. H., Matulich, K. L., Schmidt, T. M., & Handelsman, J. (2012). Fundamentals of microbial community resistance and resilience. *Frontiers in Microbiology, 3*, 417. DOI: 10.3389/fmicb.2012.00417 PMID: 23267351

Steffen, W., Richardson, K., Rockström, J., Cornell, S. E., Fetzer, I., Bennett, E. M., Biggs, R., Carpenter, S. R., de Vries, W., de Wit, C. A., Folke, C., Gerten, D., Heinke, J., Mace, G. M., Persson, L. M., Ramanathan, V., Reyers, B., & Sörlin, S. (2015). Planetary boundaries: Guiding human development on a changing planet. *Science, 347*(6223), 1259855. DOI: 10.1126/science.1259855 PMID: 25592418

Storz, G., & Hengge-Aronis, R. (Eds.). (2000). *Bacterial stress responses.* ASM Press. Print ISBN:9781119737841, Online ISBN:9781683671213

Thomas, C. M., & Nielsen, K. M. (2005). Mechanisms of, and barriers to, horizontal gene transfer between bacteria. *Nature Reviews. Microbiology, 3*(9), 711–721. DOI: 10.1038/nrmicro1234 PMID: 16138099

Wall, D. H., Bardgett, R. D., Behan-Pelletier, V., Herrick, J. E., Jones, T. H., Ritz, K., Six, J., Strong, D. R., & van der Putten, W. H. (Eds.). (2012). *Soil ecology and ecosystem services.* Oxford University Press., DOI: 10.1093/acprof:oso/9780199575923.001.0001

Webster, N. S., Negri, A. P., Botté, E. S., Laffy, P. W., Flores, F., Noonan, S., Schmidt, C., & Uthicke, S. (2016). Host-associated coral reef microbes respond to the cumulative pressures of ocean warming and ocean acidification. *Scientific Reports, 6*(1), 19324. DOI: 10.1038/srep19324 PMID: 26758800

Wittebolle, L., Marzorati, M., Clement, L., Balloi, A., Daffonchio, D., Heylen, K., De Vos, P., Verstraete, W., & Boon, N. (2009). Initial community evenness favours functionality under selective stress. *Nature*, *458*(7238), 623–626. DOI: 10.1038/nature07840 PMID: 19270679

Yoshida, S., Hiraga, K., Takehana, T., Taniguchi, I., Yamaji, H., Maeda, Y., Toyohara, K., Miyamoto, K., Kimura, Y., & Oda, K. (2016). A bacterium that degrades and assimilates poly(ethylene terephthalate). *Science*, *351*(6278), 1196–1199. DOI: 10.1126/science.aad6359 PMID: 26965627

Zettler, E. R., Mincer, T. J., & Amaral-Zettler, L. A. (2013). Life in the 'plastisphere': Microbial communities on plastic marine debris. *Environmental Science & Technology*, *47*(13), 7137–7146. DOI: 10.1021/es401288x PMID: 23745679

Zinsstag, J., Schelling, E., Waltner-Toews, D., & Tanner, M. (2011). From "one medicine" to "one health" and systemic approaches to health and well-being. *Preventive Veterinary Medicine*, *101*(3-4), 148–156. DOI: 10.1016/j.prevetmed.2010.07.003 PMID: 20832879

Chapter 9
Microbial Bioremediation of Microplastics:
Mechanisms, Challenges, and Future Prospects

Rachel Leads
http://orcid.org/0000-0002-1306-245X
Michigan State University, USA

Austin D. Gray
Virginia Polytechnic Institute and State University, USA

Ola A. Olapade
http://orcid.org/0000-0002-4505-4024
Albion College, USA

ABSTRACT

Microplastics (MPs) are ubiquitous in the environment and are defined as small (< 5 mm) heterogeneous plastic particles that vary in size, color, shape, and polymer type. These relatively new pollutants are considered contaminants of emerging concern (CECs) due to their widespread accumulation, bioavailability to aquatic and terrestrial organisms, and potential to induce toxicity in wildlife and humans. This chapter explores the main anthropogenic sources of various types of microplastic (MP) accumulation and disposal into the environment as well as potential interactions with indigenous microbial cells. The chapter also examines key degradation processes commonly used by microbes in remediating discarded MPs, in the plastisphere and in surrounding environments, and current challenges associated with

DOI: 10.4018/979-8-2600-2100-2.ch009

interactions between microbes and MPs biodegradation. Furthermore, the chapter highlights alternatives to MPs to reduce their availability and ensure degradative potentials of microbes of these pollutants in the environment.

1 INTRODUCTION TO MICROPLASTICS

Plastics represent a large group of materials that are comprised of petroleum-derived synthetic and semi-synthetic polymers (Hopewell et al., 2009; Thompson et al., 2009). The most commonly used types of plastics include polyethylene (PE), polypropylene (PP), polystyrene (PS), polyvinyl chloride (PVC), and polyethylene terephthalate (PET) (Li et al., 2021). Plastics are designed to be durable, light weight, affordable, easily manufactured, and versatile, making them extremely useful in a wide variety of applications and products. Therefore, because of their durability and stability they are recalcitrant to biodegradation, remaining in the environment for generations (Wang et al., 2020; Pilapitiya, and Ratnayake 2024). Approximately, about 50% of plastic production is used to manufacture single use plastics such as packaging and disposable consumer products (Chen et al., 2021). Synthetic textiles, such as polyester and nylon, represent the second largest portion of plastic manufacturing (Mohamadi et al., 2023). The demand for these materials continues to increase, with plastic manufacturing growing over 8% each year (Geyer et al., 2017). Between 1950 and 2018, plastic production increased from 1.3 million tons to 359 million tons (Zhang et al., 2021a). Waste management practices, infrastructure, and behavior have not kept pace with global plastic production, leading to the accumulation of plastic litter in the environment. Indeed, it is estimated that only 9% of plastic products are recycled, while 50% or more accumulate in landfills or the natural environment (Geyer et al., 2017). More recent estimates states that about 19% of plastic waste is incinerated, 22% littered, and 50% dumped in regulated landfills (OECD, 2019; Figure 1).

Figure 1. The proportion of plastic waste that is recycled, identified, or dumped into the enivronment.

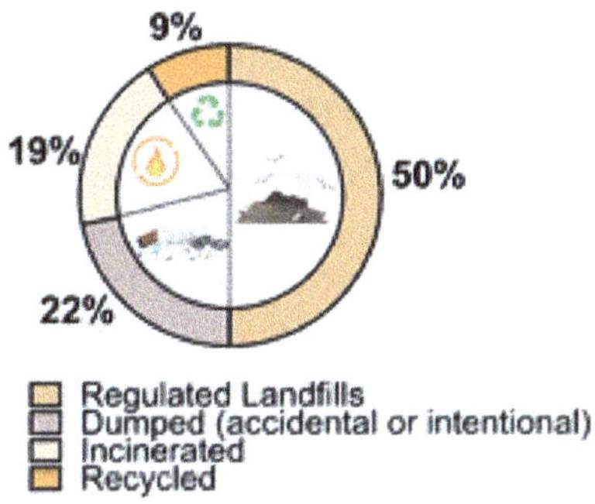

Source: Organization for Economic Cooperation and Development (OECD)

Plastic litter can be classified by size into macroplastic (>20 mm), mesoplastic (5-20 mm), and microplastic (MP) (<5 mm) (Barnes et al., 2009). Microplastics (MPs) are plastic particles generally less than 5 mm in size widely found in various environments (Kye et al., 2023; Zhai et al., 2023; Vass et al., 2024). Due to the wide range of polymers, applications, and sources, MPs are a heterogeneous mixture of plastic particles that vary in size, color, shape, and polymer type. Common MP shapes include fragments, fibers, foam, and spheres (Hidalgo-Ruz et al., 2012). These classifications can reflect the sources and degradation pathways of different MPs found in the environment. MPs are becoming ubiquitous in the biosphere, found everywhere, from the air, water, and soil as well as in organisms living in these environments, including in every extreme region of the world (Yang et al., 2023). MP pollution is so drastic and widespread that occurrences in the environment are currently estimated to be between 82 and 358 trillion pieces in the planet's oceans alone (Erikson et al., 2023), with more than 800 tons of MPs entering each year, not including those in landfills and freshwater systems (Lau et al., 2020; Yang et al., 2023). Recently, detection of MPs, comprised mostly of PE, in post-mortem human tissues and in animal guts have been widely reported (Fackelmann et al., 2023; Nihart et al., 2025).

Contaminants of emerging concern (CECs) are defined as newly identified or previously undetected chemicals that may be harmful to wildlife and human health (Wang et al., 2024). MPs are considered CECs due to their widespread distribution, bioavailability to organisms, and potential to exert toxicity in animals and humans (Weinstein et al., 2019). Compounds that are considered CECs are typically unregulated, and their risks to ecosystem and human health are not fully understood. In the environment, MPs occupy the same size fraction as plankton, sediments, and soils, making them bioavailable to a wide range of organisms such as fish and invertebrates. As a result, MPs can be ingested intentionally or accidentally by organisms and can bioaccumulate within organs and tissues across species and trophic levels (Bhatt and Chauhan, 2023; Miller et al., 2020; Gray et al., 2024). MP exposure can induce a variety of adverse effects such as gut blockage, decreased growth, inflammation, oxidative stress, gut microbiome dysbiosis, and decreased reproduction (Deng et al., 2023; Guzzetti et al., 2018; Rakib et al., 2023). In addition to physical particle toxicity, MPs can also leach and expose organisms to chemical additives (e.g., phthalates and flame retardants) and contaminants adsorbed from the environment (e.g., heavy metals and per- and polyfluoroalkyl substances [PFAS]) (Costigan et al., 2022; Godoy et al., 2020). Importantly, MPs may also serve as a vector for pathogen transmission in aquatic and terrestrial environments (Gkoutselis et al., 2021; Junaid et al., 2022). Human exposure to MPs primarily arises from ingestion of contaminated food (e.g., seafood) and inhalation of indoor air pollution (Domenech and Marcos, 2021; Sun and Wang, 2023; Yakovenko et al., 2025). Currently, the risks of MP exposure to human health are not fully understood. Given the extensive accumulation and risks to ecosystem and human health globally, there is great interest in identifying remediation and mitigation strategies for MPs in the environment, such as microbial bioremediation. Therefore, given the current ubiquitous occurrences of various MPs in natural and built environments as well as the urgent need for their effective remediation, the goal of this chapter is to critically explore the different sources, potential for their biodegradation, and eventual fate of these anthropogenic pollutants in our biosphere.

2 SOURCES OF MICROPLASTICS TO THE ENVIRONMENT

MPs can be divided into primary or secondary MPs based on their uses and sources. Primary MPs are manufactured to be small, such as plastic pre-production pellets (i.e., nurdles) or microbeads added to personal care products. Secondary MPs are those degraded from larger plastic items such as tires, synthetic textiles, and food packaging through various processes including photodegradation, erosion, and mechanical breakup (Zhang et al., 2018; Akdogan and Guven, 2019; Cheng et al., 2020). In addition, primary and secondary MPs can enter the environment from both point and non-point sources. Point sources of MP pollution are defined as identifiable, discrete sources (e.g., wastewater treatment plant [WWTP] effluent), whereas non-point sources are defined as diffuse or originating from many potential sources (e.g., runoff).

Primary MPs can be released directly into the environment from point sources such as WWTP effluent, industrial outfalls, or accidental spills (Auta et al., 2017). By contrast, secondary MPs are primarily released into the environment from diffuse, non-point sources as larger plastic items degrade over time (Weinstein et al., 2016; Gray et al., 2025). Secondary MPs, such as fibers released from synthetic textiles, can also be released from point sources such as WWTP effluent into both aquatic and terrestrial environments. While the majority of research has focused on these sources as contributors of MP accumulation in aquatic ecosystems, terrestrial environments can also accumulate MPs through the breakdown of plastic litter, atmospheric deposition, incineration, and agricultural practices such as the use of plastic mulching and land application of biosolids (Huang et al., 2020; Kumar et al., 2023; Surendran et al., 2023).

3 MICROPLASTIC DEGRADATION IN THE ENVIRONMENT

Plastic materials are generally long-lasting and persistent in the environment. However, factors such as wave action, sediment abrasion, ultraviolet (UV) radiation from sunlight, and microbial degradation can facilitate the breakdown of plastic litter (Figure 2). As described above, MPs can be produced and released into the environment as macroplastic litter breaks down through these abiotic and biotic processes. Abiotic factors refer to the physical, chemical, and environmental processes influencing degradation, whereas biotic factors refer to the biological and organismal processes. Together, these two types of degradation pathways can result in changes to the physical and/or chemical structure of the plastic through fragmentation, cracking, bond cleavage, and oxidation of polymer chains (Chamas et al., 2020). In the environment, chemical degradation typically occurs through hydrolysis (requiring water) or oxidation (requiring oxygen), each of which can be initiated by exposure to sunlight, heat, and/or microbial action (Chamas et al., 2020). Several studies have shown that discarded plastics can begin degrading and releasing MPs to the environment in as little as 2 to 8 weeks through these mechanisms (Gray et al., 2025; Weinstein et al., 2016; Weinstein et al., 2019).

Figure 2. Overview of various degradation processes that result in the formation of microplastics from macroplastics.

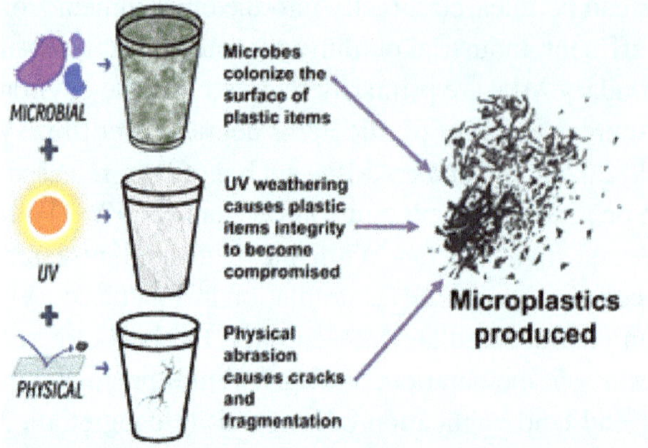

The primary abiotic processes influencing plastic degradation in the environment include mechanical stress, UV irradiation, and heat (Zhang et al., 2021). Mechanical stress and abrasion can be produced by waves, friction, collision, freezing, and thawing which can result in the physical cracking, fragmentation, and delamination of plastic materials (Weinstein et al., 2016). Photodegradation, through UV irradiation and photo-oxidation, is one of the primary processes that facilitates plastic degradation in the environment (Zhang et al., 2021). In this process, UV is absorbed by the plastic which initiates bond cleavage, chain scission, and the subsequent degradation of the polymer (Ibrahim, 2024). Thermal oxidation can also lead to polymer degradation, particularly in environments with high temperatures and/or direct sunlight (Zhang et al., 2021). This process is influenced greatly by environmental conditions, with the highest rates of thermal degradation occurring in hot environments with high moisture (Ray and Cooney, 2018). Together, these abiotic degradation mechanisms can increase the surface area and hydrophilicity of plastic particles which can then facilitate biofilm formation and microbial degradation (Ali et al., 2021). In this way, abiotic and biotic factors can act synergistically in the degradation process of plastic materials (Figures 17.2 and 17.3).

Microbial degradation is the primary biotic process facilitating the breakdown of plastic in the environment (Figure 17.3). This biodegradation process begins with the formation of a biofilm on the plastic surface, which is a collection of bacteria, fungi, viruses, archaea, algae, and protozoans that is distinct from the surrounding environment (Dhali et al., 2024). Biofilms are present on MPs in both aquatic and terrestrial environments (Ventura et al., 2024). These colonizing microorganisms can secrete intracellular and extracellular enzymes that promote the hydrolysis, breakdown, and mineralization of plastic polymers (Ali et al., 2021; Dhali et al., 2024). For over 10 years, understanding these plastic-associated microbial communities and mechanisms of biodegradation have been important areas of research, which revealed new plastic-active enzymes from various microbial phyla with plastic degradation potential such as PETases, laccases, cutinases, and MHETases (Pérez-García et al., 2025). Characterizing these processes can lead to innovative solutions for the mitigation and remediation of persistent plastic pollution. Accordingly, the interactions between microbes and MPs, mechanisms of biodegradation, and challenges and future prospects for microbial bioremediation are discussed in greater detail in the following sections.

4 MICROPLASTIC AND MICROBE INTERACTIONS IN THE ENVIRONMENT

As described above, the inappropriate discarding and mismanagement of waste has resulted in plastic debris entering the environment at rates unanticipated, where no regulation exists on a global level (Corradini et al., 2024; Valavanidis, 2024). The introduction of plastic debris into the environment has a correlative relationship with MP abundance, where high abundance of macroplastics occurs in the environment with high abundances of MPs (Lee et al., 2013). Both of these size ranges have been investigated in aquatic habitats where researchers have assessed microbial colonization on their surfaces. Thus, given rise to the term "plastisphere" which describes plastics in aquatic environments establishing microbial communities on the surface of the plastic distinct from that in the surrounding water (Sooriyakumar et al., 2022 (**Figure 3**). The term originated in 2013 by Zettler et al. (2013) when describing organisms colonizing the surface of MPs in marine debris. Since then, it has expanded to other ecosystems to describe microbial colonization of MPs.

Plastic debris can be trapped in riverbeds, debris dams, and salt marshes along its transport before reaching the ocean (Gray et al., 2025; Weinstein et al., 2016). MPs themselves are typically more buoyant and less dense, resulting in transport along greater distances than larger debris (Kumar et al., 2021). During this time, trapped plastics or free-floating MPs can become colonized by different microbes, facilitating the further degradation of plastics into MPs through biodegradation processes (Gray et al., 2025) **(Figure 4)**. Microbes colonizing the surface of plastics and MPs can change their buoyancy through biofilm development (Kaiser et al., 2017), which can ultimately impact the transport and fate of MPs in aquatic environments.

Figure 3. The primary processes involved with the microbial biodegradation of plastics

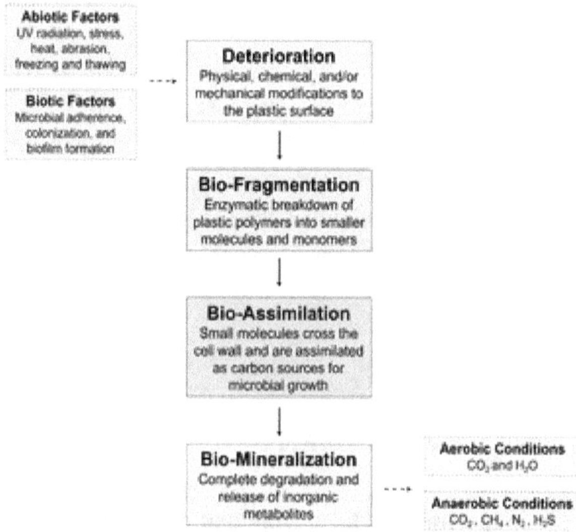

Figure 4. Scanning electron imaging of low-density polyethylene (LDPE), polyethylene terephthalate (PET), and polystyrene (PS) surfaces colonized with microbes from a low-order riverine system

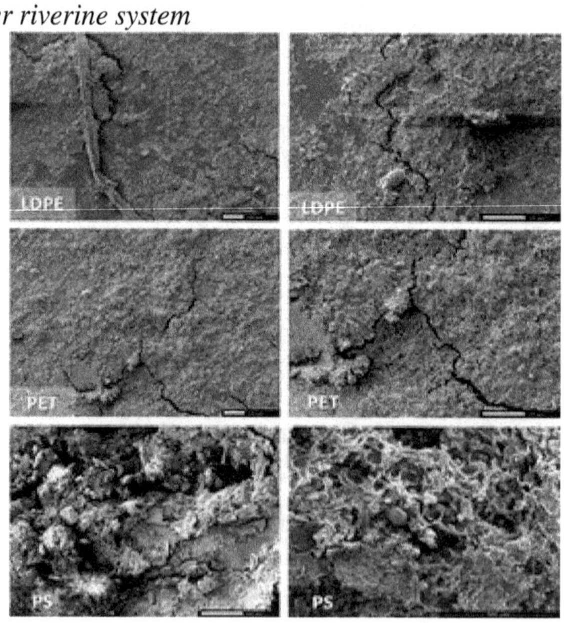

Images from Gray et al., 2025.

The factors influencing the rate of colonization of microbes onto plastics is nuanced, depending on abiotic factors, geographic location, polymer type, and additives used in the production of plastics (Rosato et al., 2020). MPs, while buoyant, are also hydrophobic, allowing for various organic molecules or matter (OM) and nutrients to adsorb to its surface. The adsorption of OM to the surface of MPs provides a resource for microbes to select for (Artham et al., 2009).

There are different types of microbes that can attach to MPs (bacteria, fungi, archaea, diatoms), producing an extracellular polymeric substance (EPS) resulting in the formation of biofilm on the surfaces of MPs (Dhali et al., 2024) (Figure 17.4). The biofilms that colonize MPs have been reported to differ compared to communities that colonize natural items (e.g., wood, stone, rock) (Oberbeckman et al., 2018). Most notably, bacterial pathogens (e.g., *Vibrio* spp.) have been shown to select for the surface of MPs versus natural items (Frere et al., 2018), highlighting a concern of MPs serving as a vector of disease. Fungal and eukaryotic pathogens have also been shown by Kettner et al., (2017; 2019) to be enriched in MP biofilms. However, there is a dearth of knowledge in comparison to bacterial pathogens.

The risk associated with pathogens selecting for MPs is tied to the fate of MPs and their inadvertent ingestion by biota, including humans, which are estimated to ingest 4.1 μg of MPs per week (Pletz, 2022). In addition to ingestion, humans are estimated to inhale between 3,200 to 68,000 MPs per day due to indoor air pollution (Yakovenko et al., 2025). These results highlight the importance of MP pollution as both an environmental and public health crisis. Currently, issues with biofilm colonization on MPs has been focused under the scope of antimicrobial resistance. In particular, antimicrobial resistance has been shown to be highly prevalent in wastewater dominated systems where unregulated prescription drugs (e.g., antibiotics) are released.

In additiona, microbes that interact with plastics aside from relationships where biofilms are considered include those specific to nutrient transformation and fixation. MPs that deposit in sediment and soils are in direct contact with microbes, where abundance is greater in these compartments versus surface waters. Because MPs are made of different chemical additives, over time they can leach into these areas in the form of dissolved organic matter, impacting microbes in close proximity and their function (Wang et al., 2025; Seeley et al., 2020). Nitrogen cycling has been the main process researchers have assessed with limited reports demonstrating that the nitrogen cycle can be impacted due to plastics and associated leachates. The impact to these processes is constrained by the environment, size, concentration, and polymer of MP (Shen et al., 2022). Specifically looking at polymers, PVC can inhibit nitrification and denitrification. Conversely, polylactic acid (PLA) and polyurethane foam both promote nitrification and denitrification (Seeley et al., 2020). Low-density polyethylene (LDPE) has been shown to impact functional genes specific to nitrification, denitrification, and ammonia oxidation (Rong et al., 2021), while both PE and PP have been reported to enhance nitrification and nitrogen fixation (Kim et al., 2023). Alterations to these processes can have global implications in regard to the nitrogen cycle, primarily due to the excess in reactive nitrogen introduced into the environment.

5 MECHANISMS OF MICROBIAL BIODEGRADATION OF MICROPLASTICS

Current available information regarding the mechanisms for microbial degradation and remediation of MPs in the environment is still very sparse and limited. Generally, microbes found in the plastisphere capable of plastic degradation—especially PE, PET, PP, PS, and PVC—have mainly been found in terrestrial environments (Zhai et al., 2023). Microbial members capable of the degradation of these plastic materials are very diverse and include members belonging to the *Achromobacter, Acinetobacter, Arthrobacter, Aspergillus, Bacillus, Comamonas, Delftia, Geobacillus, Kocuria, Micrococcus, Nesiobacter, Paenibacillus, Pseudomonas, Rahnella, Sacchramonospora, Staphylococcus, Stenotrophomonas*, *Streptococcus*, *Thermus*, and *Zalerion* genera (Orr et al., 2004; Mor and Sivan, 2008; Das et al., 2012; Kawai et al., 2014; Asmita et al., 2015; Mohan et al., 2016; Khan et al., 2017; Giacomucci et al., 2019; Chen et al., 2020).

The general processes for the microbial degradation of most MPs involves several biochemical reactions starting with their initial colonization, followed immediately by their subsequent depolymerization into small, low molecular weight molecules, assimilation as carbon sources for microbial growth, and mineralization into different small molecules (Figure 3) (e.g., Shah et al., 2008; Pathak, 2017; Zhai et al., 2023). Throughout these biochemical processes in MP degradation, various hydrolytic enzymes produced by the microbial assemblages such as hydrolases (esterase, lipase, keratinase, cutinase) and oxidoreductases (laccase, manganese peroxidase, hydroxylase and lignin peroxidases) as well as certain polymerases have been implicated (e.g., Yoon et al., 2012; Tahir et al., 2013; Mohan et al., 2016; Gan and Zhang, 2019; Amobonye et al., 2021; Zhang et al., 2021). Ecologically, it has been observed that within the complex confines of the plastisphere around plastics, the autotrophic microbial members oftenaccelerate the oxidative decomposition process, while the heterotrophic members contribute to the overall decomposition of MPs through series of activities such as biodeterioration, bio-fragmentation, bio-assimilation, and bio-mineralization (Figure 3) (Zhai et al., 2023). Biodeterioration refers to the chemical, physical, and mechanical changes that occur on the surface of plastic particles due to microbial action, while bio-fragmentation refers to the subsequent enzymatic breakdown of the plastic polymers into smaller monomers (Elahi et al., 2021). Bio-assimilation and bio-mineralization refer to the integration of carbon sources into microbial biomass and the release of inorganic metabolites such as carbon dioxide (Elahi et al., 2021). These biodegradation processes can occur in both aerobic (presence of oxygen) and anaerobic (absence of oxygen) conditions. Aerobic biodegradation occurs primarily in aquatic environments and soils and produces carbon dioxide and water during the mineralization process (Lin et al., 2022). By contrast, anaerobic biodegradation occurs primarily in sediments and landfills and can produce carbon dioxide, methane, hydrogen sulfide, and nitrogen gas, depending on the environment and electron acceptor (Lokesh et al., 2023). These processes are also species-specific, with fungal biodegradation occurring only under aerobic conditions, and bacterial biodegradation occurring in both aerobic and anaerobic conditions (Mohan et al., 2022).

Generally, bacterial members belonging to the Proteobacteria, Firmicutes, Bacteroidetes and Cyanobacteria phyla are the most dominant taxa that have been mostly implicated in the plastisphere (e.g., Oberbeckman et al., 2018; Syranidou et al., 2019; Zhai et al., 2023; Vass et al., 2024). Among five classes of the Proteobacteria, the Alpha-, Gamma- and members of the Delta-Proteobacteria are the most commonly found on MPs in various environments including populations of *Rhodobacter*, *Pseudomonas*, and *Vibrio* species (Dang et al., 2008; Zettler et al., 2013; Oberbeckman et al., 2018).

The ability for fungal species to colonize and degrade plastics has also been explored and has implicated several species including *Aspergillus*, *Cobetia*, *Fusarium*, *Exiguobacterium* and *Alcanivorax* in various environments (Ekanayaka et al., 2022; Parsaeimehr et al., 2023). Other studies have also reported the potential roles of fungi such as Crypromycota, Ascomycota, and Chytridiomycota on the plastisphere (Kettner et al., 2017; Vass et al., 2024). Recent studies have shown that fungi, through the secretion of their various extracellular enzymes, are very efficient in the degradation of several synthetic polymers in the environment (e.g., Karunarathna et al., 2025).

Phototrophic microbes such as microalgae and Cyanobacteria are recognized as common residents of the plastisphere as well as dominant colonizers of MPs where they frequently contribute nutrients for heterotrophic microbial populations (Amaral-Zettler et al., 2020; Zhai et al., 2023). Among these phototrophs, the genera *Staniera* and *Pseudomorphium* have been found abundantly on various MPs (e.g., Orr et al., 2004) while the filamentous cyanobacteria *Phormidium* is one of the dominant members that have been observed to be actively hydrocarbon degrading, especially in aquatic environments (Antic et al., 2006; Zettler et al., 2013).

6 MICROBIAL BIODEGRADATION OF PLASTIC-ASSOCIATED CHEMICALS

As described above, MPs can adsorb a number of different contaminants from the environment including heavy metals and persistent organic pollutants (POPs). These adsorbed contaminants can increase potential adverse effects for organisms because they can leach from the MPs and potentially exert additional toxic effects following ingestion. Very little research has been conducted on the impact of these plastic-associated chemicals on the plastisphere, or the ability of microorganisms to degrade them. However, a few case studies highlight the potential for microbial biodegradation of plastic-associated chemicals. For example, Sun et al. (2022) reported that biofilm development enhanced the adsorption of copper (i.e., Cu(II)) to PLA MPs through electrostatic attraction, hydrogen bonding, and surface complexation with biofilm functional groups. In a laboratory study, Rosato et al. (2020) showed that marine biofilms contributed to the biodegradation/biotransformation of polychlorinated biphenyls (PCBs) adsorbed to MPs following two weeks of incubation in anoxic marine sediments. These biodegradation processes resulted in the reductive dechlorination of adsorbed PCBs across five different types of MPs (PE, PET, PS, PP, PVC), potentially altering the toxicity (Rosato et al., 2020). In a follow-up field study, Rosato et al. (2022) determined that dehalogenating taxa preferentially recruited on PCB-contaminated MPs in marine sediments. However, they were unable to detect significant dechlorination of adsorbed PCBs in the field (Rosato et al., 2022). In a similar study, Wu et al. (2017) reported that plastic-associated microbial assemblages biotransformed adsorbed dichlorodiphenyltrichloroethane (DDT) and some polycyclic aromatic hydrocarbons (PAHs) in marine and river sediments, but had little impact on adsorbed PCBs. Together, these studies suggest that, in addition to the biodegradation of plastics, the plastisphere may also play an important role in mediating the fate and effects of other plastic-associated contaminants.

Recent research has also investigated the interaction of MPs with other emerging contaminants such as PFAS. These studies have revealed that MPs and PFAS can co-occur in aquatic and terrestrial environments and that MPs can adsorb PFAS and transfer these compounds to organisms (Freilinger et al., 2025; Parashar et al., 2023; Sobhani et al., 2021). Like Sun et al. (2022), Barhoumi et al. (2024) showed that weathering and biofilm development can increase the sorption of PFAS to MPs. To the best of our knowledge, no studies have investigated the degradation of PFAS by plastic-associated microbial assemblages. Currently, the biodegradation and biotransformation of plastic-associated chemicals by the plastisphere remains an important, yet understudied, area of research.

7 POTENTIAL FOR MICROBIAL BIOREMEDIATION OF MICROPLASTICS

Microbial bioremediation refers to the use of microorganisms, such as bacteria and fungi, to degrade and break down environmental pollutants such as MPs. These microorganisms can be naturally occurring in polluted areas or can be isolated and introduced to polluted sites (in situ) and/or collected materials (ex situ) (Mohan et al., 2022). As described above, several microbes have been implicated in the association with and subsequent breakdown of polymeric chains in MPs to their monomeric subunits using several microbial enzymes (Othman et al., 2021; Jadaun et al., 2022; Parsaeimehr et al., 2023). Harnessing these biodegradative capabilities of microbes to remediate plastic pollution is a growing area of research. Generally, investigations into microbial bioremediation involve isolating species from landfills, heavily polluted environments, or gut microbiota (Cowan et al., 2022). Species with biodegradative capabilities can be collected and isolated through enrichment of media with polymers as a carbon source, such as PE (Cowan et al., 2022). Biodegradation is generally assessed and confirmed through changes to the chemical and physical properties of the plastic such as oxidation, hydrophilicity, weight loss, molecular weight distribution, and crystallinity using methods such as Fourier-Transform Infrared Spectroscopy (FTIR), high resolution imaging, tensometer measurements, and Raman spectroscopy (Cowan et al., 2022). Similarly, microbial enzymes can also be isolated and used for bioremediation, including recent ones like PETases or PETase-like enzymes (Liu et al. 2023); laccases (Wu et al. 2025); cutinase (Sui et al. 2023), and MHETases (Knott et al. 2020). PETases as a means to hydrolyze PET dates back back to Müller et al. (2005) by use of *Thermobifida fusca*. Since then more than 250 publications have been reported on PETases (Ahituv et al., 2025). Of these publications, most have utilized PETases from *Ideonella Sakaiensis*, which was discovered in 2016 (Ermis, 2025). Its use is due to its stability at elevated temperatures. Despite the ability to enzymatically degrade PET with PETases, there is limited scaling to industry due to difficulties associated with engineering challenges (Ermis et al., 2025). Developing, refining, and standardizing these analytical techniques to measure microbial biodegradation is an important area of future research.

8 CURRENT CHALLENGES

Effectively isolating, enriching, and selecting for microorganisms and enzymes with biodegradative properties could eventually be a promising approach to the bioremediation of MPs (Mohan et al., 2022). However, several challenges persist with developing microbial bioremediation as a cost-effective and highly feasible method. Namely, scaling up microbial bioremediation from the laboratory to real-world scenarios is a prominent challenge that will require innovative solutions and investment. Because plastics are comprised of many different polymer types, identifying strains capable of degrading each polymer may be difficult and time consuming. Additionally, the additives within plastics and adsorbed contaminants may interfere with enzymes and hinder microbial degradation (Mohan et al., 2022). Microbial degradation is also a very slow process that is highly dependent on environmental conditions such as temperature and pH (Agarwal et al., 2024). For these reasons, identifying, isolating, and developing strains capable of quickly degrading various polymers and additives in situ is a significant challenge.

Fully understanding the ecological and health impacts of plastic pollution and microbial bioremediation are also prominent challenges. For example, the various microbial enzymatic processes of MP degradation within the plastisphere, especially in aquatic systems, can promote the leaching of several toxic compounds including phthalates, bisphenols, heavy metals, antioxidants, and PFAS, among others, into the environment (Sooriyakumar et al., 2022), leading to several adverse effects, including immunological, respiratory, and neurological impacts in higher organisms (e.g., Beans, 2023; Parsaeimehr et al., 2023; Yang et al., 2023). Therefore, the potential for microbes and their components to adhere to MPs to possibly change their physical and/or chemical characteristics and thereby exploit the immunological system of infected hosts is a current challenge that may influence our understanding of the pathogen-host interactions impacted by biodegradation (e.g., Yang et al., 2023). Presently, accidental inhalation or ingestion of MPs could potentially lead to various pathogen delivery that can cause systemic infections in humans (Amato-Lourenço et al., 2021; Beans, 2023), due to the concurrent presence of MPs that then hinders the immune system to respond accordingly (Wang et al., 2023). Furthermore, the process of microbial adhesion and bioremediation of persistent MPs have been associated with the spread of several multidrug resistance (MDR) genes in the environment (Pham et al., 2021). Additionally, the persistent presence of various MPs in the environment has the potential of drastically altering the occurrences, stability, and diversity of microbial assemblages in such environments (Fei et al., 2020; Yu et al., 2021). Yu et al. (2021) in their experimental study found that microplastic addition decreased the richness and diversity of bacterial populations, while those for algae on the soil surface increased significantly in wetland systems and also indicated both the direct and indirect effects of plastic residues on the plant-microbe-soil systems. These studies underscore the need to better understand the ecological effects and health impacts of MPs as well as their interactions with microbes in the environment. In addition, a thorough understanding of the ecological impacts of microbial bioremediation and introducing microorganisms with biodegradative capabilities into ecosystems is critical (Agarwal et al., 2024).

9 FUTURE PROSPECTS

Because MP pollution is a significant global environmental problem, it will require combinations of both political and scientific attention as well as targeted solutions to solve. The global rise in plastic, and subsequent MPs, has resulted in significant challenges faced in various environments due to their over-production, overuse, and relatively recalcitrant and durable properties making them unsustainable. Therefore, in order to efficiently regulate and mitigate MP contamination in our environment, a number of sustainable approaches must be effectively implemented, including solutions involving microbial bioremediation. Several advancements have been developed to overcome some of the challenges associated with microbial bioremediation. For example, molecular techniques such as CRISPR-Cas9 and enzyme engineering have helped create genetically modified organisms (GMOs) and improve enzymes capable of degrading plastics (Agarwal et al., 2024; Mohan et al., 2022). However, special considerations need to be taken regarding the use or introduction of GMOs into the environment. Similarly, multi-omics approaches such as genome sequencing, metagenomics, transcriptomics, metabolomics, and proteomics have been used to identify enzymes, genes, and metabolic pathways involved with plastic and additive degradation in a number of different bacterial species (Dangi et al., 2019; Shah et al., 2024). These techniques can help optimize the use of microbes in bioremediation. Ex situ bioremediation, such as composting, is another promising technique that can provide ideal environmental conditions and provide supplementation that optimizes the biodegradation process (Shahnawaz et al., 2019).

In addition to microbial bioremediation, the use of more biodegradable plastics as alternatives to non-biodegradable ones should be promoted and enhanced globally. These are plastic materials including PLA, polybutylene succinate (PBS), poly(butylene adipate-co-terephthalate) (PBAT), and polyhydroxyalkanoate (PHA) usually produced from renewable raw materials and from partial fossil resources (Kim et al., 2021) and as recently as the year 2020, were estimated to be at about 1.23 million tons of global production, which accounts for roughly 60% of the 2.11 million tons of total bioplastic production (e.g., Shin et al., 2021). More importantly, the use of biodegradable plastics may further enhance and support the proliferation of indigenous bacterial populations of polluted sites (Li et al., 2022). Additionally, it is imperative that we emphasize proper waste management and disposal for plastic materials whether synthetic, semi-synthetic, or biodegradable. Reducing waste that enters the environment is crucial towards mitigating the issue of MP pollution.

In addition to actively promoting the use of more biodegradable plastic materials, concerted efforts should also be made to further understand and support various microorganisms in light of their bioremediation capabilities, given that this process of MP removal is more cost-efficient and environmentally friendly as compared to the chemical and physical remediation approaches that are energy consuming and costly (Delangiz et al., 2022).

CONFLICT OF INTEREST STATEMENT

The authors declare no known conflicts of interest.

ACKNOWLEDGEMENTS

The authors thank the editors and reviewers for their contributions to this chapter.

REFERENCES

Agarwal, T., Atray, N., & Sharma, J. G. (2024). A critical examination of advanced approaches in green chemistry: Microbial bioremediation strategies for sustainable mitigation of plastic pollution. *Future Journal of Pharmaceutical Sciences, 10*(1), 78. DOI: 10.1186/s43094-024-00645-x

Ahituv, N., Freund, D., Mireles, R., & Noda-García, L. (2025). The diversity of PET degrading enzymes: A systematic review of sequence, structure, and function. *Protein Science : a Publication of the Protein Society, 34*(10), e70282. DOI: 10.1002/pro.70282 PMID: 40944418

Akdogan, Z. & Guven, B. (2019). Microplastics in the environment: a critical review of current understanding and identification of future research needs. *Environmental Pollution*, 254:113011. DOI: . 2029.113011DOI: 10.1016/j.envpol

Ali, S. S., Elsamahy, T., Koutra, E., Kornaros, M., El-Sheekh, M., Abdelkarim, E. A., Zhu, D., & Sun, J. (2021). Degradation of conventional plastic wastes in the environment: A review on current status of knowledge and future perspectives of disposal. *The Science of the Total Environment, 771*, 144719. DOI: 10.1016/j.scitotenv.2020.144719 PMID: 33548729

Amaral-Zettler, L. A., Zettler, E. R., & Mincer, T. J. (2020). Ecology of the plastisphere. *Nature Reviews. Microbiology, 18*(3), 139–151. DOI: 10.1038/s41579-019-0308-0 PMID: 31937947

Amato-Lourenço, L. F., Carvalho-Oliveira, R., Junior, G. R., Galvao, L. D. S., Ando, R. A., & Mauad, T. (2021). Presence of airborne microplastics in human lung tissue. *Journal of Hazardous Materials, 416*, 126124. DOI: 10.1016/j.jhazmat.2021.126124 PMID: 34492918

Amobonye, A., Bhagwat, P., Singh, S., & Pillai, S. (2021). Plastic biodegradation: Frontline microbes and their enzymes. *The Science of the Total Environment, 759*, 143526. DOI: 10.1016/j.scitotenv.2020.143536 PMID: 33190901

Antic, M. P., Jovancicevic, B. S., Ilic, M., Vrvic, M. M., & Schwarzbauer, J. (2006). Petroleum pollutant degradation by surface water microorganisms. *Environmental Science and Pollution Research International, 13*(5), 320–327. DOI: 10.1065/espr2006.03.296 PMID: 17067026

Artham, T., Sudhakar, M., Venkatesan, R., Nair, C. M., Murty, K. V. G. K., & Doble, M. (2009). Biofouling and stability of synthetic polymers in sea water. *International Biodeterioration & Biodegradation, 63*(7), 884–890. DOI: 10.1016/j.ibiod.2009.03.003

Asmita, K., Shubhamsingh, T., & Tejashree, S. (2015). Isolation of plastic degrading microorganisms from soil samples collected at various locations in Mumbai, India. *International Journal of Environmental Research, 4*, 77–85.

Auta, H. S., Emenike, C. U., & Fauziah, S. H. (2017). Distribution and importance of microplastics in the marine environment: A review of the sources, fate, effects, and potential solutions. *Environment International, 102*, 165–176. DOI: 10.1016/j.envint.2017.02.013 PMID: 28284818

Barhoumi, B., Metian, M., Alonso-Hernández, C. M., Oberhaensli, F., Mourgkogiannis, N., Karapanagioti, H. K., Bersuder, P., & Tolosa, I. (2024). Insight into the effect of natural aging of polystyrene microplastics on the sorption of legacy and emerging per-and polyfluorinated alkyl substances in seawater. *Heliyon, 10*(23), e40490. Advance online publication. DOI: 10.1016/j.heliyon.2024.e40490 PMID: 39654741

Barnes, D. K., Galgani, F., Thompson, R. C., & Barlaz, M. (2009, July 27). Accumulation and fragmentation of plastic debris in global environments. *Philosophical Transactions of the Royal Society of London. Series B, Biological Sciences, 364*(1526), 1985–1998. DOI: 10.1098/rstb.2008.0205 PMID: 19528051

Beans, C. (2023). Are microplastics spreading infectious disease? *Proceedings of the National Academy of Sciences of the United States of America, 120*(31), e2311253120. Advance online publication. DOI: 10.1073/pnas.2311253120 PMID: 37494399

Bhatt, V., & Chauhan, J. S. (2023). Microplastic in freshwater ecosystem: Bioaccumulation, trophic transfer, and biomagnification. *Environmental Science and Pollution Research International, 30*(4), 9389–9400. DOI: 10.1007/s11356-022-24529-w PMID: 36508090

Chamas, A., Moon, H., Zheng, J., Qiu, Y., Tabassum, T., Jang, J. H., Abu-Omar, M., Scott, S. L., & Suh, S. (2020). Degradation rates of plastics in the environment. *ACS Sustainable Chemistry & Engineering, 8*(9), 3494–3511. DOI: 10.1021/acssuschemeng.9b06635

Chen, Y., Awasthi, A. K., Wei, F., Tan, Q., & Li, J. (2021). Single-use plastics: Production, usage, disposal, and adverse impacts. *The Science of the Total Environment, 752*, 141772. DOI: 10.1016/j.scitotenv.2020.141772 PMID: 32892042

Chen, Z., Zhao, W., Xing, R., Xie, S., Yang, X., Cui, P., Lü, J., Liao, H., Yu, Z., Wang, S., & Zhou, S. (2020). Enhanced in situ biodegradation of microplastics in sewage sludge using hyperthermophylic composting technology. *Journal of Hazardous Materials, 384*, 121271. DOI: 10.1016/j.jhazmat.2019.121271 PMID: 31611021

Cheng, J., Jacquin, J., Conan, P., Pujo-Pay, M., Barbe, V., George, M., Fabre, P., Bruzaud, S., Ter Halle, A., Meistertzheim, A.-L., & Ghiglione, J.-F. (2020). Relative influence of plastic debris size and shape, chemical composition and phytoplankton bacteria interactions in driving seawater plastisphere abundance, diversity and activity. *Frontiers in Microbiology, 11*, 610231. DOI: 10.3389/fmicb.2020.610231 PMID: 33519764

Corradini, F., Melillán, E., Schoebitz, M., Carrasco, R., & Jiménez, M. (2024). Assessing on-site plastics fragmentation: Linking macroplastics litter to microplastics pollution in vegetable fields. *The Science of the Total Environment, 955*, 177168. DOI: 10.1016/j.scitotenv.2024.177168 PMID: 39461513

Costigan, E., Collins, A., Hatinoglu, M. D., Bhagat, K., MacRae, J., Perreault, F., & Apul, O. (2022). Adsorption of organic pollutants by microplastics: Overview of a dissonant literature. *Journal of Hazardous Materials Advances, 6*, 100091. DOI: 10.1016/j.hazadv.2022.100091

Cowan, A. R., Costanzo, C. M., Benham, R., Loveridge, E. J., & Moody, S. C. (2022). Fungal bioremediation of polyethylene: Challenges and perspectives. *Journal of Applied Microbiology, 132*(1), 78–89. DOI: 10.1111/jam.15203 PMID: 34218487

Dang, H., Li, T., Chen, M., & Hang, G. (2008). Cross-ocean distribution of Rhodobacteriales bacteria as primary surface colonizers in temperate coastal marine waters. *Applied and Environmental Microbiology, 74*(1), 52–60. DOI: 10.1128/AEM.01400-07 PMID: 17965206

Dangi, A. K., Sharma, B., Hill, R. T., & Shukla, P. (2019). Bioremediation through microbes: Systems biology and metabolic engineering approach. *Critical Reviews in Biotechnology, 39*(1), 79–98. DOI: 10.1080/07388551.2018.1500997 PMID: 30198342

Das, G., Bordoloi, N.K., Rai, S.K., Mukherjee, A.K., & Karak, N. (2012). Biodegradable and biocompatible epoxidized vegetable oil modified thermostable poly (vinyl chloride): Thermal and performance characteristics post biodegradation with Pseudomonas aeruginosa and Achromobacter sp. *Journal of Hazardous Materials,* 209-210: 434-442. DOI: . 2012. 01. 043DOI: 10.1016/j.jhazmat

Delangiz, N., Aliyar, S., Pashapoor, N., Nobaharan, K., Lajayer, B. A., & Rodríguez-Couto, S. (2022). Can polymer-degrading microorganisms solve the bottleneck of plastics' environmental challenges? *Chemosphere, 294*, 13709. DOI: 10.1016/j.chemosphere.2022.133709 PMID: 35074325

Deng, Y., Wu, J., Chen, J., & Kang, K. (2023). Overview of microplastic pollution and its influence on the health of organisms. *Journal of Environmental Science and Health. Part A, Toxic/Hazardous Substances & Environmental Engineering, 58*(4), 412–422. DOI: 10.1080/10934529.2023.2190715 PMID: 36942439

Dhali, S. L., Parida, D., Kumar, B., & Bala, K. (2024). Recent trends in microbial and enzymatic plastic degradation: A solution for plastic pollution predicaments. *Biotechnology for Sustainable Materials, 1*(1), 11. DOI: 10.1186/s44316-024-00011-0

Domenech, J., & Marcos, R. (2021). Pathways of human exposure to microplastics, and estimation of the total burden. *Current Opinion in Food Science, 39*, 144–151. DOI: 10.1016/j.cofs.2021.01.004

Editorial, N. M. (2025). Microplastic research must consider microbes. *Nature Microbiology, 10*(3), 603. Advance online publication. DOI: 10.1038/s41564-025-01960-6 PMID: 40044874

Ekanayaka, A. H., Tibpromma, S., Dai, D., Xu, R., Suwannarach, N., Stephenson, S. L., Dao, C., & Karunarathna S. C. (2022). A review of the fungi that degrade plastic. *Journal of Fungi* 8,8: 772. DOI: DOI: 10.3390/jof8080772

Elahi, A., Bukhari, D. A., Shamim, S., & Rehman, A. (2021). Plastics degradation by microbes: A sustainable approach. *Journal of King Saud University. Science, 33*(6), 101538. DOI: 10.1016/j.jksus.2021.101538

Ermis, H. (2025). A mini-review on the role of PETase in polyethylene terephthalate degradation. *Reviews in Environmental Science and Biotechnology, 24*(3), 545–555. DOI: 10.1007/s11157-025-09737-3

Fackelmann, G., Pham, C. K., Rodriguez, Y., Mallory, M. L., Provencher, J. F., Baak, J. E., & Sommer, S. (2023). Current levels of microplastic pollution impacts wild seabird gut microbiomes. *Nature Ecology & Evolution, 7*(5), 698–706. DOI: 10.1038/s41559-023-02013-z PMID: 36973353

Fei, Y., Huang, S., Zhang, H., Tong, Y., Wen, D., Xia, X., Wang, H., Luo, Y., & Barceló, D. (2020). Response of soil enzyme activities and bacterial communities to the accumulation of microplastics in an acid cropped soil. *The Science of the Total Environment, 707*, 135634. DOI: 10.1016/j.scitotenv.2019.135634 PMID: 31761364

Freilinger, J., Kappacher, C., Huter, K., Hofer, T. S., Back, J. O., Huck, C. W., & Bakry, R. (2025). Interactions between perfluorinated alkyl substances (PFAS) and microplastics (MPs): Findings from an extensive investigation. *Journal of Hazardous Materials Advances, 18*, 100740. DOI: 10.1016/j.hazadv.2025.100740

Frere, L., Maignien, L., Chalopin, M., Huvet, A., Rinnert, E., Morrison, H., Kerninon, S., Cassone, A. L., Lambert, C., Reveillaud, J., & Paul-Pont, I. (2018). Microplastic bacterial communities in the Bay of Brest: Influence of polymer type and size. *Environmental Pollution, 242*, 614–625. DOI: 10.1016/j.envpol.2018.07.023 PMID: 30014939

Gan, Z., & Zhang, H. (2019). PMBD: A comprehensive plastics microbial biodegradation database. *Database : The Journal of Biological Databases and Curation, 2019*, baz119. Advance online publication. DOI: 10.1093/database/baz119 PMID: 31738435

Geyer, R., Jambeck, J. R., & Law, K. L. (2017, July 19). Production, use, and fate of all plastics ever made. *Science Advances, 3*(7), e1700782. DOI: 10.1126/sciadv.1700782 PMID: 28776036

Giacomucci, L., Raddadi, N., Soccio, M., Lotti, N., & Fava, F. (2019). Polyvinylchloride biodegradation by *Pseudomonas citronellolis* and *Bacillus flexus*. *New Biotechnology, 52*, 35–41. DOI: 10.1016/j.nbt.2019.04.005 PMID: 31026607

Gkoutselis, G., Rohrbach, S., Harjes, J., Obst, M., Brachmann, A., Horn, M. A., & Rambold, G. (2021). Microplastics accumulate fungal pathogens in terrestrial ecosystems. *Scientific Reports, 11*(1), 13214. DOI: 10.1038/s41598-021-92405-7 PMID: 34267241

Godoy, V., Martín-Lara, M. A., Calero, M., & Blázquez, G. (2020). The relevance of interaction of chemicals/pollutants and microplastic samples as route for transporting contaminants. *Process Safety and Environmental Protection, 138*, 312–323. DOI: 10.1016/j.psep.2020.03.033

Gray, A., Gore, B., Gaesser, M., Rodriguez Sequeira, L., Thibodeau, T., Montgomery, A., Purvis, S., Ouimet, K., Dura, T., & Mayer, K. (2025). Insights into the in-situ degradation and fragmentation of macroplastics in a low-order riverine system. *Environmental Toxicology and Chemistry, 44*(7), 2017–2028. DOI: 10.1093/etojnl/vgaf116 PMID: 40316505

Gray, A., Mayer, K., Gore, B., Gaesser, M., & Ferguson, N. (2024). Microplastic burden in native (*Cambarus appalachiensis*) and non-native (*Faxonius cristavarius*) crayfish along semi-rural and urban streams in southwest Virginia, USA. *Environmental Research, 258*, 119494. DOI: 10.1016/j.envres.2024.119494 PMID: 38936498

Guzzetti, E., Sureda, A., Tejada, S., & Faggio, C. (2018). Microplastic in marine organisms: Environmental and toxicological effects. *Environmental Toxicology and Pharmacology, 64*, 164–171. DOI: 10.1016/j.etap.2018.10.009 PMID: 30412862

Hidalgo-Ruz, V., Gutow, L., Thompson, R. C., & Thiel, M. (2012). Microplastics in the marine environment: A review of the methods used for identification and quantification. *Environmental Science & Technology, 46*(6), 3060–3075. DOI: 10.1021/es2031505 PMID: 22321064

Hopewell, J., Dvorak, R., & Kosior, E. (2009). Plastics recycling: Challenges and opportunities. *Philosophical Transactions of the Royal Society of London. Series B, Biological Sciences, 364*(1526), 2115–2126. DOI: 10.1098/rstb.2008.0311 PMID: 19528059

Huang, Y., Liu, Q., Jia, W., Yan, C., & Wang, J. (2020). Agricultural plastic mulching as a source of microplastics in the terrestrial environment. *Environmental Pollution, 260*, 114096. DOI: 10.1016/j.envpol.2020.114096 PMID: 32041035

Ibrahim, A. M. (2024). Plastics: Photodegradations and mechanisms. In *Microplastics and Pollutants: Interactions, Degradations and Mechanisms* (pp. 25-49). Cham: Springer Nature Switzerland. DOI: DOI: 10.1007/978-3-031-54565-8_2

Jadaun, J. S., Bansal, S., Sonthalia, A., Rai, K. A., & Singh, S. P. (2022). Biodegradation of plastics for sustainable environment. *Bioresource Technology, 126697*, 126697. Advance online publication. DOI: 10.1016/j.biortech.2022.126697

Junaid, M., Siddiqui, J. A., Sadaf, M., Liu, S., & Wang, J. (2022). Enrichment and dissemination of bacterial pathogens by microplastics in the aquatic environment. *The Science of the Total Environment, 830*, 154720. DOI: 10.1016/j.scitotenv.2022.154720 PMID: 35337880

Kaiser, D., Kowalski, N., & Waniek, J. J. (2017). Effects of biofouling on the sinking behavior of microplastics. *Environmental Research Letters, 12*(12), 124003. DOI: 10.1088/1748-9326/aa8e8b

Karunarathna, B. S., Gajasinghe, S. T., Diwankara, U. T., Hunupolagama, M. D., & Chen, X. (2025). Fungal-mediated biodegradation of synthetic plastic: Mechanisms, enzymes, and strategies for enhanced plastic waste management. *New Zealand Journal of Botany*, •••, 1–27. DOI: 10.1080/0028825X.2025.2522207

Kawai, F., Oda, M., Tamashiro, T., Waku, T., Tanaka, N., Yamamoto, M., Mizushima, H., Miyakawa, T., & Tanokura, M. (2014). A novel Ca^{2+} activated, thermostabilized polyesterase capable of hydrolyzing polyethylene terephthalate from *Saccharomonospora viridis* AHK190. *Applied Microbiology and Biotechnology*, *98*(24), 10053–10064. DOI: 10.1007/s00253-014-5860-y PMID: 24929560

Kettner, M. T., Oberbeckmann, S., Labrenz, M., & Grossart, H. P. (2019). The eukaryotic life on microplastics in brackish ecosystems. *Frontiers in Microbiology*, *10*, e00538. DOI: 10.3389/fmicb.2019.00538 PMID: 30949147

Kettner, M. T., Rojas-Jimenez, K., Oberbeckmann, S., Labrenz, M., & Grossart, H. P. (2017). Microplastics alter composition of fungal communities in aquatic ecosystems. *Environmental Microbiology*, *19*(11), 4447–4459. DOI: 10.1111/1462-2920.13891 PMID: 28805294

Khan, S., Nadir, S., Shah, Z. U., Shah, A. A., Karunarathna, S. C., Xu, J., Khan, A., Munir, S., & Hasan, F. (2017). Biodegradation of polyester polyurethane by *Aspergillus tubingensis*. *Environmental Pollution*, *225*, 469–480. DOI: 10.1016/j.envpol.2017.03.012 PMID: 28318785

Kim, H., Jeon, H., Shin, G., Lee, M., Jegal, J., Hwang, S. Y., Oh, D. X., Koo, J. M., Eom, Y., & Park, J. (2021). Biodegradable nanocomposite of poly(ester-*co*-carbonate) and cellulose nanocrystals for tough tear-resistant disposable bags. *Green Chemistry*, *23*(6), 2293–2299. DOI: 10.1039/D0GC04072J

Kim, K., Song, I. G., Yoon, H., & Park, J. W. (2023). Sub-micron microplastics affect nitrogen cycling by altering microbial abundance and activities in a soil-legume system. *Journal of Hazardous Materials*, *460*, 132504. DOI: 10.1016/j.jhazmat.2023.132504 PMID: 37703725

Kumar, A., Mishra, S., Pandey, R., Yu, Z. G., Kumar, M., Khoo, K. S., Thakur, T. K., & Show, P. L. (2023). Microplastics in terrestrial ecosystems: Un-ignorable impacts on soil characterises, nutrient storage and its cycling. *Trends in Analytical Chemistry*, *158*, 116869. DOI: 10.1016/j.trac.2022.116869

Kumar, R., Sharma, P., Verma, A., Jha, P. K., Singh, P., Gupta, P. K., Chandra, R., & Prasad, P. V. (2021). Effect of physical characteristics and hydrodynamic conditions on transport and deposition of microplastics in riverine ecosystem. *Water (Basel)*, *13*(19), 2710. DOI: 10.3390/w13192710

Kye, H., Kim, J., Ju, S., Lee, J., Lim, C., & Yoon, Y. (2023). Microplastics in water systems: A review of their impacts on the environment and their potential hazards. *Heliyon*, *9*(3), e14359. DOI: 10.1016/j.heliyon.2023.e14359 PMID: 36950574

Lau, W. W. Y., Shiran, Y., Bailey, R. M., Cook, E., Stuchtey, M. R., Keskella, J., Velis, C. A., Godfrey, L., Boucher, J., Murphy, M. B., Thompson, R. C., Jankowska, E., Castillo, A. C., Pilditch, T. D., Dixon, B., Koerselman, L., Kosior, E., Favoino, E., Gutberlet, J., & Palardy, J. E. (2020). Evaluating scenarios toward zero plastic pollution. *Science*, *369*(6510), 1455–1461. DOI: 10.1126/science.aba9475 PMID: 32703909

Lee, J., Hong, S., Song, Y. K., Hong, S. H., Jang, Y. C., Jang, M., Heo, N. W., Han, G. M., Lee, M. J., Kang, D., & Shim, W. J. (2013). Relationships among the abundances of plastic debris in different size classes on beaches in South Korea. *Marine Pollution Bulletin*, *77*(1-2), 349–354. DOI: 10.1016/j.marpolbul.2013.08.013 PMID: 24054782

Li, C., Cui, Q., Li, Y., Zhang, K., Lu, X., & Yong, Z. (2022). Effect of different additions of low-density polyethylene and microplastics polyadipate/butylene terephthalate on soil bacterial community structure. *Journal of Hazardous Materials*. 429:10, 128353. DOI: DOI: 10.1016/j.jhazmat.2022.128353

Li, P., Wang, X., Su, M., Zou, X., Duan, L., & Zhang, H. (2021). Characteristics of plastic pollution in the environment: A review. *Bulletin of Environmental Contamination and Toxicology*, *107*(4), 577–584. DOI: 10.1007/s00128-020-02820-1 PMID: 32166334

Lin, Z., Jin, T., Zou, T., Xu, L., Xi, B., Xu, D., He, J., Xiong, L., Tang, C., Peng, J., Zhou, Y., & Fei, J. (2022). Current progress on plastic/microplastic degradation: Fact influences and mechanism. *Environmental Pollution*, *304*, 119159. DOI: 10.1016/j.envpol.2022.119159 PMID: 35304177

Liu, F., Wang, T., Yang, W., Zhang, Y., Gong, Y., Fan, X., Wang, G., Lu, Z., & Wang, J. (2023). Current advances in the structural biology and molecular engineering of PETase. *Frontiers in Bioengineering and Biotechnology*, *11*, 1263996. DOI: 10.3389/fbioe.2023.1263996 PMID: 37795175

Lokesh, P., Shobika, R., Omer, S., Reddy, M., Saravanan, P., Rajeshkannan, R., Saravanan, V., & Venkatkumar, S. (2023). Bioremediation of plastics by the help of microbial tool: A way for control of plastic pollution. *Sustainable Chemistry for the Environment*, *3*, 100027. DOI: 10.1016/j.scenv.2023.100027

Miller, M. E., Hamann, M., & Kroon, F. J. (2020). Bioaccumulation and biomagnification of microplastics in marine organisms: A review and meta-analysis of current data. *PLoS One*, *15*(10), e0240792. DOI: 10.1371/journal.pone.0240792 PMID: 33064755

Mohamadi, S., Madadi, R., Rakib, M. R. J., De-la-Torre, G. E., & Idris, A. M. (2023). Abundance and characterization of personal protective equipment (PPE) polluting Kish Island, *Persian Gulf. The Science of the Total Environment*, *854*, 158678. DOI: 10.1016/j.scitotenv.2022.158678 PMID: 36099950

Mohan, A. J., Sekhar, V. C., Bhaskar, T., & Nampoothiri, K. M. (2016). Microbial assisted high impact polystyrene (HIPS) degradation. *Bioresource Technology*, *213*, 204–207. DOI: 10.1016/j.biortech.2016.03.021 PMID: 26993201

Mohan, L., Cherian, E., & Joshi, T. J. (2022). Emerging issues and challenges for plastic bioremediation. In *Microbes and Microbial Biotechnology for Green Remediation* (pp. 589–600). Elsevier., DOI: 10.1016/B978-0-323-90452-0.00010-4

Mor, R., & Sivan, A. (2008). Biofilm formation and partial biodegradation of polystyrene by the actinomycete *Rhodococcus ruber*: Biodegradation of polystyrene. *Biodegradation*, *19*(6), 851–858. DOI: 10.1007/s10532-008-9188-0 PMID: 18401686

Müller, R. J., Schrader, H., Profe, J., Dresler, K., & Deckwer, W. D. (2005). Enzymatic degradation of poly (ethylene terephthalate): Rapid hydrolyse using a hydrolase from T. fusca. *Macromolecular Rapid Communications*, *26*(17), 1400–1405. DOI: 10.1002/marc.200500410

Nihart, A. J., Garcia, M. A., Hyek, E. E., Liu, R., Olewine, M., Kingston, J. D., Castillo, E. F., Gullapalli, R. R., Howard, T., Bleske, B., Scott, J., Gonzalez-Estrella, J., Gross, J. M., Spilde, M., Adolphi, N. L., Gallego, D. F., Jarrell, H. S., Dvorscak, G., Zuluaga-Ruiz, M. E., & Campen, M. J. (2025). Bioaccumulation of plastics in decent human brains. *Nature Medicine*, *31*(4), 1114–1119. DOI: 10.1038/s41591-024-03453-1 PMID: 39901044

Oberbeckman, S., Kreikemeyer, B., & Labrenz, M. (2018). Environmental factors support the function of specific bacterial assemblages on microplastics. *Frontiers in Microbiology*, *8*, 2709. DOI: 10.3389/fmicb.2017.02709 PMID: 29403454

Orr, I. G., Hadar, Y., & Sivan, A. (2004). Colonization, biofilm formation and biodegradation of polyethylene by a strain of *Rhodococcus ruber. Applied Microbiology and Biotechnology*, *65*, 97–104. 101007.s00253-004-1584-8 PMID: 15221232

Othman, A. R., Hasan, H. A., Muhamad, M. H., Ismail, N. I., & Abdullah, S. R. S. (2021). Microbial degradation of microplastics by enzymatic processes: A review. *Environmental Chemistry Letters*, *19*(4), 3057–3073. DOI: 10.1007/s10311-021-01197-9

Parashar, N., Mahanty, B., & Hait, S. (2023). Microplastics as carriers of per-and polyfluoroalkyl substances (PFAS) in aquatic environment: Interactions and ecotoxicological effects. *Water Emerging Contaminants & Nanoplastics*, *2*(7), 15. Advance online publication. DOI: 10.20517/wecn.2023.25

Parsaeimehr, A., Miller, C. M., & Ozbay, G. (2023). Microplastics and their interactions with microbiota. *Heliyon*, *9*(4), e15104. DOI: 10.1016/j.heliyon.2023.e15104 PMID: 37089279

Pathak, V., & Navneet, . (2017). Review on the current status of polymer degradation: A microbial approach. *Bioresources and Bioprocessing*, *4*(1), 15. DOI: 10.1186/s40643-017-0145-9

Pérez-García, P., Sass, K., Wongwattanarat, S., Amann, J., Feuerriegel, G., Neumann, T., Base, N., Schmitz, L. S., Dierkes, R. F., Gurschke, M. F., Wypych, A., Bounabi, H., de Divitiis, M., Vollstedt, C., & Streit, W. R. (2025). Microbial plastic degradation: Enzymes, pathways, challenges, and perspectives. *Microbiology and Molecular Biology Reviews : MMBR*, *89*(4), e00087–e24. DOI: 10.1128/mmbr.00087-24 PMID: 40970732

Pham, D. N., Clark, L., & Li, M. (2021). Microplastics as hubs enriching antibiotic-resistant bacteria and pathogens in municipal activated sludge. *Journal of Hazardous Materials Letters*, *2*, 100014. DOI: 10.1016/j.hazl.2021.100014

Pilapitiya, P. G. C. N. T., & Ratnayake, A. S. (2024). The world of plastic waste: A review. *Cleaner Materials*, *11*, 100220. Advance online publication. DOI: 10.1016/j.clema.2024.100220

Pletz, M. (2022). Ingested microplastics: Do humans eat one credit card per week? *Journal of Hazardous Materials Letters*, *3*, 100071. DOI: 10.1016/j.hazl.2022.100071

Rakib, M. R. J., Sarker, A., Ram, K., Uddin, M. G., Walker, T. R., Chowdhury, T., Uddin, J., Khandaker, M. U., Rahman, M. M., & Idris, A. M. (2023). Microplastic toxicity in aquatic organisms and aquatic ecosystems: A review. *Water, Air, and Soil Pollution*, *234*(1), 52. DOI: 10.1007/s11270-023-06062-9

Ray, S., & Cooney, R. P. (2018). Thermal degradation of polymer and polymer composites. In *Handbook of Environmental Degradation of Materials* (pp. 185–206). William Andrew publishing., DOI: 10.1016/B978-0-323-52472-8.00009-5

Rong, L., Zhao, L., Zhao, L., Cheng, Z., Yao, Y., Yuan, C., Wang, L., & Sun, H. (2021). LDPE microplastics affect soil microbial communities and nitrogen cycling. *The Science of the Total Environment*, *773*, 145640. DOI: 10.1016/j.scitotenv.2021.145640 PMID: 33582358

Rosato, A., Barone, M., Negroni, A., Brigidi, P., Fava, F., Biagi, E., Candela, M., & Zanaroli, G. (2022). Bacterial colonization dynamics of different microplastic types in an anoxic salt marsh sediment and impact of adsorbed polychlorinated biphenyls on the plastisphere. *Environmental Pollution*, *315*, 120411. DOI: 10.1016/j.envpol.2022.120411 PMID: 36240963

Rosato, A., Barone, M., Negroni, A., Brigidi, P., Fava, F., Xu, P., Candela, M., & Zanaroli, G. (2020). Microbial colonization of different microplastic types and biotransformation of sorbed PCBs by a marine anaerobic bacterial community. *The Science of the Total Environment*, *705*, 135790. DOI: 10.1016/j.scitotenv.2019.135790 PMID: 31972939

Seeley, M. E., Song, B., Passie, R., & Hale, R. C. (2020). Microplastics affect sedimentary microbial communities and nitrogen cycling. *Nature Communications*, *11*(1), 2372. DOI: 10.1038/s41467-020-16235-3 PMID: 32398678

Shah, A. A., Hasan, F., Hameed, A., & Ahmed, S. (2008). Biological degradation of plastics: A comprehensive review. *Biotechnology Advances*, *26*(3), 246–265. DOI: 10.1016/j.biotechadv.2007.12.005 PMID: 18337047

Shah, B. A., Malhotra, H., Papade, S. E., Dhamale, T., Ingale, O. P., Kasarlawar, S. T., & Phale, P. S. (2024). Microbial degradation of contaminants of emerging concern: Metabolic, genetic and omics insights for enhanced bioremediation. *Frontiers in Bioengineering and Biotechnology*, *12*, 1470522. DOI: 10.3389/fbioe.2024.1470522 PMID: 39364263

Shahnawaz, M., Sangale, M. K., & Ade, A. B. (2019). Ex situ bioremediation technology for plastic degradation. In *Bioremediation Technology for Plastic Waste* (pp. 77–83). Springer Singapore., DOI: 10.1007/978-981-13-7492-0_7

Shen, M., Song, B., Zhou, C., Almatrafi, E., Hu, T., Zeng, G., & Zhang, Y. (2022). Recent advances in impacts of microplastics on nitrogen cycling in the environment: A review. *The Science of the Total Environment*, *815*, 152740. DOI: 10.1016/j.scitotenv.2021.152740 PMID: 34974017

Shin, G., Park, S. A., Koo, J. M., Kim, M., Lee, M., Jegal, J., Hwang, S. Y., Park, J., & Oh, D. X. (2021). A micro-spray-based high-throughput screening system for bioplastic-degrading microorganisms. *Green Chemistry*, *23*(15), 5429–5436. rsc.li/greenchem. DOI: 10.1039/D1GC01916C

Sobhani, Z., Fang, C., Naidu, R., & Megharaj, M. (2021). Microplastics as a vector of toxic chemicals in soil: Enhanced uptake of perfluorooctane sulfonate and perfluorooctanoic acid by earthworms through sorption and reproductive toxicity. *Environmental Technology & Innovation*, *22*, 101476. DOI: 10.1016/j.eti.2021.101476

Sooriyakumar, P., Bolan, N., Kumar, M., Singh, L., Yu, Y., Li, Y., Weralupitiya, C., Vithanage, M., Ramamayaka, S., Sarkar, B., Wang, F., Gleeson, D. B., Zhang, D., Kirkham, M. B., Rinklebe, J., & Siddique, K. H. M. (2022). Biofilm formation and its implications on the properties and fate of microplastics in aquatic environments: A review. *Journal of Hazardous Materials Advances, 100077*, 100077. Advance online publication. DOI: 10.1016/j.hazadv.2022.100077

Sui, B., Wang, T., Fang, J., Hou, Z., Shu, T., Lu, Z., Liu, F., & Zhu, Y. (2023). Recent advances in the biodegradation of polyethylene terephthalate with cutinase-like enzymes. *Frontiers in Microbiology, 14*, 1265139. DOI: 10.3389/fmicb.2023.1265139 PMID: 37849919

Sun, A., & Wang, W. X. (2023). Human exposure to microplastics and its associated health risks. *Environment & Health, 1*(3), 139–149. DOI: 10.1021/envhealth.3c00053 PMID: 39473618

Sun, Y., Wang, X., Xia, S., & Zhao, J. (2022). Cu (II) adsorption on poly (lactic acid) microplastics: Significance of microbial colonization and degradation. *Chemical Engineering Journal, 429*, 132306. DOI: 10.1016/j.cej.2021.132306

Surendran, U., Jayakumar, M., Raja, P., Gopinath, G., & Chellam, P. V. (2023). Microplastics in terrestrial ecosystem: Sources and migration in soil environment. *Chemosphere, 318*, 137946. DOI: 10.1016/j.chemosphere.2023.137946 PMID: 36708782

Syranidou, E., Karkanorachaki, K., Amorotti, F., Avgeropoulos, A., Kolvenbach, B., Zhou, N., Fava, F., Corvini, P., & Kalogerakis, N. (2019). Biodegradation of mixture of plastic films by tailored marine consortia. *Journal of Hazardous Materials, 375*, 33–42. DOI: 10.1016/j.jhazmat.2019.04.078 PMID: 31039462

Tahir, L., Ali, M. I., Zia, M., Atiq, N., Hasan, F., & Ahmed, S. (2013). Production and characterization of esterase in *Lantinus tigrinus* for degradation of polystyrene. *Polish Journal of Microbiology, 62*(1), 101–108. DOI: 10.33073/pjm-2013-015 PMID: 23829086

Thompson, R. R., Moore, C. J., vom Saal, F. S., & Swan, S. H. (2009). Plastics, the environment and human health: Current consensus and future trends. *Philosophical Transactions of the Royal Society of London. Series B, Biological Sciences, 364*(1526), 2153–2166. DOI: 10.1098/rstb.2009.0053 PMID: 19528062

Valavanidis, A. (2024). Plastic pollution, unsustainable environmental problem? Industry and academic research must co-operate to sort out the pervasive global problem of plastic waste. *Scientific Review (Singapore), 1*, 1–42.

Vass, M., Ramasamy, K. P., & Andersson, A. (2024). Microbial hitchhikers on microplastics: The exchange of aquatic microbes across distinct aquatic habitats. *Environmental Microbiology*, *26*(4), e16618. Advance online publication. DOI: 10.1111/1462-2920.16618 PMID: 38561820

Ventura, E., Marín, A., Gámez-Pérez, J., & Cabedo, L. (2024). Recent advances in the relationships between biofilms and microplastics in natural environments. *World Journal of Microbiology & Biotechnology*, *40*(7), 220. DOI: 10.1007/s11274-024-04021-y PMID: 38809290

Wang, C., Wu, W., Pang, Z., Liu, J., Qiu, J., Luan, T., Deng, J, & Fang, Z. (2023). Polystyrene microplastics significantly facilitate influenza A virus infection of host cells. *Journal of Hazardous Materials*, 15(446):130617. DOI: . 130617DOI: 10.1016/j.jharmat.2022

Wang, C., Zhao, L., Lim, M. K., Chen, W.-Q., & Sutherland, J. W. (2020). Structure of the global plastic waste trade network and the impact of China's import ban. Resource Conservatory. *Resources, Conservation and Recycling*, *153*, 104591. DOI: 10.1016/j.resconrec.2019.104591

Wang, F., Xiang, L., Leung, K. S. Y., Elsner, M., Zhang, Y., & Guo, Y.. (2024). Emerging contaminants: A one health perspective. *Innovation (Cambridge (Mass.))*, *5*(4), 100612. Advance online publication. DOI: 10.1016/j.xinn.2024.100612 PMID: 38756954

Wang, M., Ma, L., Zhang, D., Yang, F., Yang, X., Wang, B., Li, X., Ge, C., Luo, J., & Zhang, Y. (2025). Dissolved organic matter derived from long-term photodegradation of plastics alters microbial methane conversion in mangrove sediments. *Bioresource Technology*, *132677*, 132677. Advance online publication. DOI: 10.1016/j.biortech.2025.132677 PMID: 40378974

Weinstein, J. E., Crocker, B. K., & Gray, A. D. (2016). From macroplastic to microplastic: Degradation of high-density polyethylene, polypropylene, and polystyrene in a salt marsh habitat. *Environmental Toxicology and Chemistry*, *35*(7), 1632–1640. DOI: 10.1002/etc.3432 PMID: 26992845

Weinstein, J. E., Viado, H. F.Jr, Leads, R. R., Deloughry, E., Schandera, L., Liddy, K., & Scott, G. I. (2019). Microplastics: A global water pollution problem. *Encyclopedia of Water. Science, Technology & Society*, •••, 1–15. DOI: 10.1002/9781119300762.wsts0203

Wu, C. C., Bao, L. J., Liu, L. Y., Shi, L., Tao, S., & Zeng, E. Y. (2017). Impact of polymer colonization on the fate of organic contaminants in sediment. *Environmental Science & Technology*, *51*(18), 10555–10561. DOI: 10.1021/acs.est.7b03310 PMID: 28825800

Wu, L., Wang, X., Li, A., Cui, H., & Li, X. (2025). Discovery of a polyethylene biodegradable laccase from *Acinetobacter dijkshoorniae*. *AIChE Journal*. 71(10):e70005. DOI:. 70005WU ET AL . 14 of 14DOI: 10.1002/aic

Yakovenko, N., Pérez-Serrano, L., Segur, T., Hagelskjaer, O., Margenat, H., Le Roux, G., & Sonke, J. E. (2025). Human exposure to PM10 microplastics in indoor air. *PLoS One*, *20*(7), e0328011. DOI: 10.1371/journal.pone.0328011 PMID: 40737229

Yang, W., Li, Y., & Boraschi, D. (2023). Association between microorganisms and microplastics: How does it change the host-pathogen interactions and subsequent immune system? *International Journal of Molecular Sciences*, *24*(4), 4065. DOI: 10.3390/ijms24044065 PMID: 36835476

Yoon, M. G., Jeon, H. J., & Kim, N. M. (2012). Biodegradation of polyethylene by a soil bacterium and *AlkB* cloned recombinant cell. *Journal of Bioremediation & Biodegradation*, *3*(4), 145. DOI: 10.4172/2155-6199.1000145

Yu, H., Qi, W., Cao, X., Hu, J., Li, Y., Peng, J., Hu, C., & Qu, J. (2021). Microplastic residues in wetland ecosystems: Do they truly threaten the plant-microbe-soil system? *Environment International*, *156*, 106708. DOI: 10.1016/j.envint.2021.106708 PMID: 34153891

Zettler, E. R., Mincer, T. J., & Amaral-Zettler, L. A. (2013). Life in the "plastisphere": Microbial communities on plastic marine debris. *Environmental Science & Technology*, *47*(13), 7137–7146. DOI: 10.1021/es401288x PMID: 23745679

Zhai, X., Zhang, X.-H., & You, M. (2023). Microbial colonization and degradation of microplastics in the plastisphere: A review. *Frontiers in Microbiology*, *14*, 1227308. DOI: 10.3389/fmicb.2023.1127308 PMID: 36876073

Zhang, K., Hamidian, A. H., Tubić, A., Zhang, Y., Fang, J. K., Wu, C., & Lam, P. K. (2021a). Understanding plastic degradation and microplastic formation in the environment: A review. *Environmental Pollution*, *274*, 116554. DOI: 10.1016/j.envpol.2021.116554 PMID: 33529891

Zhang, K., Shi, H., Peng, J., Xiong, X., Wu, C., & Lam, P. K. S. (2018). Microplastic pollution in China's inland water systems: A review of findings, methods, characteristics, effects and management. *The Science of the Total Environment*, *630*, 1641–1653. DOI: 10.1016/j.scitotenv.2018.02.300 PMID: 29554780

Zhang, X., Li, Y., Ouyang, D., Lei, J., Tan, Q., Xie, L., Li, Z., Liu, T., Xiao, Y., Farooq, T. H., Wu, X., Chen, L., & Yan, W. (2021b). Systematic review of interactions between microplastics and microorganisms in the soil environment. *Journal of Hazardous Materials*, *418*, 126288. DOI: 10.1016/j.jhazmat.2021.126288 PMID: 34102358

Compilation of References

Abdelhafeez, I. A., El-Tohamy, S. A., Abdul-Malik, M. A., Abdel-Raheem, S. A., & El-Dars, F. M. S. (2022). A review on green remediation techniques for hydrocarbons and heavy metals contaminated soil. *Curr. Chem. Lett*, *11*(1), 43–62. DOI: 10.5267/j.ccl.2021.9.006

Abdel-Shafy, H. I., & Mansour, M. S. (2016). A review on polycyclic aromatic hydrocarbons: source, environmental impact, effect on human health and remediation. *Egyptian journal of petroleum, 25*(1), 107-123.

Abdi, G., Jain, M., Barwant, M., Tendulkar, R., Tendulkar, M., Tariq, M., & Amir, A. (2024). Unveiling the dynamic role of bioinformatics in automation for efficient and accurate data processing and interpretation. In *Advances in bioinformatics* (pp. 279–319). Springer. DOI: 10.1007/978-981-99-8401-5_15

Acosta, M., Galleguillos, P. A., Guajardo, M., & Demergasso, C. (2017). *Microbial survey on industrial bioleaching heap by high-throughput 16S sequencing and metagenomics analysis*. Solid. DOI: 10.4028/www.scientific.net/SSP.262.219

Adams, G. O., Fufeyin, P. T., Okoro, S. E., & Ehinomen, I. (2015). Bioremediation, biostimulation and bioaugmention: A review. *International Journal of Environmental Bioremediation & Biodegradation*, *3*(1), 28–39. DOI: 10.12691/ijebb-3-1-5

Adebayo, A. A., Enagbonma, B. J., & Babalola, O. O. (2025). Comparative metagenomics on community structure and diversity of rhizomicrobiome associated with monoculture and soybean precedent carrot. *Scientific Reports*, *15*(1), 28161. DOI: 10.1038/s41598-025-13605-z PMID: 40750823

Adebusoye, S. A. (2017). Biological degradation of 4-chlorobenzoic acid by a PCB-metabolizing bacterium through a pathway not involving chlorocatechol. *Biodegradation*, *28*(10), 37–51. DOI: 10.1007/s10532-016-9776-3 PMID: 27766437

Adebusoye, S. A., Ilori, M. O., Picardal, F. W., & Amund, O. O. (2008). Characterization of multiple novel aerobic Polychlorinated Biphenyl (PCB)-utilizing bacterial strains ndigenous to Tropical African Soils. *Biodegradation, 19*(1), 145–159. DOI: 10.1007/s10532-007-9122-x PMID: 17534725

Agarwal, T., Atray, N., & Sharma, J. G. (2024). A critical examination of advanced approaches in green chemistry: Microbial bioremediation strategies for sustainable mitigation of plastic pollution. *Future Journal of Pharmaceutical Sciences, 10*(1), 78. DOI: 10.1186/s43094-024-00645-x

Aguilera, A., de Diego-Castilla, G., Osuna, S., Bardera, R., Sor Mendi, S., & Blanco, Y.. (2018). Microbial ecology in the atmosphere: the last extreme environment. In Najjari, A., Cherif, A., Sghaier, H., & Ouzari, H. I. (Eds.), *Extremophilic Microbes and Metabolites-Diversity, Bioprospecting and Biotechnological Applications*. IntechOpen., DOI: 10.5772/intechopen.81650

Agustinho, D. P., Fu, Y., Menon, V. K., Metcalf, G. A., Treangen, T. J., & Sedlazeck, F. J. (2024). Unveiling microbial diversity: Harnessing long-read sequencing technology. *Nature Methods, 21*(6), 954–966. DOI: 10.1038/s41592-024-02262-1 PMID: 38689099

Ahituv, N., Freund, D., Mireles, R., & Noda-García, L. (2025). The diversity of PET degrading enzymes: A systematic review of sequence, structure, and function. *Protein Science : a Publication of the Protein Society, 34*(10), e70282. DOI: 10.1002/pro.70282 PMID: 40944418

Ahmed, F. (2024). Genomics and bioinformatics: Integrating data for better genetic insights. *Frontiers in Biotechnology and Genetics, 1*(2), 126–146. DOI: 10.71465/fbg74

Ahmed, T., Scholz, M., Al-Faraj, F., & Niaz, W. (2016). Water-related impacts of climate change on agriculture and subsequently on public health: A review for generalists with particular reference to Pakistan. *International Journal of Environmental Research and Public Health, 13*(11), 1051. DOI: 10.3390/ijerph13111051 PMID: 27801802

Ahsan, Z., Kalsoom, U., Bhatti, H. N., Aftab, K., Khalid, N., & Bilal, M. (2021). Enzyme-assisted bioremediation approach for synthetic dyes and polycyclic aromatic hydrocarbons degradation. *Journal of Basic Microbiology, 61*(11), 960–981. DOI: 10.1002/jobm.202100218 PMID: 34608659

Aimar, F. (2023). Limiting soil sealing and depaving: Local actions for regenerating public spaces to build green infrastructures. In: Giudice, B., Novarina, G., Voghera, A. (eds), *Green infrastructure*. The urban book series. Springer, Cham. DOI: 10.1007/978-3-031-28772-5_11

Akanmu, A. M., Lawal, I. B., Ibrahim, S. L., van Marle-Köster, E., & Hassen, A. (2025b). Metagenomic data from the rumen of South African Mutton Merino sheep supplemented with crude or encapsulated Acacia tannin extracts. *BMC Genomic Data*, *26*(1), 86. DOI: 10.1186/s12863-025-01378-3 PMID: 41239201

Akanmu, A. O., Asemoloye, M. D., Marchisio, M. A., & Babalola, O. O. (2024). Adoption of CRISPR-Cas for crop production: Present status and future prospects. *PeerJ*, *12*, e17402. DOI: 10.7717/peerj.17402 PMID: 38860212

Akdogan, Z. & Guven, B. (2019). Microplastics in the environment: a critical review of current understanding and identification of future research needs. *Environmental Pollution*, 254:113011. DOI: . 2029.113011DOI: 10.1016/j.envpol

Akhtar, N., & Mannan, M. A. U. (2020). Mycoremediation: Expunging environmental pollutants. *Biotechnology Reports (Amsterdam, Netherlands)*, *26*, e00452. DOI: 10.1016/j.btre.2020.e00452 PMID: 32617263

Al Raish, S. M., Sourani, O. M., & Abu-Elsaoud, A. M. (2025). Plant growth-promoting microorganisms as biocontrol agents: Mechanisms, challenges, and future prospects. *Applied Microbiology (Basel, Switzerland)*, *5*(2), 44. DOI: 10.3390/applmicrobiol5020044

Alabbosh, K. F. (2025). Leveraging the potential of environmental microorganisms: An extensive examination of their capacity in tackling global and local environmental and ecological challenges. *Polish Journal of Environmental Studies*, *34*(1), 1–11. DOI: 10.15244/pjoes/185701

Alford, K. L., & Kumar, N. (2021). Pulmonary Health Effects of Indoor Volatile Organic Compounds—A Meta-Analysis. *International Journal of Environmental Research and Public Health*, *18*(4), 1578. DOI: 10.3390/ijerph18041578 PMID: 33562372

Ali, M., Song, X., Ding, D., Wang, Q., Zhang, Z., & Tang, Z. (2022). Bioremediation of PAHs and heavy metals co-contaminated soils: Challenges and enhancement strategies. *Environmental Pollution*, *295*, 118686. DOI: 10.1016/j.envpol.2021.118686 PMID: 34920044

Ali, Q., Khan, A. R., Raza, W., Bilal, M. S., Khalid, S., Ayaz, M., Khan, A., & Mundra, S. (2025). Mechanisms of Microbial VOC-Mediated communication in plant ecosystems and agricultural applications. *Journal of Sustainable Agriculture and Environment*, *4*(1), e70044. Advance online publication. DOI: 10.1002/sae2.70044

Ali, S. S., Elsamahy, T., Koutra, E., Kornaros, M., El-Sheekh, M., Abdelkarim, E. A., Zhu, D., & Sun, J. (2021). Degradation of conventional plastic wastes in the environment: A review on current status of knowledge and future perspectives of disposal. *The Science of the Total Environment*, *771*, 144719. DOI: 10.1016/j.scitotenv.2020.144719 PMID: 33548729

Allison, S. D., & Martiny, J. B. H. (2008). Resistance, resilience, and redundancy in microbial communities. *Proceedings of the National Academy of Sciences of the United States of America*, *105*(Supplement 1), 11512–11519. DOI: 10.1073/pnas.0801925105 PMID: 18695234

Alotaibi, F., Hijri, M., & St-Arnaud, M. (2021). Overview of approaches to improve rhizoremediation of petroleum hydrocarbon-contaminated soils. *Applied Microbiology (Basel, Switzerland)*, *1*(2), 329–351. DOI: 10.3390/applmicrobiol1020023

Alpaidze, L., & Pace, R. (2021). Ecosystem services provided by urban forests in the southern Caucasus region: A modeling study in Tbilisi, Georgia. *Climate (Basel)*, *9*(11), 157. DOI: 10.3390/cli9110157

Al-Shaarani, A. A. Q. A., & Pecoraro, L. (2024). A review of pathogenic airborne fungi and bacteria: Unveiling occurrence, sources, and profound human health implication. *Frontiers in Microbiology*, *15*, 1428415. DOI: 10.3389/fmicb.2024.1428415 PMID: 39364169

Álvarez-Castillo, E., Felix, M., Bengoechea, C., & Guerrero, A. (2021). Proteins from agri-food industrial biowastes or co-products and their applications as green materials. *Foods*, *10*(5), 981. DOI: 10.3390/foods10050981 PMID: 33947093

Alvarez-Cohen, L., & Speitel, G. E. (2001). Cometabolic biodegradation of chlorinated solvents. *Environmental Health Perspectives*, *109*(Suppl 2), 225–238.

Amaral-Zettler, L. A., Zettler, E. R., & Mincer, T. J. (2020). Ecology of the plastisphere. *Nature Reviews. Microbiology*, *18*(3), 139–151. DOI: 10.1038/s41579-019-0308-0 PMID: 31937947

Amato-Lourenço, L. F., Carvalho-Oliveira, R., Junior, G. R., Galvao, L. D. S., Ando, R. A., & Mauad, T. (2021). Presence of airborne microplastics in human lung tissue. *Journal of Hazardous Materials*, *416*, 126124. DOI: 10.1016/j.jhazmat.2021.126124 PMID: 34492918

Aminov, R. I. (2011). Horizontal gene exchange in environmental microbiota. *Frontiers in Microbiology*, *2*, 158. DOI: 10.3389/fmicb.2011.00158 PMID: 21845185

Amoako, T. N., & Dzogbefia, V. P. (2011). Effect of fermentation on the quality and detoxification of cassava mash. *International Food Research Journal*, *18*(2), 695–699.

Amobonye, A., Bhagwat, P., Singh, S., & Pillai, S. (2021). Plastic biodegradation: Frontline microbes and their enzymes. *The Science of the Total Environment*, *759*, 143526. DOI: 10.1016/j.scitotenv.2020.143536 PMID: 33190901

Amund, O. O. (1996). Utilization and degradation of an ester-based synthetic lubricant by *Acinetobacter lwoffi*. *Biodegradation*, *7*(2), 91–95. DOI: 10.1007/BF00114621 PMID: 8882802

Amund, O. O., & Adebiyi, A. G. (1991). Effect of viscosity on the biodegradability of automotive lubricating oils. *Tribology International*, *24*(4), 235–237. DOI: 10.1016/0301-679X(91)90049-F

Andorfer, M. C., King-Roberts, D. T., Imrich, C. N., Brotheridge, B. G., & Drennan, C. L. (2023). Development of an *in vitro* method for activation of X-succinate synthases for fumarate hydroalkylation. *iScience*, *26*(6), 106902. DOI: 10.1016/j.isci.2023.106902 PMID: 37283811

Angon, P. B., Islam, M. S., Das, A., Anjum, N., Poudel, A., & Suchi, S. A. (2024). Sources, effects and present perspectives of heavy metals contamination: Soil, plants and human food chain. *Heliyon*, *10*(7), e28357. DOI: 10.1016/j.heliyon.2024.e28357 PMID: 38590838

Ansari, T. U., Valsan, A. E., Ojha, N., Ravikrishna, R., Narasimhan, B., & Gunthe, S. S. (2015). Model simulations of fungal spore distribution over the Indian region. *Atmospheric Environment*, *122*, 552–560. DOI: 10.1016/j.atmosenv.2015.10.020

Antic, M. P., Jovancicevic, B. S., Ilic, M., Vrvic, M. M., & Schwarzbauer, J. (2006). Petroleum pollutant degradation by surface water microorganisms. *Environmental Science and Pollution Research International*, *13*(5), 320–327. DOI: 10.1065/espr2006.03.296 PMID: 17067026

Anyansi, C., Straub, T. J., Manson, A. L., Earl, A. M., & Abeel, T. (2020). Computational methods for strain-level microbial detection in colony and metagenome sequencing data. *Frontiers in Microbiology*, *11*, 1925. DOI: 10.3389/fmicb.2020.01925 PMID: 33013732

Aparicio, J. D., Raimondo, E. E., Saez, J. M., Costa-Gutierrez, S. B., Alvarez, A., Benimeli, C. S., & Polti, M. A. (2022). The current approach to soil remediation: A review of physicochemical and biological technologies, and the potential of their strategic combination. *Journal of Environmental Chemical Engineering, 10*(2), 107141. DOI: 10.1016/j.jece.2022.107141

Aplakidou, E., Vergoulidis, N., Chasapi, M., Venetsianou, N. K., Kokoli, M., Panagiotopoulou, E., Iliopoulos, I., Karatzas, E., Pafilis, E., Georgakopoulos-Soares, I., Kyrpides, N. C., Pavlopoulos, G. A., & Baltoumas, F. A. (2024). Visualizing metagenomic and metatranscriptomic data: A comprehensive review. *Computational and Structural Biotechnology Journal, 23*, 2011–2033. DOI: 10.1016/j.csbj.2024.04.060 PMID: 38765606

Archer, S. D. J., & Pointing, S. B. (2020). Publisher Correction: Anthropogenic impact on the atmospheric microbiome. *Nature Microbiology, 5*(5), 776. DOI: 10.1038/s41564-020-0680-6 PMID: 32099094

Arıkan, M., & Atabay, B. (2024). Construction of protein sequence databases for metaproteomics: A review of the current tools and databases. *Journal of Proteome Research, 23*(12), 5250–5262. DOI: 10.1021/acs.jproteome.4c00665 PMID: 39449618

Arora, J., Kumari, A., Ranjan, A., Rajput, V. D., Shende, S., Prazdnova, E. V. E., & Jindal, T. (2024). Microbial adaptation in different extreme environmental conditions and its usefulness in differently polluted soil. In *Extremophiles for Sustainable Agriculture and Soil Health Improvement* (pp. 47–62). Springer Nature Switzerland. DOI: 10.1007/978-3-031-70203-7_3

Artham, T., Sudhakar, M., Venkatesan, R., Nair, C. M., Murty, K. V. G. K., & Doble, M. (2009). Biofouling and stability of synthetic polymers in sea water. *International Biodeterioration & Biodegradation, 63*(7), 884–890. DOI: 10.1016/j.ibiod.2009.03.003

Arthur-Holmes, F., Busia, K. A., Yakovleva, N., & Vazquez-Brust, D. A. (2022). Artisanal and small-scale mining methods and the Sustainable Development Goal 6: Perceived implications for clean water supply. *Environmental Science & Policy, 137*, 205–215. DOI: 10.1016/j.envsci.2022.08.017

Asemoloye, M. D., Ahmad, R., & Jonathan, S. G. (2017). Synergistic action of rhizospheric fungi with Megathyrsus maximus root speeds up hydrocarbon degradation kinetics in oil polluted soil. *Chemosphere, 187*, 1–10. DOI: 10.1016/j.chemosphere.2017.07.158 PMID: 28787637

Asina, F., Brzonova, I., Kozliak, E., Kubátová, A., & Ji, Y. (2017). Microbial treatment of industrial lignin: Successes, problems and challenges. *Renewable & Sustainable Energy Reviews*, *77*, 1179–1205. DOI: 10.1016/j.rser.2017.03.098

Asmita, K., Shubhamsingh, T., & Tejashree, S. (2015). Isolation of plastic degrading microorganisms from soil samples collected at various locations in Mumbai, India. *International Journal of Environmental Research*, *4*, 77–85.

Auta, H. S., Emenike, C. U., & Fauziah, S. H. (2017). Distribution and importance of microplastics in the marine environment: A review of the sources, fate, effects, and potential solutions. *Environment International*, *102*, 165–176. DOI: 10.1016/j.envint.2017.02.013 PMID: 28284818

Aza, P., & Camarero, S. (2023). Fungal laccases: Fundamentals, engineering and classification update. *Biomolecules*, *13*(12), 1716. DOI: 10.3390/biom13121716 PMID: 38136587

Azhar, U., Ahmad, H., Shafqat, H., Babar, M., Munir, H. M. S., Sagir, M., & Khoo, K. S. (2022). Remediation techniques for elimination of heavy metal pollutants from soil: A review. *Environmental Research*, *214*, 113918. DOI: 10.1016/j.envres.2022.113918 PMID: 35926577

Babalola, O. O., Adebayo, A. A., & Enagbonma, B. J. (2025). Shotgun metagenomics dataset of the core rhizo-microbiome of monoculture and soybean-precedent carrot. *BMC Genomic Data*, *26*(1), 26. DOI: 10.1186/s12863-025-01320-7 PMID: 40221653

Babalola, O. O., Akanmu, A. O., & Ayangbenro, A. S. (2024). Draft genome sequence of Bacillus velezensis strains AOA1 and AKS2, the potential plant growth-promoting rhizobacteria. *Microbiology Resource Announcements*, *13*(4), e00877–e23. DOI: 10.1128/mra.00877-23 PMID: 38411072

Babalola, O. O., & Enagbonma, B. J. (2025). Dataset of shotgun metagenomic evaluation of Sorghum bicolor rhizosphere microbiome in soils preceded by Glycine max. *Data in Brief*, *58*, 111270. DOI: 10.1016/j.dib.2025.111270 PMID: 39906131

Babalola, O. O., Fadiji, A. E., Enagbonma, B. J., Alori, E. T., Ayilara, M. S., & Ayangbenro, A. S. (2020). The nexus between plant and plant microbiome: Revelation of the networking strategies. *Frontiers in Microbiology*, *11*, 548037. DOI: 10.3389/fmicb.2020.548037 PMID: 33013781

Babalola, O. O., Molefe, R. R., & Amoo, A. E. (2021a). Metagenome assembly and metagenome-assembled genome sequences from the rhizosphere of maize plants in Mafikeng, South Africa. *Microbiology Resource Announcements*, *10*(2), e01300–e01320. DOI: 10.1128/MRA.00954-20 PMID: 33632860

Babalola, O. O., Molefe, R. R., & Amoo, A. E. (2021b). Revealing the active microbiome connected with the rhizosphere soil of maize plants in Ventersdorp, South Africa. *Biodiversity Data Journal*, *9*, e60245. DOI: 10.3897/BDJ.9.e60245 PMID: 33688297

Bala, S., Garg, D., Thirumalesh, B. V., Sharma, M., Sridhar, K., Inbaraj, B. S., & Tripathi, M. (2022). Recent Strategies for Bioremediation of Emerging Pollutants: A review for a Green and Sustainable environment. *Toxics*, *10*(8), 484. DOI: 10.3390/toxics10080484 PMID: 36006163

Bálint, M., Bahram, M., Eren, A. M., Faust, K., Fuhrman, J. A., Lindahl, B., O'Hara, R. B., Öpik, M., Sogin, M. L., Unterseher, M., & Tedersoo, L. (2016). Millions of reads, thousands of taxa: Microbial community structure and associations analyzed via marker genes. *FEMS Microbiology Reviews*, *40*(5), 686–700. DOI: 10.1093/femsre/fuw017 PMID: 27358393

Bardgett, R. D., & van der Putten, W. H. (2014). Belowground biodiversity and ecosystem functioning. *Nature*, *515*(7528), 505–511. DOI: 10.1038/nature13855 PMID: 25428498

Barhoumi, B., Metian, M., Alonso-Hernández, C. M., Oberhaensli, F., Mourgkogiannis, N., Karapanagioti, H. K., Bersuder, P., & Tolosa, I. (2024). Insight into the effect of natural aging of polystyrene microplastics on the sorption of legacy and emerging per-and polyfluorinated alkyl substances in seawater. *Heliyon*, *10*(23), e40490. Advance online publication. DOI: 10.1016/j.heliyon.2024.e40490 PMID: 39654741

Barla, R. J., Gupta, S., & Raghuvanshi, S. (2024). Sustainable synergistic approach to chemolithotrophs—Supported bioremediation of wastewater and flue gas. *Scientific Reports*, *14*(1), 16529. Advance online publication. DOI: 10.1038/s41598-024-67053-2 PMID: 39019921

Barnard, R. L., Osborne, C. A., & Firestone, M. K. (2013). Responses of soil bacterial and fungal communities to extreme desiccation and rewetting. *The ISME Journal*, *7*(11), 2229–2241. DOI: 10.1038/ismej.2013.104 PMID: 23823489

Barnes, D. K., Galgani, F., Thompson, R. C., & Barlaz, M. (2009, July 27). Accumulation and fragmentation of plastic debris in global environments. *Philosophical Transactions of the Royal Society of London. Series B, Biological Sciences*, *364*(1526), 1985–1998. DOI: 10.1098/rstb.2008.0205 PMID: 19528051

Bashiardes, S., Zilberman-Schapira, G., & Elinav, E. (2016a). Use of metatranscriptomics in microbiome research. Bioinformatics and Biology Insights, 10, BBI.S34610.

Bashiardes, S., Zilberman-Schapira, G., & Elinav, E. (2016b). Use of metatranscriptomics in microbiome research. Bioinformatics and Biology Insights, 10, BBI.S34610.

Bashir, I., Lone, F. A., Bhat, R. A., Mir, S. A., Dar, Z. A., & Dar, S. A. (2020). Concerns and threats of contamination on aquatic ecosystems. In *Bioremediation and biotechnology: Sustainable approaches to pollution degradation* (pp. 1–26). Springer International Publishing. DOI: 10.1007/978-3-030-35691-0_1

Bauer, H., Giebl, H., Hitzenberger, R., Kasper-Giebl, A., Reischl, G., Zibuschka, F., & Puxbaum, H. (2003). Airborne bacteria as cloud condensation nuclei. *Journal of Geophysical Research*, *108*(D21), 2003JD003545. Advance online publication. DOI: 10.1029/2003JD003545

Bautista-Cruz, A., Aquino-Bolaños, T., Hernández-Canseco, J., & Quiñones-Aguilar, E. E. (2024). Cellulolytic aerobic bacteria isolated from agricultural and forest soils: An overview. *Biology (Basel)*, *13*(2), 102. DOI: 10.3390/biology13020102 PMID: 38392320

Beans, C. (2023). Are microplastics spreading infectious disease? *Proceedings of the National Academy of Sciences of the United States of America*, *120*(31), e2311253120. Advance online publication. DOI: 10.1073/pnas.2311253120 PMID: 37494399

Beattie, G., Edlund, A., Esiobu, N., Gilbert, J., Nicolaisen, M., Jansson, J., Jensen, P., Keiluweit, M., Lennon, J. T., Martiny, J., Minnis, V., Newman, D., Peixoto, R., Schadt, C., & van-der Meer, J. (2025). Soil microbiome interventions for carbon sequestration and climate mitigation. *mSystems*, *10*(1), e01129–e24. DOI: 10.1128/msystems.01129-24 PMID: 39692482

Behzad, H., Gojobori, T., & Mineta, K. (2015). Challenges and opportunities of airborne metagenomics. *Genome Biology and Evolution*, *7*(5), 1216–1226. DOI: 10.1093/gbe/evv064 PMID: 25953766

Benites-Alfaro, E., Lizarzaburu Aguinaga, D., Cabrera Carranza, C., Jave Nakayo, J., Suca-Apaza, G. R., Atuncar Yrribari, J. F., Lopez Bulnes, J., & Munoz-Ccuro, F. (2025). Biological method with *Pseudomonas Putida* for Chromium VI reduction in chromium plating process wastewater. *Chemical Engineering Transactions*, *117*, 427–432. DOI: 10.3303/CET25117072

Berendonk, T. U., Manaia, C. M., Merlin, C., Fatta-Kassinos, D., Cytryn, E., Walsh, F., Bürgmann, H., Sørum, H., Norström, M., Pons, M. N., Kreuzinger, N., Huovinen, P., Stefani, S., Schwartz, T., Kisand, V., Baquero, F., & Martinez, J. L. (2015). Tackling antibiotic resistance: The environmental framework. *Nature Reviews. Microbiology*, *13*(5), 310–317. DOI: 10.1038/nrmicro3439 PMID: 25817583

Berkinbayeva, A., Saulebekkyzy, S., Kenzhaliyev, B., Sailed, K., Yessengaziyev, A., Nurtazina, N., Karim, D., & Birlikzhan, Y. (2025). Sodium percarbonate for eco-efficient cyanide detoxification in gold mining tailings. *Metals*, *15*(10), 1162. DOI: 10.3390/met15101162

Berry, O., Briand, E., Bagot, A., Chaigne, M., Meslet-Cladière, L., Wang, J., Grovel, O., Jansen, J. J., Ruiz, N., & du Pont, T. R. (2021). Deciphering microbiome impacts on fungal-microalgal interaction in a marine environment using metabolomics. *bioRxiv*. DOI: 10.1101/2021.05.27.445989

Bhadrecha, P., Singh, S., & Dwibedi, V. (2023). 'A plant's major strength in rhizosphere': The plant growth promoting rhizobacteria. *Archives of Microbiology*, *205*(5), 165. DOI: 10.1007/s00203-023-03502-2 PMID: 37012531

Bhatt, V., & Chauhan, J. S. (2023). Microplastic in freshwater ecosystem: Bioaccumulation, trophic transfer, and biomagnification. *Environmental Science and Pollution Research International*, *30*(4), 9389–9400. DOI: 10.1007/s11356-022-24529-w PMID: 36508090

Bher, A., Mayekar, P. C., Auras, R. A., & Schvezov, C. E. (2022). Biodegradation of biodegradable polymers in mesophilic aerobic environments. *International Journal of Molecular Sciences*, *23*(20), 12165. DOI: 10.3390/ijms232012165 PMID: 36293023

Bialik, O. M., Jarochowska, E., & Grossowicz, M. (2021). Ordination analysis in sedimentology, geochemistry and palaeoenvironment—Background, current trends and recommendations. *The Depositional Record: a Journal of Biological, Physical and Geochemical Sedimentary Processes*, *7*(3), 541–563. DOI: 10.1002/dep2.161

Bochenek, A. D., Klemm, K. and Woźna, M. (2022). Assessment of effectiveness of selected adaptation actions to climate change: The example of the new centre of Lodz. *Budownictwo i Architektura 21*(4), 025-042. .DOI: 10.35784/bud-arch.3150

Bokade, P., Gaur, V. K., Tripathi, V., Bobate, S., Manickam, N., & Bajaj, A. (2023). Bacterial remediation of pesticide polluted soils: Exploring the feasibility of site restoration. *Journal of Hazardous Materials*, *441*, 129906. DOI: 10.1016/j.jhazmat.2022.129906 PMID: 36088882

Bonfá, M. R., Grossman, M. J., Mellado, E., & Durrant, L. R. (2011). Biodegradation of aromatic hydrocarbons by Haloarchaea and their use for the reduction of the chemical oxygen demand of hypersaline petroleum produced water. *Chemosphere*, *84*(11), 1671–1676. DOI: 10.1016/j.chemosphere.2011.05.005 PMID: 21621813

Bonnet, M., Lagier, J. C., Raoult, D., & Khelaifia, S. (2020). Bacterial culture through selective and non-selective conditions: The evolution of culture media in clinical microbiology. *New Microbes and New Infections*, *34*, 100622. DOI: 10.1016/j.nmni.2019.100622 PMID: 31956419

Bonoli, A., Conte, A., Maglionico, M., & Stojkov, I. (2013). Green roofs for sustainable water management in urban areas. *Environmental Engineering and Management Journal*, *12*(S11), 153–156.

Boro, M., Sannyasi, S., Chettri, D., & Verma, A. K. (2022). Microorganisms in biological control strategies to manage microbial plant pathogens: A review. *Archives of Microbiology*, *204*(11), 666. DOI: 10.1007/s00203-022-03279-w PMID: 36214917

Borozan, A. B., Misca, C. D., Morar, A., Obistioiu, D., Raba, D. N., Pirvulescu, L., Caba, I. L., Alexa, E., Poiana, M. A., Bordean, D. M., & Petcu, C. D. (2021). Soil pollution with heavy metals and bioremediation methods. *AgroLife Scientific Journal*, *10*(1), 54–66. DOI: 10.17930/AGL202115

Bosch, A., Gkogka, E., Le Guyader, F. S., Loisy-Hamon, F., Lee, A., Van Lieshout, L., Marthi, B., Myrmel, M., Sansom, A., Schultz, A. C., Winkler, A., Zuber, S., & Phister, T. (2018). Foodborne viruses: Detection, risk assessment, and control options in food processing. *International Journal of Food Microbiology*, *285*, 110–128. DOI: 10.1016/j.ijfoodmicro.2018.06.001 PMID: 30075465

Bouzouidja, R., Leconte, F., Kiss, M., Pierret, M., Pruvot, C., Détriché, S., Berthelot, J. C., & Rousseau, D. (2021). Experimental comparative study between conventional and green parking lots: Analysis of subsurface thermal behavior under warm and dry summer conditions. *Atmosphere (Basel)*, *12*(8), 994. DOI: 10.3390/atmos12080994

Bowers, R. M., Clements, N., Emerson, J. B., Wiedinmyer, C., Hannigan, M. P., & Fierer, N. (2013). Seasonal variability in bacterial and fungal diversity of the near-surface atmosphere. *Environmental Science & Technology*, *47*(21), 12097–12106. DOI: 10.1021/es402970s PMID: 24083487

Bowers, R. M., Kyrpides, N. C., Stepanauskas, R., Harmon-Smith, M., Doud, D., Reddy, T., & Schulz, F. (2017). Minimum information about a metagenome-assembled genome (MIMAG) of bacteria and archaea. *Nature Biotechnology*, *35*(8), 725–731. DOI: 10.1038/nbt.3893 PMID: 28787424

Bowers, R. M., Lauber, C. L., Wiedinmyer, C., Hamady, M., Hallar, A. G., Fall, R., Knight, R., & Fierer, N. (2009). Characterization of airborne microbial communities at a High-Elevation site and their potential to act as atmospheric ice nuclei. *Applied and Environmental Microbiology*, *75*(15), 5121–5130. DOI: 10.1128/AEM.00447-09 PMID: 19502432

Bowman, J. S., & Deming, J. W. (2014). Alkane hydroxylase genes in psychrophile genomes and the potential for cold active catalysis. *BMC Genomics*, *15*(1), 1120. DOI: 10.1186/1471-2164-15-1120 PMID: 25515036

Boyd, D. R., Sharma, N. D., McMurray, B., Haughey, S. A., Allen, C. C. R., Hamilton, J. T. G., McRoberts, W. C., More O'Ferrall, R. A., Nikodinovic-Runic, J., Coulombel, L. A., & O'Connor, K. E. (2011). Bacterial dioxygenase- and monooxygenase-catalysed sulfoxidation of benzo[b]thiophenes. *Organic & Biomolecular Chemistry*, *10*(4), 782–790. DOI: 10.1039/C1OB06678A PMID: 22134441

Boyd, D. R., Sharma, N. D., Stevenson, P. J., Hoering, P., Allen, C. C., & Dansette, P. M. (2022). Monooxygenase-and dioxygenase-catalyzed oxidative dearomatization of thiophenes by sulfoxidation, cis-dihydroxylation and epoxidation. *International Journal of Molecular Sciences*, *23*(2), 909. DOI: 10.3390/ijms23020909 PMID: 35055091

Bradford, M. A., Davies, C. A., Frey, S. D., Maddox, T. R., Melillo, J. M., Mohan, J. E., Reynolds, J. F., Treseder, K. K., & Wallenstein, M. D. (2008). Thermal adaptation of soil microbial respiration to elevated temperature. *Ecology Letters*, *11*(12), 1316–1327. DOI: 10.1111/j.1461-0248.2008.01251.x PMID: 19046360

Bretzel, F., Vannucchi, F., Pezzarossa, B., Paraskevopoulou, A., & Romano, D. (2024). Establishing wildflower meadows in anthropogenic landscapes. *Frontiers in Horticulture*, *2*, 1248785. DOI: 10.3389/fhort.2023.1248785

Briffa, J., Sinagra, E., & Blundell, R. (2020). Heavy metal pollution in the environment and their toxicological effects on humans. *Heliyon*, *6*(9), e04691. DOI: 10.1016/j.heliyon.2020.e04691 PMID: 32964150

Bruins, M. R., Kapil, S., & Oehme, F. W. (2000). Microbial resistance to metals in the environment. *Ecotoxicology and Environmental Safety*, *45*(3), 198–207. DOI: 10.1006/eesa.1999.1860 PMID: 10702338

Buraimoh, O. M., Amund, O. O., & Ilori, M. O. (2015a). Kraft lignin degradation by autochthonous *Streptomyces* strains isolated from a tropical lagoon ecosystem. *Journal of Microbiology, Biotechnology and Food Sciences*, *5*(3), 248253. DOI: 10.15414/jmbfs.2015/16.5.3.248-253

Buraimoh, O. M., Ilori, M. O., Amund, O. O., Michel, F. C.Jr, & Grewal, S. K. (2015b). Assessment of bacterial degradation of lignocellulosic residues (sawdust) in a tropical estuarine microcosm using improvised floating raft equipment. *International Biodeterioration & Biodegradation*, *104*, 186–193. DOI: 10.1016/j.ibiod.2015.06.010

Burmølle, M., Johnsen, K., Al-Soud, W. A., Hansen, L. H., & Sørensen, S. J. (2009). The presence of embedded bacterial pure cultures in agar plates stimulate the culturability of soil bacteria. *Journal of Microbiological Methods*, *79*(2), 166–173. DOI: 10.1016/j.mimet.2009.08.006 PMID: 19699242

Byrne, A. M., Kukor, J. J., & Olsen, R. H. (1995). Sequence analysis of the gene cluster encoding toluene-3-monooxygenase from Pseudomonas pickettii PKO1. *Gene*, *154*(1), 65–70. DOI: 10.1016/0378-1119(94)00844-I PMID: 7867951

Cai, Z., Wang, J., Ma, J., Zhu, X., Cai, J., & Yang, G. (2015). Anaerobic degradation pathway of the novel chiral insecticide paichongding and its impact on bacterial communities in soils. *Journal of Agricultural and Food Chemistry*, *63*(32), 7151–7160. DOI: 10.1021/acs.jafc.5b02645 PMID: 26216485

Calfapietra, C. (2020). Nature-based solutions for microclimate regulation and air quality. In: *Nature-based solutions: State of the art in EU-funded projects*. DOI: 10.2777/383904

Callaghan, A. V. (2013). Enzymes involved in the anaerobic biodegradation of hydrocarbons. *Frontiers in Microbiology*, *4*, 89. DOI: 10.3389/fmicb.2013.00089 PMID: 23717304

Cambiaghi, A., Ferrario, M., & Masseroli, M. (2017). Analysis of metabolomic data: Tools, current strategies and future challenges for omics data integration. *Briefings in Bioinformatics*, *18*(3), 498–510. PMID: 27075479

Canizares-Gonzalez, R., Benitez, E., & Ogunseitan, O. A. (2011). Molecular analyses of beta-glucosidase diversity and function in soil. *European Journal of Soil Biology*, *47*(1), 1–8. DOI: 10.1016/j.ejsobi.2010.11.002

Cao, L., Men, Q., Zhang, Z., Yue, H., Cui, S., Huang, X., Zhang, Y., Wang, J., Chen, M., & Li, H. (2024). Significance of Volatile Organic Compounds to Secondary Pollution Formation and Health Risks Observed during a Summer Campaign in an Industrial Urban Area. *Toxics*, *12*(1), 34. DOI: 10.3390/toxics12010034 PMID: 38250990

Cardinale, B. J., Duffy, J. E., Gonzalez, A., Hooper, D. U., Perrings, C., Venail, P., Narwani, A., Mace, G. M., Tilman, D., Wardle, D. A., Kinzig, A. P., Daily, G. C., Loreau, M., Grace, J. B., Larigauderie, A., Srivastava, D. S., & Naeem, S. (2012). Biodiversity loss and its impact on humanity. *Nature*, *486*(7401), 59–67. DOI: 10.1038/nature11148 PMID: 22678280

Carini, P., Marsden, P. J., Leff, J. W., Morgan, E. E., Strickland, M. S., & Fierer, N. (2020). Effects of spatial variability and relic DNA removal on the detection of temporal dynamics in soil microbial communities. *mBio*, *11*(3), e02776–e19. DOI: 10.1128/mBio.02776-19 PMID: 31964728

Cases, I., & de Lorenzo, V. (2005). Genetically modified organisms for the environment: Stories of success and failure and what we have learned from them. *International Microbiology*, *8*(3), 213–222. https://digital.csic.es/bitstream/10261/1986/1/Cases.pdf PMID: 16200500

Castro-Severyn, J., Remonsellez, F., Valenzuela, S. L., Salinas, C., Fortt, J., Aguilar, P., Pardo-Esté, C., Dorador, C., Quatrini, R., Molina, F., Aguayo, D., Castro-Nallar, E., & Saavedra, C. P. (2017). Comparative genomics analysis of a new Exiguobacterium strain from Salar de Huasco reveals a repertoire of stress-related genes and arsenic resistance. *Frontiers in Microbiology*, *8*, 456. DOI: 10.3389/fmicb.2017.00456 PMID: 28377753

Čaušević, S., Dubey, M., Morales, M., Salazar, G., Sentchilo, V., Carraro, N., Ruscheweyh, H. J., Sunagawa, S., & van der Meer, J. R. (2024). Niche availability and competitive loss by facilitation control proliferation of bacterial strains intended for soil microbiome interventions. *Nature Communications*, *15*(1), 2557. DOI: 10.1038/s41467-024-46933-1 PMID: 38519488

Causin, V. (2015). Polymers: An overview. In Polymers on the crime scene: Forensic analysis of polymeric trace evidence (pp. 9–52).

Cavicchioli, R., Ripple, W. J., Timmis, K. N., Azam, F., Bakken, L. R., Baylis, M., Behrenfeld, M. J., Boetius, A., Boyd, P. W., Classen, A. T., Crowther, T. W., Danovaro, R., Foreman, C. M., Huisman, J., Hutchins, D. A., Jansson, J. K., Karl, D. M., Koskella, B., Mark Welch, D. B., & Webster, N. S. (2019). Scientists' warning to humanity: Microorganisms and climate change. *Nature Reviews. Microbiology*, *17*(9), 569–586. DOI: 10.1038/s41579-019-0222-5 PMID: 31213707

CBD (Convention on Biological Diversity) (2010). Aichi Biodiversity Targets of the Strategic Plan 2011–2020.

Chamas, A., Moon, H., Zheng, J., Qiu, Y., Tabassum, T., Jang, J. H., Abu-Omar, M., Scott, S. L., & Suh, S. (2020). Degradation rates of plastics in the environment. *ACS Sustainable Chemistry & Engineering*, *8*(9), 3494–3511. DOI: 10.1021/acssuschemeng.9b06635

Chang, J., Peng, P., DiSpirito, A. A., & Semrau, J. D. (2022). Variable inhibition of nitrous oxide reduction in denitrifying bacteria by different forms of methanobactin. *Applied and Environmental Microbiology*, *88*(7), e02346-21. Advance online publication. DOI: 10.1128/aem.02346-21 PMID: 35285718

Chaudhary, D. K., & Kim, J. (2019). New insights into bioremediation strategies for oil-contaminated soil in cold environments. *International Biodeterioration & Biodegradation*, *142*, 58–72. DOI: 10.1016/j.ibiod.2019.05.001

Chaudhary, D. K., Park, J., Kim, P., Ok, Y. S., & Hong, Y. (2023). Enrichment cultivation of VOC-degrading bacteria using diffusion bioreactor and development of bacterial-immobilized biochar for VOC bioremediation. *Environmental Pollution*, *320*, 121089. DOI: 10.1016/j.envpol.2023.121089 PMID: 36669717

Chaudhary, P., Padhi, S. K., Pattanaik, L., & Dikshit, P. K. (2025). Recent advances in treating volatile organic compounds from waste gas streams: The potential of anoxic/anaerobic bioreactors, parameters affecting their performance and future perspectives. *Biomass Conversion and Biorefinery*, *15*(11), 16383–16408. DOI: 10.1007/s13399-025-06635-0

Chávez-Vergara, B., Solleiro-Rebolledo, E., López-Martínez, R., Beltrán-Paz, O., Ceniceros-Gómez, Á. E., & Yañez-Mendoza, G. (2025). The release of arsenic is a hidden risk during the in-situ decomposition of landed sargassum litter. *Aquatic Botany*, *199*, 103884. DOI: 10.1016/j.aquabot.2025.103884

Chawla, H., Singh, S. K., & Haritash, A. K. (2024). Reversing the damage: Ecological restoration of polluted water bodies affected by pollutants due to anthropogenic activities. *Environmental Science and Pollution Research International*, *31*(1), 127–143. DOI: 10.1007/s11356-023-31295-w PMID: 38044406

Chen, B. S., & Wu, C. C. (2013). Systems biology as an integrated platform for bioinformatics, systems synthetic biology, and systems metabolic engineering. *Cells*, *2*(4), 635–688. DOI: 10.3390/cells2040635 PMID: 24709875

Cheng, J., Jacquin, J., Conan, P., Pujo-Pay, M., Barbe, V., George, M., Fabre, P., Bruzaud, S., Ter Halle, A., Meistertzheim, A.-L., & Ghiglione, J.-F. (2020). Relative influence of plastic debris size and shape, chemical composition and phytoplankton bacteria interactions in driving seawater plastisphere abundance, diversity and activity. *Frontiers in Microbiology*, *11*, 610231. DOI: 10.3389/fmicb.2020.610231 PMID: 33519764

Cheng, M., Chen, D., Parales, R. E., & Jiang, J. (2022). Oxygenases as powerful weapons in the microbial degradation of pesticides. *Annual Review of Microbiology*, *76*(1), 325–348. DOI: 10.1146/annurev-micro-041320-091758 PMID: 35650666

Cheng, Y., He, H., Yang, C., Zeng, G., Li, X., Chen, H., & Yu, G. (2016). Challenges and solutions for biofiltration of hydrophobic volatile organic compounds. *Biotechnology Advances*, *34*(6), 1091–1102. DOI: 10.1016/j.biotechadv.2016.06.007 PMID: 27374790

Chen, J., Lin, L., Tu, Q., Peng, Q., Wang, X., Liang, C., Zhou, J., & Yu, X. (2024). Metagenomic-based discovery and comparison of the lignin degrading potential of microbiomes in aquatic and terrestrial ecosystems via the LCdb database. *Molecular Ecology Resources*, *24*(5), e13950. DOI: 10.1111/1755-0998.13950 PMID: 38567644

Chen, L., Zhou, M., Wang, J., Zhang, Z., Duan, C., Wang, X., Zhao, S., Bai, X., Li, Z., Li, Z., & Fang, L. (2022). A global meta-analysis of heavy metal (loid) s pollution in soils near copper mines: Evaluation of pollution level and probabilistic health risks. *The Science of the Total Environment*, *835*, 155441. DOI: 10.1016/j.scitotenv.2022.155441 PMID: 35469881

Chen, M., Li, Q., Liu, C., Meng, E., & Zhang, B. (2025). Microbial degradation of lignocellulose for sustainable biomass utilization and future research perspectives. *Sustainability (Basel)*, *17*(9), 4223. DOI: 10.3390/su17094223

Chen, S. (2025). Fastp 1.0: An ultra-fast all-round tool for FASTQ data quality control and preprocessing. *iMeta*, *4*(1), e70078. DOI: 10.1002/imt2.70078 PMID: 41112039

Chen, W., Zeng, F., Liu, W., Bu, J., Hu, G., Xie, S., Yao, H., Zhou, H., Qi, S., & Huang, H. (2021). Organochlorine pesticides in karst soil: Levels, distribution, and source diagnosis. *International Journal of Environmental Research and Public Health*, *18*(21), 11589. DOI: 10.3390/ijerph182111589 PMID: 34770109

Chen, X., Kumari, D., & Achal, V. (2020). A review on Airborne Microbes: The Characteristics of Sources, Pathogenicity and Geography. *Atmosphere (Basel)*, *11*(9), 919. DOI: 10.3390/atmos11090919

Chen, X., Zhao, Y., Zhang, C., Zhang, D., Yao, C., Meng, Q., Zhao, R., & Wei, Z. (2020). Speciation, toxicity mechanism and remediation ways of heavy metals during composting: A novel theoretical microbial remediation method is proposed. *Journal of Environmental Management*, *272*, 111109. DOI: 10.1016/j.jenvman.2020.111109 PMID: 32854897

Chen, Y., Awasthi, A. K., Wei, F., Tan, Q., & Li, J. (2021). Single-use plastics: Production, usage, disposal, and adverse impacts. *The Science of the Total Environment*, *752*, 141772. DOI: 10.1016/j.scitotenv.2020.141772 PMID: 32892042

Chen, Y., Li, E.-M., & Xu, L.-Y. (2022). Guide to metabolomics analysis: A bioinformatics workflow. *Metabolites, 12*(4), 357. DOI: 10.3390/metabo12040357 PMID: 35448542

Chen, Y., Wang, Z., Fang, Y., Wang, G., Zhou, F., Yu, J., Chi, R., & Xiao, C. (2025). Occurrence and microbial remediation of polycyclic aromatic hydrocarbons and heavy metals pollution in soils. *World Journal of Microbiology & Biotechnology, 41*(8), 280. DOI: 10.1007/s11274-025-04498-1 PMID: 40719956

Chen, Z., Zhao, W., Xing, R., Xie, S., Yang, X., Cui, P., Lü, J., Liao, H., Yu, Z., Wang, S., & Zhou, S. (2020). Enhanced in situ biodegradation of microplastics in sewage sludge using hyperthermophylic composting technology. *Journal of Hazardous Materials, 384*, 121271. DOI: 10.1016/j.jhazmat.2019.121271 PMID: 31611021

Chern, E. C., Tsai, D. W., & Ogunseitan, O. A. (2007). Deposition of glomalin-related soil proteins and sequestered toxic metals into watersheds. *Environmental Science & Technology, 41*(10), 3566–3572. DOI: 10.1021/es0628598 PMID: 17547179

Chia, X. K., Hadibarata, T., Jusoh, M. N. H., Sutiknowati, L. I., Tan, I. S., & Foo, H. C. Y. (2024). Extremophilic Microbes for Environmental Bioremediation: A Review. *Biointerface Research in Applied Chemistry, 14*(5), 103. DOI: 10.33263/BRIAC145.103

Chia, X. K., Hadibarata, T., Kristanti, R. A., Jusoh, M. N. H., Tan, I. S., & Foo, H. C. Y. (2024). The function of microbial enzymes in breaking down soil contaminated with pesticides: A review. *Bioprocess and Biosystems Engineering, 47*(5), 597–620. DOI: 10.1007/s00449-024-02978-6 PMID: 38456898

Chikere, C. B., Okpokwasili, G. C., & Chikere, B. O. (2011). Monitoring of microbial hydrocarbon remediation in the soil. 3. *Biotechnology, 1*(3), 117–138. DOI: 10.1007/s13205-011-0014-8 PMID: 22611524

Chikere, C., & Ekwuabu, C. (2014). Culture-dependent characterization of hydrocarbon-utilizing bacteria in selected crude oil impacted sites in Bodo, Ogoniland, Nigeria. *African Journal of Environmental Science and Technology, 8*(6), 401–406. DOI: 10.5897/AJEST2014.1707

Chivian, D., Jungbluth, S. P., Dehal, P. S., Wood-Charlson, E. M., Canon, R. S., Allen, B. H., Clark, M. M., Gu, T., Land, M. L., Price, G. A., Riehl, W. J., Sneddon, M. W., Sutormin, R., Zhang, Q., Cottingham, R. W., Henry, C. S., & Arkin, A. P. (2023). Metagenome-assembled genome extraction and analysis from microbiomes using KBase. *Nature Protocols, 18*(1), 208–238. DOI: 10.1038/s41596-022-00747-x PMID: 36376589

Chklovski, A., Parks, D. H., Woodcroft, B. J., & Tyson, G. W. (2023). CheckM2: A rapid, scalable and accurate tool for assessing microbial genome quality using machine learning. *Nature Methods*, *20*(8), 1203–1212. DOI: 10.1038/s41592-023-01940-w PMID: 37500759

Christen, P., Vega, A., Casalot, L., Simon, G., & Auria, R. (2012). Kinetics of aerobic phenol biodegradation by the acidophilic and hyperthermophilic archaeon Sulfolobus solfataricus 98/2. *Biochemical Engineering Journal*, *62*, 56–61. DOI: 10.1016/j.bej.2011.12.012

Chukwuma, O. B., Rafatullah, M., Tajarudin, H. A., & Ismail, N. (2021). Review on bacterial contribution to lignocellulose breakdown into useful bio-products. *International Journal of Environmental Research and Public Health*, *18*(11), 6001. DOI: 10.3390/ijerph18116001 PMID: 34204975

Colman, B. P., Arnaout, C. L., Anciaux, S., Gunsch, C. K., Hochella, M. F.Jr, Kim, B., Lowry, G. V., McGill, B. M., Reinsch, B. C., Richardson, C. J., Unrine, J. M., Wright, J. P., Yin, L., & Bernhardt, E. S. (2013). Low concentrations of silver nanoparticles in biosolids cause adverse ecosystem responses under realistic field scenario. *PLoS One*, *8*(2), e57189. DOI: 10.1371/journal.pone.0057189 PMID: 23468930

Cornu, J. Y., Huguenot, D., Jézéquel, K., Lollier, M., & Lebeau, T. (2017). Bioremediation of copper-contaminated soils by bacteria. *World Journal of Microbiology & Biotechnology*, *33*(2), 26. DOI: 10.1007/s11274-016-2191-4 PMID: 28044274

Corradini, F., Melillán, E., Schoebitz, M., Carrasco, R., & Jiménez, M. (2024). Assessing on-site plastics fragmentation: Linking macroplastics litter to microplastics pollution in vegetable fields. *The Science of the Total Environment*, *955*, 177168. DOI: 10.1016/j.scitotenv.2024.177168 PMID: 39461513

Costa, A. S., Romao, L. P. C., Aranjo, B. R., Lucas, S. C. O., Maciel, S. T. A., Wisniewski, A.Jr, & Alexandre, M. R. (2012). Environmental strategies to remove volatile fractions (BTEX) from petroleum industry wastewater using biomass. *Bioresource Technology*, *105*, 31–39. DOI: 10.1016/j.biortech.2011.11.096 PMID: 22178487

Costanza, R., d'Arge, R., de Groot, R., Farber, S., Grasso, M., Hannon, B., Limburg, K., Naeem, S., O'Neill, R. V., Paruelo, J., Raskin, R. G., Sutton, P., & van den Belt, M. (1997). The value of the world's ecosystem services and natural capital. *Nature*, *387*(6630), 253–260. DOI: 10.1038/387253a0

Costigan, E., Collins, A., Hatinoglu, M. D., Bhagat, K., MacRae, J., Perreault, F., & Apul, O. (2022). Adsorption of organic pollutants by microplastics: Overview of a dissonant literature. *Journal of Hazardous Materials Advances*, *6*, 100091. DOI: 10.1016/j.hazadv.2022.100091

Cowan, A. R., Costanzo, C. M., Benham, R., Loveridge, E. J., & Moody, S. C. (2022). Fungal bioremediation of polyethylene: Challenges and perspectives. *Journal of Applied Microbiology*, *132*(1), 78–89. DOI: 10.1111/jam.15203 PMID: 34218487

Cowan, D. A., Ramond, J. B., Makhalanyane, T. P., & De Maayer, P. (2015). Metagenomics of extreme environments. *Current Opinion in Microbiology*, *25*, 97–102. DOI: 10.1016/j.mib.2015.05.005 PMID: 26048196

Coyte, K. Z., Schluter, J., & Foster, K. R. (2015). The ecology of the microbiome: Networks, competition, and stability. *Science*, *350*(6261), 663–666. DOI: 10.1126/science.aad2602 PMID: 26542567

Creer, S., Deiner, K., Frey, S., Porazinska, D., Taberlet, P., Thomas, W. K., Potter, C., & Bik, H. M. (2016). The ecologist's field guide to sequence-based identification of biodiversity. *Methods in Ecology and Evolution*, *7*(9), 1008–1018. DOI: 10.1111/2041-210X.12574

Cristiano, E., Urru, S., Farris, S., Ruggiu, D., Deidda, R., & Viola, F. (2020). Analysis of potential benefits on flood mitigation of a CAM green roof in Mediterranean urban areas. *Building and Environment*, *183*, 107179. DOI: 10.1016/j.buildenv.2020.107179

Cundell, A. M. (2018). Microbial ecology of the human skin. *Microbial Ecology*, *76*(1), 113–120. DOI: 10.1007/s00248-016-0789-6 PMID: 27245597

Dakal, T. C., Sharma, A., & Singh, R. (2025). Artificial intelligence and deep learning in microbial genomics: Applications and challenges. *Computational and Structural Biotechnology Journal*, *23*, 1059–1078. DOI: 10.1016/j.csbj.2025.01.014

Dang, H., Li, T., Chen, M., & Hang, G. (2008). Cross-ocean distribution of Rhodobacteriales bacteria as primary surface colonizers in temperate coastal marine waters. *Applied and Environmental Microbiology*, *74*(1), 52–60. DOI: 10.1128/AEM.01400-07 PMID: 17965206

Dangi, A. K., Sharma, B., Hill, R. T., & Shukla, P. (2019). Bioremediation through microbes: Systems biology and metabolic engineering approach. *Critical Reviews in Biotechnology*, *39*(1), 79–98. DOI: 10.1080/07388551.2018.1500997 PMID: 30198342

Danso, D., Schmeisser, C., Chow, J., Zimmermann, W., Wei, R., Leggewie, C., Li, X., Hazen, T., & Streit, W. R. (2018). New insights into the function and global distribution of polyethylene terephthalate (PET)-degrading bacteria and enzymes in marine and terrestrial metagenomes. *Applied and Environmental Microbiology*, *84*(8), e02773–e17. DOI: 10.1128/AEM.02773-17 PMID: 29427431

Das, G., Bordoloi, N.K., Rai, S.K., Mukherjee, A.K., & Karak, N. (2012). Biodegradable and biocompatible epoxidized vegetable oil modified thermostable poly (vinyl chloride): Thermal and performance characteristics post biodegradation with Pseudomonas aeruginosa and Achromobacter sp. *Journal of Hazardous Materials*, 209-210: 434-442. DOI: . 2012. 01. 043DOI: 10.1016/j.jhazmat

Dash, R. R., Gaur, A., & Balomajumder, C. (2009). Cyanide in industrial wastewaters and its removal: A review on biotreatment. *Journal of Hazardous Materials*, *163*(1), 1–11. DOI: 10.1016/j.jhazmat.2008.06.051 PMID: 18657360

Das, N., & Chandran, P. (2011). Microbial degradation of petroleum hydrocarbon contaminants: An overview. *Biotechnology Research International*, *2011*(1), 941810. DOI: 10.4061/2011/941810 PMID: 21350672

Das, P. P., Singh, K. R., Nagpure, G., Mansoori, A., Singh, R. P., Ghazi, I. A., Kumar, A., & Singh, J. (2022). Plant-soil-microbes: A tripartite interaction for nutrient acquisition and better plant growth for sustainable agricultural practices. *Environmental Research*, *214*, 113821. DOI: 10.1016/j.envres.2022.113821 PMID: 35810815

David, E., & Niculescu, V. (2021). Volatile Organic compounds (VOCs) as environmental pollutants: Occurrence and mitigation using nanomaterials. *International Journal of Environmental Research and Public Health*, *18*(24), 13147. DOI: 10.3390/ijerph182413147 PMID: 34948756

Davies, J., & Davies, D. (2010). Origins and evolution of antibiotic resistance. *Microbiology and Molecular Biology Reviews : MMBR*, *74*(3), 417–433. DOI: 10.1128/MMBR.00016-10 PMID: 20805405

De Filippo, C., Ramazzotti, M., Fontana, P., & Cavalieri, D. (2012). Bioinformatic approaches for functional annotation and pathway inference in metagenomics data. *Briefings in Bioinformatics*, *13*(6), 696–710. DOI: 10.1093/bib/bbs070 PMID: 23175748

de Oliveira, A. R., de Toledo Rós, B., Jardim, R., Kotowski, N., de Barros, A., Pereira, R. H. G., Almeida, N. F., & Dávila, A. M. R. (2024). A comparative genomics study of the microbiome and freshwater resistome in Southern Pantanal. *Frontiers in Genetics*, *15*, 1352801. DOI: 10.3389/fgene.2024.1352801 PMID: 38699231

De Souza, M. L., Seffernick, J., Martinez, B., Sadowsky, M. J., & Wackett, L. P. (1998). The atrazine catabolism genes atzABC are widespread and highly conserved. *Journal of Bacteriology, 180*(7), 1951–1954. DOI: 10.1128/JB.180.7.1951-1954.1998 PMID: 9537398

de Vries, F. T., Liiri, M. E., Bjørnlund, L., Bowker, M. A., Christensen, S., Setälä, H. M., & Bardgett, R. D. (2012). Soil bacterial networks are less stable under drought than fungal networks. *Nature Communications*, *3*, 2409. DOI: 10.1038/s41467-018-05516-7 PMID: 30072764

Debeljak, P., & Baltar, F. (2023). Fungal diversity and community composition across ecosystems. *Journal of Fungi (Basel, Switzerland)*, *9*(5), 510. DOI: 10.3390/jof9050510 PMID: 37233221

Decena, B. C.Jr, & dela Cruz, T. E. E. (2024). Detection of changes in soil microbial community physiological profiles in relation to forest types and presence of antibiotics using BIOLOG EcoPlate. *Indian Journal of Microbiology*, *64*(3), 773–779. DOI: 10.1007/s12088-024-01294-7 PMID: 39011008

Delangiz, N., Aliyar, S., Pashapoor, N., Nobaharan, K., Lajayer, B. A., & Rodríguez-Couto, S. (2022). Can polymer-degrading microorganisms solve the bottleneck of plastics' environmental challenges? *Chemosphere*, *294*, 13709. DOI: 10.1016/j.chemosphere.2022.133709 PMID: 35074325

Delgado-Baquerizo, M., Oliverio, A. M., Brewer, T. E., Benavent-González, A., Eldridge, D. J., Bardgett, R. D., Maestre, F. T., Singh, B. K., & Fierer, N. (2018). A global atlas of the dominant bacteria found in soil. *Science*, *359*(6373), 320–325. DOI: 10.1126/science.aap9516 PMID: 29348236

Demirbaga, Ü., Aujla, G. S., Jindal, A., & Kalyon, O. (2024). Big data analytics in bioinformatics. In *Big data analytics: Theory, techniques, platforms, and applications* (pp. 265–284). Springer. DOI: 10.1007/978-3-031-55639-5_12

Deng, Y., Wu, J., Chen, J., & Kang, K. (2023). Overview of microplastic pollution and its influence on the health of organisms. *Journal of Environmental Science and Health. Part A, Toxic/Hazardous Substances & Environmental Engineering*, *58*(4), 412–422. DOI: 10.1080/10934529.2023.2190715 PMID: 36942439

Dennett, G. V., & Blamey, J. M. (2016). A new thermophilic nitrilase from an antarctic hyperthermophilic microorganism. *Frontiers in Bioengineering and Biotechnology*, *4*, 5. DOI: 10.3389/fbioe.2016.00005 PMID: 26973832

Derakhshan Nejad, Z., Jung, M. C., & Kim, K. H. (2018). Remediation of soils contaminated with heavy metals with an emphasis on immobilization technology. *Environmental Geochemistry and Health, 40*(3), 927–953. DOI: 10.1007/s10653-017-9964-z PMID: 28447234

Desa, S., Saidin, S., Azeyanty, J. A., Alobaidi, K. H., Al-Taie, B. S., Allawi, M. Y., Ahmad-Kamil, E., Adibah, A., & Al-Obaidi, J. R. (2023). Plant-environment interactions: Proteomics, metabolomics and genetic engineering perspective. In *New frontiers in plant-environment interactions: Innovative technologies and developments* (pp. 15–51). Springer. DOI: 10.1007/978-3-031-43729-8_2

Desidery, L., & Lanotte, M. (2022). Polymers and plastics: Types, properties, and manufacturing. In *Plastic waste for sustainable asphalt roads* (pp. 3–28). Woodhead Publishing. DOI: 10.1016/B978-0-323-85789-5.00001-0

Després, V. R., Huffman, J. A., Burrows, S. M., Hoose, C., Safatov, A., Buryak, G., Fröhlich-Nowoisky, J., Elbert, W., Andreae, M. O., Pöschl, U., & Jaenicke, R. (2012). Primary biological aerosol particles in the atmosphere: A review. Tellus B. *Tellus. Series B, Chemical and Physical Meteorology, 64*(1), 15598. DOI: 10.3402/tellusb.v64i0.15598

Dey, S., Rout, A. K., Ghosh, K., Jana, A. K., & Behera, B. K. (2024). Microbial ecology in microplastics: Impact on aquatic ecosystems and bioremediation. In *Current trends in fisheries biotechnology* (pp. 79–93). Springer Nature Singapore. DOI: 10.1007/978-981-97-3165-7_6

Dey, S., Veerendra, G. T. N., Babu, P. A., Manoj, A. P., & Nagarjuna, K. (2024). Degradation of plastics waste and its effects on biological ecosystems: A scientific analysis and comprehensive review. *Biomedical Materials & Devices, 2*(1), 70–112. DOI: 10.1007/s44174-023-00085-w

Dhali, S. L., Parida, D., Kumar, B., & Bala, K. (2024). Recent trends in microbial and enzymatic plastic degradation: A solution for plastic pollution predicaments. *Biotechnology for Sustainable Materials, 1*(1), 11. DOI: 10.1186/s44316-024-00011-0

Dhanraj, N. D., Hatha, A. M., & Jisha, M. S. (2022). Biodegradation of petroleum based and bio-based plastics: Approaches to increase the rate of biodegradation. *Archives of Microbiology, 204*(5), 258. DOI: 10.1007/s00203-022-02883-0 PMID: 35419707

Di Genova, B. M., & Tonelli, R. R. (2016). Infection strategies of intestinal parasite pathogens and host cell responses. *Frontiers in Microbiology, 7*, 256. DOI: 10.3389/fmicb.2016.00256 PMID: 26973630

Díaz-Rodríguez, A. M., Parra Cota, F. I., Cira Chávez, L. A., García Ortega, L. F., Estrada Alvarado, M. I., Santoyo, G., & de los Santos-Villalobos, S. (2025). Microbial inoculants in sustainable agriculture: Advancements, challenges, and future directions. *Plants*, *14*(2), 191. DOI: 10.3390/plants14020191 PMID: 39861545

Dinsdale, E. A., Edwards, R. A., Hall, D., Angly, F., Breitbart, M., Brulc, J. M., Furlan, M., Desnues, C., Haynes, M., Li, L., McDaniel, L., Moran, M. A., Nelson, K. E., Nilsson, C., Olson, R., Paul, J., Brito, B. R., Ruan, Y., Swan, B. K., & Rohwer, F. (2008). Functional metagenomic profiling of nine biomes. *Nature*, *452*(7187), 629–632. DOI: 10.1038/nature06810 PMID: 18337718

Diyaolu, C. O., & Folarin, I. O. (2024). The role of biodiversity in agricultural resilience: Protecting ecosystem services for sustainable food production. *International Journal of Research Publication and Reviews*, *5*(3), 1560–1573. DOI: 10.55248/gengpi.5.1024.2741

Dlamini, S. P., Akanmu, A. O., & Babalola, O. O. (2023). Variations in the functional diversity of root microbiome of healthy and Northern corn leaf blight infected maize (Zea mays L.). *Spanish Journal of Soil Science*, *13*, 10964. DOI: 10.3389/sjss.2023.10964

Domenech, J., & Marcos, R. (2021). Pathways of human exposure to microplastics, and estimation of the total burden. *Current Opinion in Food Science*, *39*, 144–151. DOI: 10.1016/j.cofs.2021.01.004

Don, A., Seidel, F., Leifeld, J., Kätterer, T., Martin, M., Pellerin, S., Emde, D., Seitz, D., & Chenu, C. (2023). Carbon sequestration in soils and climate change mitigation-Definitions and pitfalls. *Global Change Biology*, *29*(12), e16983. DOI: 10.1111/gcb.16983 PMID: 37905459

Donati, G. F. A., Bolliger, J., Psomas, A., Maurer, M., & Bach, P. M. (2022). Reconciling cities with nature: Identifying local blue-green infrastructure interventions for regional biodiversity enhancement. *Journal of Environmental Management*, *316*, 115254. DOI: 10.1016/j.jenvman.2022.115254 PMID: 35576714

Doney, S. C., Fabry, V. J., Feely, R. A., & Kleypas, J. A. (2009). Ocean acidification: The other CO_2 problem. *Annual Review of Marine Science*, *1*(1), 169–192. DOI: 10.1146/annurev.marine.010908.163834 PMID: 21141034

Dong, C., Tiwari, A., Anisha, G. S., Chen, C., Singh, A., Haldar, D., Patel, A. K., & Singhania, R. R. (2023). Laccase: A potential biocatalyst for pollutant degradation. *Environmental Pollution*, *319*, 120999. DOI: 10.1016/j.envpol.2023.120999 PMID: 36608728

Dong, X., Chu, Y., Tong, Z., Sun, M., Meng, D., Yi, X., Feng, Y., Xiao, R., & Chen, H. (2024). Mechanisms of adsorption and functionalization of biochar for pesticides: A review. *Ecotoxicology and Environmental Safety*, *272*, 116019. DOI: 10.1016/j.ecoenv.2024.116019 PMID: 38295734

Doolittle, W. F., Boucher, Y., Nesbø, C. L., Douady, C. J., Andersson, J. O., & Roger, A. J. (2003). How big is the iceberg of which organellar genes in nuclear genomes are but the tip? *Philosophical Transactions of the Royal Society of London. Series B, Biological Sciences*, *358*(1429), 39–58. DOI: 10.1098/rstb.2002.1185 PMID: 12594917

Dorst, H., van der Jagt, A., Raven, R., & Runhaar, H. (2019). Urban greening through nature-based solutions: Key characteristics of an emerging concept. *Sustainable Cities and Society*, *49*, 101620. DOI: 10.1016/j.scs.2019.101620

Douglass, J. F., Radosevich, M., & Tuovinen, O. H. (2017). Microbial attenuation of atrazine in agricultural soils: Biometer assays, bacterial taxonomic diversity, and catabolic genes. *Chemosphere*, *176*, 352–360. DOI: 10.1016/j.chemosphere.2017.02.102 PMID: 28273542

Doytchinov, V. V., & Dimov, S. G. (2022). Microbial community composition of the Antarctic ecosystems: Review of the bacteria, fungi, and archaea identified through an NGS-based metagenomics approach. *Life (Chicago, Ill.)*, *12*(6), 916. DOI: 10.3390/life12060916 PMID: 35743947

Dris, R., Imhof, H., Sanchez, W., Gasperi, J., Galgani, F., Tassin, B., & Laforsch, C. (2015). Beyond the ocean: Contamination of freshwater ecosystems with (micro-) plastic particles. *Environmental Chemistry*, *12*(5), 539–550. DOI: 10.1071/EN14172

Dueck, H., Eberwine, J., & Kim, J. (2016). Variation is function: Are single cell differences functionally important? Testing the hypothesis that single cell variation is required for aggregate function. *BioEssays*, *38*(2), 172–180. DOI: 10.1002/bies.201500124 PMID: 26625861

Eberwein, M., Hellmold, N., Frank, R., Deobald, D., & Adrian, L. (2024). Reductive dehalogenase of Dehalococcoides mccartyi strain CBDB1 reduces cobalt-containing metal complexes enabling anodic respiration. *Frontiers in Microbiology*, *15*, 1457014. DOI: 10.3389/fmicb.2024.1457014 PMID: 39507334

Editorial, N. M. (2025). Microplastic research must consider microbes. *Nature Microbiology*, *10*(3), 603. Advance online publication. DOI: 10.1038/s41564-025-01960-6 PMID: 40044874

Egbe, C. C., Oyetibo, G. O., & Ilori, M. O. (2021). Ecological impact of organochlorine pesticides consortium on autochthonous microbial community in agricultural soil. *Ecotoxicology and Environmental Safety*, *207*, 111319. DOI: 10.1016/j.ecoenv.2020.111319 PMID: 32947214

Ejaz, M., Zhao, B., Wang, X., Bashir, S., Haider, F. U., Aslam, Z., & Mustafa, A. (2021). Isolation and characterization of oil-degrading Enterobacter sp. from naturally hydrocarbon-contaminated soils and their potential use against the bioremediation of crude oil. *Applied Sciences (Basel, Switzerland)*, *11*(8), 3504. DOI: 10.3390/app11083504

Ejiohuo, O., Onyeaka, H., Akinsemolu, A., Nwabor, O. F., Siyanbola, K. F., Tamasiga, P., & Al-Sharify, Z. T. (2025). Ensuring water purity: Mitigating environmental risks and safeguarding human health. Water Biology and Security, 4(2), Article 100341.

Ekanayaka, A. H., Tibpromma, S., Dai, D., Xu, R., Suwannarach, N., Stephenson, S. L., Dao, C., & Karunarathna S. C. (2022). A review of the fungi that degrade plastic. *Journal of Fungi* 8,8: 772. DOI: DOI: 10.3390/jof8080772

Elahi, A., Bukhari, D. A., Shamim, S., & Rehman, A. (2021). Plastics degradation by microbes: A sustainable approach. *Journal of King Saud University. Science*, *33*(6), 101538. DOI: 10.1016/j.jksus.2021.101538

El-Sharkawy, R. M., Khairy, M., Abbas, M. H. H., Zaki, M. E. A., & El-Hadary, A. E. (2024). Innovative optimization for enhancing Pb^{2+} biosorption from aqueous solutions using *Bacillus subtilis*. *Frontiers in Microbiology*, *15*, 1384639. DOI: 10.3389/fmicb.2024.1384639 PMID: 39176280

Enagbonma, B. J., Amoo, A. E., & Babalola, O. O. (2021). Biopedturbation by termites affects respiration profiles of microbial communities from termite mound soils. *Journal of Soil Science and Plant Nutrition*, *21*(4), 2115–2123. DOI: 10.1007/s42729-021-00507-y

Enagbonma, B. J., Aremu, B. R., & Babalola, O. O. (2019). Profiling the functional diversity of termite mound soil bacteria as revealed by shotgun sequencing. *Genes*, *10*(9), 637. DOI: 10.3390/genes10090637 PMID: 31450818

Enagbonma, B. J., & Babalola, O. O. (2022). Metagenomics shows that termite activities influence the diversity and composition of soil invertebrates in termite mound soils. *Applied and Environmental Soil Science*, *2022*, 7111775. DOI: 10.1155/2022/7111775

Enagbonma, B. J., & Babalola, O. O. (2023). Metagenomics reveals the microbiome multifunctionalities of environmental importance from termite mound soils. *Bioinformatics and Biology Insights*, *17*, 11779322231184025. Advance online publication. DOI: 10.1177/11779322231184025 PMID: 37424707

Enagbonma, B. J., Cromwell, S., Amoo, A. E., & Babalola, O. O. (2024a). Unveiling the mycobiome of ecological importance from termitarium soils. *Annals of Microbiology*, *74*(1), 34. DOI: 10.1186/s13213-024-01779-2

Enagbonma, B. J., Fadiji, A. E., & Babalola, O. O. (2024b). Anthropogenic fertilization influences a shift in barley rhizosphere microbial communities. *PeerJ*, *12*, e17303. DOI: 10.7717/peerj.17303 PMID: 39006020

Enagbonma, B. J., Imade, E. E., Babalola, O. O., Modise, D. M., Rigobelo, E. C., & Cruz, C. (2025a). Exploring plant-microbe interactions in extreme environments: Lessons from arid and desert ecosystems. In *Microbial allies* (pp. 49–65). Springer. DOI: 10.1007/978-3-031-90530-8_3

Enagbonma, B. J., Modise, D. M., & Babalola, O. O. (2025b). Effects of legume–cereal rotation on sorghum rhizosphere microbial community structure and nitrogen-cycling functions. *MicrobiologyOpen*, *14*(2), e70085. DOI: 10.1002/mbo3.70085 PMID: 41121668

Engels, J. M., & Ebert, A. W. (2021). A critical review of the current global ex situ conservation system for plant agrobiodiversity. II. Strengths and weaknesses of the current system and recommendations for its improvement. *Plants*, *10*(9), 1904. DOI: 10.3390/plants10091904 PMID: 34579439

Ermis, H. (2025). A mini-review on the role of PETase in polyethylene terephthalate degradation. *Reviews in Environmental Science and Biotechnology*, *24*(3), 545–555. DOI: 10.1007/s11157-025-09737-3

Ervens, B., Amato, P., Aregahegn, K., Joly, M., Khaled, A., Labed-Veydert, T., Mathonat, F., Nuñez López, L., Péguilhan, R., & Zhang, M. (2025). Ideas and perspectives: Microorganisms in the air through the lenses of atmospheric chemistry and microphysics. *Biogeosciences*, *22*(1), 243–256. DOI: 10.5194/bg-22-243-2025

European Commission. (2013). Green infrastructure (GI): Enhancing Europe's natural capital. In: *Communication from the Commission to the European Parliament, the European Economic Council, the European Economic and Social Committee and the Committee of the Regions*. DOI: 10.1017/CBO9781107415324.004

Evans, S. E., & Wallenstein, M. D. (2014). Climate change alters ecological strategies of soil bacteria. *Ecology Letters*, *17*(2), 155–164. DOI: 10.1111/ele.12206 PMID: 24261594

Eze, C. N., Anyanwu, B. N., & Onwurah, I. N. E. (2020). Biodegradation of cyanide in cassava mill effluent using indigenous bacterial isolates. *Environmental Monitoring and Assessment*, *192*(6), 389. PMID: 32447581

Fackelmann, G., Pham, C. K., Rodriguez, Y., Mallory, M. L., Provencher, J. F., Baak, J. E., & Sommer, S. (2023). Current levels of microplastic pollution impacts wild seabird gut microbiomes. *Nature Ecology & Evolution*, *7*(5), 698–706. DOI: 10.1038/s41559-023-02013-z PMID: 36973353

Fadiji, A. E., Ayangbenro, A. S., Akanmu, A. O., & Babalola, O. O. (2023). Draft genome sequence of Acinetobacter sp. AYS6, a potential plant growth-promoting endophyte. *Microbiology Resource Announcements*, *12*(10), e00464–e23. DOI: 10.1128/MRA.00464-23 PMID: 37737613

Fadiji, A. E., Lanrewaju, A. A., Akanmu, A. O., Ayangbenro, A. S., & Babalola, O. O. (2026). The relevant role of soil microorganisms in ameliorating abiotic stress in crop plants. In Santoyo, G. (Ed.), *Soil microorganisms for plant growth promotion and soil health* (pp. 421–456). Elsevier., DOI: 10.1016/B978-0-443-34055-0.00022-9

Falkowski, P. G. (2015). Biogeochemical controls and feedbacks on ocean primary production. *Science*, *350*(6262), 763–767. PMID: 9660741

Fan, B.-L., Chen, L.-H., Chen, L.-L., & Guo, H. (2025). Integrative multi-omics approaches for identifying and characterizing biological elements in crop traits: Current progress and future prospects. *International Journal of Molecular Sciences*, *26*(3), 1466. DOI: 10.3390/ijms26041466 PMID: 40003933

Fang, X., Liu, Y., Xiao, J., Ma, C., & Huang, Y. (2023). GC–MS and LC-MS/MS metabolomics revealed dynamic changes of volatile and non-volatile compounds during withering process of black tea. *Food Chemistry*, *410*, 135396. DOI: 10.1016/j.foodchem.2023.135396 PMID: 36634561

Farhangi-Abriz, S., & Torabian, S. (2017). Antioxidant enzyme and osmotic adjustment changes in bean seedlings as affected by biochar under salt stress. *Ecotoxicology and Environmental Safety*, *137*, 64–70. DOI: 10.1016/j.ecoenv.2016.11.029 PMID: 27915144

Faroon, O., Jones, D., & De Rosa, C. (2000). Effects of polychlorinated biphenyls on the nervous system. *Toxicology and Industrial Health*, *16*(7-8), 305–333. DOI: 10.1177/074823370001600708 PMID: 11693948

Farrance, C. E. (2019). Identification of microorganisms. In *Pharmaceutical microbiological quality assurance and control: Practical guide for non-sterile manufacturing* (pp. 265–328). Wiley. DOI: 10.1002/9781119356196.ch9

Farrington, J. W. (2014). Oil pollution in the marine environment II: Fates and effects of oil spills. *Environment, 56*(4), 16–31. DOI: 10.1080/00139157.2014.922382

Faust, K., & Raes, J. (2012). Microbial interactions: From networks to models. *Nature Reviews. Microbiology, 10*(8), 538–550. DOI: 10.1038/nrmicro2832 PMID: 22796884

Fei, Y., Huang, S., Zhang, H., Tong, Y., Wen, D., Xia, X., Wang, H., Luo, Y., & Barceló, D. (2020). Response of soil enzyme activities and bacterial communities to the accumulation of microplastics in an acid cropped soil. *The Science of the Total Environment, 707*, 135634. DOI: 10.1016/j.scitotenv.2019.135634 PMID: 31761364

Feng, Y. Y., Hou, L. C., Ping, N. X., Ling, T. D., & Kyo, C. I. (2004). Development of mariculture and its impacts in Chinese coastal waters. *Reviews in Fish Biology and Fisheries, 14*(1), 1–10. DOI: 10.1007/s11160-004-3539-7

Fenik, J., Tankiewicz, M., & Biziuk, M. (2011). Properties and determination of pesticides in fruits and vegetables. *Trends in Analytical Chemistry, 30*(6), 814–826. DOI: 10.1016/j.trac.2011.02.008

Fernández-Braña, A., Salgado, L., Gallego, J. L. R., Afif, E., Boente, C., & Forján, R. (2023). Phytoremediation potential depends on the degree of soil pollution: A case study in an urban brownfield. *Environmental Science and Pollution Research International, 30*(25), 67708–67719. DOI: 10.1007/s11356-023-26968-5 PMID: 37118389

Ferrocino, I., Bianchi, D., Decastelli, L., Cocolin, L., & Rantsiou, K. (2023). The need for an integrated multi-omics approach in microbiome science in the genomic era. *Comprehensive Reviews in Food Science and Food Safety, 22*(3), 2169–2192. DOI: 10.1111/1541-4337.13184

Fierer, N., Leff, J. W., Adams, B. J., Nielsen, U. N., Bates, S. T., Lauber, C. L., Owens, S., Gilbert, J. A., Wall, D. H., & Caporaso, J. G. (2012). Cross-biome metagenomic analyses of soil microbial communities and their functional attributes. *Proceedings of the National Academy of Sciences of the United States of America, 109*(52), 21390–21395. DOI: 10.1073/pnas.1215210110 PMID: 23236140

Focardi, S., Corsi, I., & Franchi, E. (2005). Safety issues and sustainable development of European aquaculture: New tools for environmentally sound aquaculture. *Aquaculture International, 13*(1), 3–17. DOI: 10.1007/s10499-004-9036-0

Freilinger, J., Kappacher, C., Huter, K., Hofer, T. S., Back, J. O., Huck, C. W., & Bakry, R. (2025). Interactions between perfluorinated alkyl substances (PFAS) and microplastics (MPs): Findings from an extensive investigation. *Journal of Hazardous Materials Advances*, *18*, 100740. DOI: 10.1016/j.hazadv.2025.100740

Frere, L., Maignien, L., Chalopin, M., Huvet, A., Rinnert, E., Morrison, H., Kerninon, S., Cassone, A. L., Lambert, C., Reveillaud, J., & Paul-Pont, I. (2018). Microplastic bacterial communities in the Bay of Brest: Influence of polymer type and size. *Environmental Pollution*, *242*, 614–625. DOI: 10.1016/j.envpol.2018.07.023 PMID: 30014939

Froelich, B. A., & Noble, R. T. (2014). Factors affecting the uptake and retention of Vibrio vulnificus in oysters. *Applied and Environmental Microbiology*, *80*(24), 7454–7459. DOI: 10.1128/AEM.02042-14 PMID: 25261513

Fröhlich-Nowoisky, J., Kampf, C. J., Weber, B., Huffman, J. A., Pöhlker, C., Andreae, M. O., Lang-Yona, N., Burrows, S. M., Gunthe, S. S., Elbert, W., Su, H., Hoor, P., Thines, E., Hoffmann, T., Després, V. R., & Pöschl, U. (2016). Bioaerosols in the Earth system: Climate, health, and ecosystem interactions. *Atmospheric Research*, *182*, 346–376. DOI: 10.1016/j.atmosres.2016.07.018

Fujita, H., Ushio, M., Suzuki, K., Abe, M. S., Yamamichi, M., Okazaki, Y., Canarini, A., Hayashi, I., Fukushima, K., Fukuda, S., Kiers, E. T., & Toju, H. (2023). Metagenomic analysis of ecological niche overlap and community collapse in microbiome dynamics. Frontiers in Microbiology, 14, Article 1234567. (Note: Volume 14 - 2023; exact article number may vary by platform.)

Ganguly, P., Mandal, J., Sengupta, S., & Sinha, B. (2023). Basics of environment, pollution, and bioremediation techniques. In *Apple Academic Press eBooks* (pp. 1–17). https://doi.org/DOI: 10.1201/9781003412236-1

Gan, Z., & Zhang, H. (2019). PMBD: A comprehensive plastics microbial biodegradation database. *Database : The Journal of Biological Databases and Curation*, *2019*, baz119. Advance online publication. DOI: 10.1093/database/baz119 PMID: 31738435

Gao, J., Han, H., Gao, C., Wang, Y., Dong, B., & Xu, Z. (2023). Organic amendments for in situ immobilization of heavy metals in soil: A review. *Chemosphere*, *335*, 139088. DOI: 10.1016/j.chemosphere.2023.139088 PMID: 37268229

Gasparetto, J., Magenta, L., Sinisi, F., Zotti, S., Costantini, A., Martelli, B., Giangregorio, T., & Pippucci, T. (2024, March). Computational Genomics platform: A cloud-enabled approach. In *Proceedings of the International Symposium on Grids and Clouds (ISGC2024)* (Vol. 24, p. 29).

Gawai, D. (2018). Role of fungi as biocontrol agents for the control of plant diseases in sustainable agriculture. In *Fungi and their role in sustainable development: current perspectives* (pp. 283–291). Springer. DOI: 10.1007/978-981-13-0393-7_16

Gerba, C. P. (2009). Environmentally transmitted pathogens. In *Environmental microbiology* (pp. 445–484). Academic Press. DOI: 10.1016/B978-0-12-370519-8.00022-5

Gerber, R., Smit, N. J., Van Vuren, J. H., Nakayama, S. M., Yohannes, Y. B., Ikenaka, Y., Ishizuka, M., & Wepener, V. (2016). Bioaccumulation and human health risk assessment of DDT and other organochlorine pesticides in an apex aquatic predator from a premier conservation area. *The Science of the Total Environment, 550*, 522–533. DOI: 10.1016/j.scitotenv.2016.01.129 PMID: 26845188

Ge, S., Sun, S., Xu, H., Cheng, Q., & Ren, Z. (2025). Deep learning in single-cell and spatial transcriptomics data analysis: Advances and challenges from a data science perspective. *Briefings in Bioinformatics, 26*(2), bbaf136. DOI: 10.1093/bib/bbaf136 PMID: 40185158

Geyer, R., Jambeck, J. R., & Law, K. L. (2017, July 19). Production, use, and fate of all plastics ever made. *Science Advances, 3*(7), e1700782. DOI: 10.1126/sciadv.1700782 PMID: 28776036

Ghiasian, M., Akhavan, S. A., Amoozegar, M., Saadatmand, S., & Shavandi, M. (2017). Bacterial diversity determination using culture-dependent and culture-independent methods.

Ghosal, D., Omelchenko, M. V., Gaidamakova, E. K., Matrosova, V. Y., Vasilenko, A., Venkateswaran, A., & Daly, M. J. (2005). How radiation kills cells: Survival of Deinococcus radiodurans and Shewanella oneidensis under oxidative stress. *FEMS Microbiology Reviews, 29*(2), 361–375. PMID: 15808748

Giacomucci, L., Raddadi, N., Soccio, M., Lotti, N., & Fava, F. (2019). Polyvinylchloride biodegradation by *Pseudomonas citronellolis* and *Bacillus flexus. New Biotechnology, 52*, 35–41. DOI: 10.1016/j.nbt.2019.04.005 PMID: 31026607

Gieg, L. M., Fowler, S. J., & Berdugo-Clavijo, C. (2014). Syntrophic biodegradation of hydrocarbon contaminants. *Current Opinion in Biotechnology, 27*, 21–29. DOI: 10.1016/j.copbio.2013.09.002 PMID: 24863893

Giller, K. E., Witter, E., & McGrath, S. P. (1998). Toxicity of heavy metals to microorganisms and microbial processes in agricultural soils: A review. *Soil Biology & Biochemistry, 30*(10-11), 1389–1414. DOI: 10.1016/S0038-0717(97)00270-8

Gkoutselis, G., Rohrbach, S., Harjes, J., Obst, M., Brachmann, A., Horn, M. A., & Rambold, G. (2021). Microplastics accumulate fungal pathogens in terrestrial ecosystems. *Scientific Reports*, *11*(1), 13214. DOI: 10.1038/s41598-021-92405-7 PMID: 34267241

Glick, B. R. (2012). Plant growth-promoting bacteria: Mechanisms and applications. *Scientifica*, *2012*(1), 963401. PMID: 24278762

Godoy, V., Martín-Lara, M. A., Calero, M., & Blázquez, G. (2020). The relevance of interaction of chemicals/pollutants and microplastic samples as route for transporting contaminants. *Process Safety and Environmental Protection*, *138*, 312–323. DOI: 10.1016/j.psep.2020.03.033

Gondal, M. N., Shah, S. U. R., Chinnaiyan, A. M., & Cieslik, M. (2024). A systematic overview of single-cell transcriptomics databases, their use cases, and limitations. *Frontiers in Bioinformatics*, *4*, 1417428. DOI: 10.3389/fbinf.2024.1417428 PMID: 39040140

Gonzalez, J. M., Santana, M. M., Gomez, E. J., & Delgado, J. A. (2023). Soil thermophiles and their extracellular enzymes: A set of capabilities able to provide significant services and risks. *Microorganisms*, *11*(7), 1650. DOI: 10.3390/microorganisms11071650 PMID: 37512823

González-Martín, J., Cantera, S., Muñoz, R., & Lebrero, R. (2023). Indoor air VOCs biofiltration by bioactive coating packed bed bioreactors. *Journal of Environmental Management*, *349*, 119362. DOI: 10.1016/j.jenvman.2023.119362 PMID: 37897901

Gouda, S., Kerry, R. G., Das, G., Paramithiotis, S., Shin, H. S., & Patra, J. K. (2018). Revitalization of plant growth promoting rhizobacteria for sustainable development in agriculture. *Microbiological Research*, *206*, 131–140. DOI: 10.1016/j.micres.2017.08.016 PMID: 29146250

Grant, K. R. (2022). Next-generation amplicon sequencing: A cost-effective method for exploring microbial biodiversity. In Kumar, A., Choudhury, B., Dayanandan, S., & Khan, M. L. (Eds.), *Molecular genetics and genomics tools in biodiversity conservation* (pp. 203–236). Springer Nature Singapore. DOI: 10.1007/978-981-16-6005-4_10

Gray, A., Gore, B., Gaesser, M., Rodriguez Sequeira, L., Thibodeau, T., Montgomery, A., Purvis, S., Ouimet, K., Dura, T., & Mayer, K. (2025). Insights into the in-situ degradation and fragmentation of macroplastics in a low-order riverine system. *Environmental Toxicology and Chemistry*, *44*(7), 2017–2028. DOI: 10.1093/etojnl/vgaf116 PMID: 40316505

Gray, A., Mayer, K., Gore, B., Gaesser, M., & Ferguson, N. (2024). Microplastic burden in native (*Cambarus appalachiensis*) and non-native (*Faxonius cristavarius*) crayfish along semi-rural and urban streams in southwest Virginia, USA. *Environmental Research, 258*, 119494. DOI: 10.1016/j.envres.2024.119494 PMID: 38936498

Griffiths, B. S., & Philippot, L. (2013). Insights into the resistance and resilience of the soil microbial community. *FEMS Microbiology Reviews, 37*(2), 112–129. DOI: 10.1111/j.1574-6976.2012.00343.x PMID: 22568555

Grigson, S. R., Bouras, G., Dutilh, B. E., Olson, R. D., & Edwards, R. A. (2025). Computational function prediction of bacteria and phage proteins. *Microbiology and Molecular Biology Reviews : MMBR, 89*(1), e00022–e00025. DOI: 10.1128/mmbr.00022-25 PMID: 40824055

Grujcic, V., Taylor, G. T., & Foster, R. A. (2022). One cell at a time: Advances in single-cell methods and instrumentation for discovery in aquatic microbiology. Frontiers in Microbiology, 13, Article 876543. (Note: Volume 13 - 2022.)

Guidi Nissim, W., Castiglione, S., Guarino, F., Pastore, M. C., & Labra, M. (2023). Beyond cleansing: Ecosystem services related to phytoremediation. *Plants, 12*(5), 1031. DOI: 10.3390/plants12051031 PMID: 36903892

Gu, J.-D. (2021). On enrichment culturing and transferring technique. *Applied Environmental Biotechnology, 6*(1), 1–5. DOI: 10.26789/AEB.2021.01.001

Gupta, P., Samant, K., & Sahu, A. (2012). Isolation of cellulose-degrading bacteria and determination of their cellulolytic potential. *International Journal of Microbiology, 2012*, 1–5. DOI: 10.1155/2012/578925 PMID: 22315612

Guzzetti, E., Sureda, A., Tejada, S., & Faggio, C. (2018). Microplastic in marine organisms: Environmental and toxicological effects. *Environmental Toxicology and Pharmacology, 64*, 164–171. DOI: 10.1016/j.etap.2018.10.009 PMID: 30412862

Haines, S. R., Hall, E. C., Marciniak, K., Misztal, P. K., Goldstein, A. H., Adams, R. I., & Dannemiller, K. C. (2021). Microbial growth and volatile organic compound (VOC) emissions from carpet and drywall under elevated relative humidity conditions. *Microbiome, 9*(1), 209. Advance online publication. DOI: 10.1186/s40168-021-01158-y PMID: 34666813

Hale, R. C., Seeley, M. E., La Guardia, M. J., Mai, L., & Zeng, E. Y. (2020). A global perspective on microplastics. Journal of Geophysical Research: Oceans, 125(1), Article e2018JC014719.

Hamid, B., Bashir, Z., Yatoo, A. M., Mohiddin, F., Majeed, N., Bansal, M., & Alfaifi, M. Y. (2022). Cold-active enzymes and their potential industrial applications review. *Molecules (Basel, Switzerland)*, *27*(18), 5885. DOI: 10.3390/molecules27185885 PMID: 36144621

Hansen, R., & Pauleit, S. (2014). From multifunctionality to multiple ecosystem services? A conceptual framework for multifunctionality in green infrastructure planning for urban areas. *Ambio*, *43*(4), 516–529. DOI: 10.1007/s13280-014-0510-2 PMID: 24740622

Haque, S., Srivastava, N., Pal, D. B., Alkhanani, M. F., Almalki, A. H., Areeshi, M. Y., Naidu, R., & Gupta, V. K. (2022). Functional microbiome strategies for the bioremediation of petroleum-hydrocarbon and heavy metal contaminated soils: A review. *The Science of the Total Environment*, *833*, 155222. DOI: 10.1016/j.scitotenv.2022.155222 PMID: 35421499

Harayama, S., Kasai, Y., & Hara, A. (2004). Microbial communities in oil-contaminated seawater. *Current Opinion in Biotechnology*, *15*(3), 205–214. DOI: 10.1016/j.copbio.2004.04.002 PMID: 15193328

Harnisch, F., Deutzmann, J. S., Boto, S. T., & Rosenbaum, M. A. (2024). Microbial electrosynthesis: Opportunities for microbial pure cultures. *Trends in Biotechnology*, *42*(11), 1035–1047. DOI: 10.1016/j.tibtech.2024.02.004 PMID: 38431514

Harper, D. M. (1992). Eutrophication of freshwaters.

Harwood, C. S., & Parales, R. E. (1996). The β-ketoadipate pathway and the biology of aromatic compound degradation. *Annual Review of Microbiology*, *50*, 553–590. DOI: 10.1146/annurev.micro.50.1.553 PMID: 8905091

Hausdorf, B. (2025). Changing views on speciation. *Molecular Ecology*, *34*(24), e70175. DOI: 10.1111/mec.70175 PMID: 41236164

Hazaimeh, M., Kanaan, B. M., AlFaleh, F. A., Elhaig, M. M., Khamaiseh, E. I., Zia, Q., Alaidarous, M., Seth, C. S., Alsowayeh, N., & Ahmad, F. (2024). Biodegradation of petroleum hydrocarbons using a novel bacterial strain isolated from hydrocarbons contaminated soil of Saudi Arabia. *Biocatalysis and Agricultural Biotechnology*, *57*, 103074. DOI: 10.1016/j.bcab.2024.103074

Heeley-Hill, A. C., Grange, S. K., Ward, M. W., Lewis, A. C., Owen, N., Jordan, C., Hodgson, G., & Adamson, G. (2021). Frequency of use of household products containing VOCs and indoor atmospheric concentrations in homes. *Environmental Science. Processes & Impacts*, *23*(5), 699–713. DOI: 10.1039/D0EM00504E PMID: 34037627

Heidrich, V., & Beule, L. (2022). Are short-read amplicons suitable for the prediction of microbiome functional potential? A critical perspective. *iMeta, 1*(3), e38. DOI: 10.1002/imt2.38 PMID: 38868716

Hemme, C. L., Deng, Y., Gentry, T. J., Fields, M. W., Wu, L., Barua, S., Barry, K., Tringe, S. G., Watson, D. B., He, Z., Hazen, T. C., Tiedje, J. M., Rubin, E. M., & Zhou, J. (2010). Metagenomic insights into evolution of a heavy metal-contaminated groundwater microbial community. *The ISME Journal, 4*(5), 660–672. DOI: 10.1038/ismej.2009.154 PMID: 20182523

He, Q., Wang, Z., Liu, H., Xu, P., Duan, R., Xu, C., Chen, J., & Wei, M. (2024). Influence of terrestrial and marine air mass on the constituents and intermixing of bioaerosols over a coastal atmosphere. *Atmospheric Chemistry and Physics, 24*(22), 12775–12792. DOI: 10.5194/acp-24-12775-2024

Hibbert, K., & Ogunseitan, O. A. (2014). Risks of toxic ash from artisanal mining of discarded cellphones. *Journal of Hazardous Materials, 278*, 1–7. DOI: 10.1016/j.jhazmat.2014.05.089 PMID: 24937657

Hickey, W. J. (2021). Biodegradation of environmental pollutants. In *Elsevier eBooks* (pp. 581–605). https://doi.org/DOI: 10.1016/B978-0-12-820202-9.00021-6

Hidalgo-Ruz, V., Gutow, L., Thompson, R. C., & Thiel, M. (2012). Microplastics in the marine environment: A review of the methods used for identification and quantification. *Environmental Science & Technology, 46*(6), 3060–3075. DOI: 10.1021/es2031505 PMID: 22321064

Hildebrand, G. A., Honeker, L. K., Freire-Zapata, V., Ayala-Ortiz, C., Rajakaruna, S., Fudyma, J., Daber, L. E., AminiTabrizi, R., Chu, R. L., Toyoda, J., Flowers, S. E., Hoyt, D. W., Hamdan, R., Gil-Loaiza, J., Shi, L., Dippold, M. A., Ladd, S. N., Werner, C., Meredith, L. K., & Tfaily, M. M. (2023). Uncovering the dominant role of root metabolism in shaping rhizosphere metabolome under drought in tropical rainforest plants. *The Science of the Total Environment, 899*, 165689. DOI: 10.1016/j.scitotenv.2023.165689 PMID: 37481084

Hinsu, A., Dumadiya, A., Joshi, A., Kotadiya, R., Andharia, K., Koringa, P., & Kothari, R. (2021). To culture or not to culture: A snapshot of culture-dependent and culture-independent bacterial diversity from peanut rhizosphere. *PeerJ, 9*, e12035. DOI: 10.7717/peerj.12035 PMID: 34557347

Hiroshi, S. (2024). Controlling genetic engineering in Environmental biotechnology: Potentials and challenges. *Advancements in Genetic Engineering, 13*(02). Advance online publication. DOI: 10.35841/2169-0111.24.13.264

Hlihor, R. M., & Cozma, P. (2023). Microbial bioremediation of environmental pollution. *Processes (Basel, Switzerland)*, *11*(5), 1543. DOI: 10.3390/pr11051543

Holling, C. S. (1973). Resilience and stability of ecological systems. *Annual Review of Ecology and Systematics*, *4*(1), 1–23. DOI: 10.1146/annurev.es.04.110173.000245

Hommel, R. K. (1990). Formation and phylogenetic role of biosurfactants. *Journal of Applied Microbiology*, *89*(1), 158–219.

Hong, J., Boussetta, N., Enderlin, G., Merlier, F., & Grimi, N. (2022). Degradation of residual herbicide atrazine in agri-food and washing water. *Food*, *11*(16), 2416. DOI: 10.3390/foods11162416 PMID: 36010414

Hopewell, J., Dvorak, R., & Kosior, E. (2009). Plastics recycling: Challenges and opportunities. *Philosophical Transactions of the Royal Society of London. Series B, Biological Sciences*, *364*(1526), 2115–2126. DOI: 10.1098/rstb.2008.0311 PMID: 19528059

Hosokawa, M., & Nishikawa, Y. (2024). Tools for microbial single-cell genomics for obtaining uncultured microbial genomes. *Biophysical Reviews*, *16*(1), 69–77. DOI: 10.1007/s12551-023-01124-y PMID: 38495448

Hou, D., Al-Tabbaa, A., O'Connor, D., Hu, Q., Zhu, Y. G., Wang, L., Kirkwood, N., Ok, Y. S., Tsang, D. C. W., Bolan, N. S., & Rinklebe, J. (2023). Sustainable remediation and redevelopment of brownfield sites. *Nature Reviews. Earth & Environment*, *4*(4), 271–286. DOI: 10.1038/s43017-023-00404-1

Houmenou, C. T., Sokhna, C., Fenollar, F., & Mediannikov, O. (2025). Advancements and challenges in bioinformatics tools for microbial genomics in the last decade: Toward the smart integration of bioinformatics tools, digital resources, and emerging technologies for the analysis of complex biological data. *Infection, Genetics and Evolution*, *136*, 105859. DOI: 10.1016/j.meegid.2025.105859 PMID: 41297621

Huang, Y., Liu, Q., Jia, W., Yan, C., & Wang, J. (2020). Agricultural plastic mulching as a source of microplastics in the terrestrial environment. *Environmental Pollution*, *260*, 114096. DOI: 10.1016/j.envpol.2020.114096 PMID: 32041035

Huang, Y., Sheth, R. U., Zhao, S., Cohen, L. A., Dabaghi, K., Moody, T., Sun, Y., Ricaurte, D., Richardson, M., Velez-Cortes, F., Blazejewski, T., Kaufman, A., Ronda, C., & Wang, H. H. (2023). High-throughput microbial culturomics using automation and machine learning. *Nature Biotechnology*, *41*(10), 1424–1433. DOI: 10.1038/s41587-023-01674-2 PMID: 36805559

Hu, F., Wang, P., Li, Y., Ling, J., Ruan, Y., Yu, J., & Zhang, L. (2023). Bioremediation of environmental organic pollutants by Pseudomonas aeruginosa: Mechanisms, methods and challenges. *Environmental Research, 239*, 117211. DOI: 10.1016/j.envres.2023.117211 PMID: 37778604

Ibanez-Lligona, M., Colomer-Castell, S., Gonzalez-Sanchez, A., Gregori, J., Campos, C., Garcia-Cehic, D., Andres, C., Pinana, M., Pumarola, T., Rodriguez-Frias, F., Antón, A., & Quer, J. (2023). Bioinformatic tools for NGS-based metagenomics to improve the clinical diagnosis of emerging, re-emerging and new viruses. *Viruses, 15*(3), 587. DOI: 10.3390/v15020587 PMID: 36851800

Ibrahim, A. M. (2024). Plastics: Photodegradations and mechanisms. In *Microplastics and Pollutants: Interactions, Degradations and Mechanisms* (pp. 25-49). Cham: Springer Nature Switzerland. DOI: DOI: 10.1007/978-3-031-54565-8_2

Icer, M. A., Özbay, S., Ağagündüz, D., Kelle, B., Bartkiene, E., Rocha, J. M. F., & Ozogul, F. (2023). The impacts of acidophilic lactic acid bacteria on food and human health: A review of the current knowledge. *Foods, 12*(15), 2965. DOI: 10.3390/foods12152965 PMID: 37569234

Ilori, M. O., & Obayori, O. S. (2023). Introduction to aeromicrobiology. Ilori MO, Obayori OS. Salam LB (eds), Aeromicrobiology. First Edition, Publisher: Academic Press, an imprint of Elsevier, London, United Kingdom, pp. 1–16. https://doi.org/ DOI: 10.1016/B978-0-323-96122-6.00005-0

Intergovernmental Panel on Climate Change. (2022). *Climate change 2022: Impacts, adaptation, and vulnerability. Contribution of Working Group II to the Sixth Assessment Report of the Intergovernmental Panel on Climate Change*. Cambridge University Press., DOI: 10.1017/9781009325844

Iqbal, S., Xu, J., Allen, S. D., Khan, S., Nadir, S., Arif, M. S., & Yasmeen, T. (2020). Unraveling consequences of soil micro-and nano-plastic pollution on soil-plant system: Implications for nitrogen (N) cycling and soil microbial activity. *Chemosphere, 260*, 127578. DOI: 10.1016/j.chemosphere.2020.127578 PMID: 32683024

Irfana, S., Ahmad Farooq, L., Moieza, A., Aneesul Mohammad, M., & Asmat, R. (2016). Biofilters in Mitigation of Odour Pollution - A Review. *Nature Environment and Pollution Technology, 15*(4), 1177–1185.

Irwin, J. A. (2020). Overview of extremophiles and their food and medical applications. In *Physiological and biotechnological aspects of extremophiles* (pp. 65–87). Academic Press. DOI: 10.1016/B978-0-12-818322-9.00006-X

Isha, N., Ali, S., Khalid, A., Naseer, I. A., Raza, H., & Chang, Y. (2024). Bioremediation of smog: Current trends and future perspectives. *Processes (Basel, Switzerland)*, *12*(10), 2266. DOI: 10.3390/pr12102266

Ishii, S., Suzuki, S., Norden-Krichmar, T. M., Tenney, A., Chain, P. S. G., Scholz, M. B., Nealson, K. H., & Bretschger, O. (2013). A novel metatranscriptomic approach to identify gene expression dynamics during extracellular electron transfer. *Nature Communications*, *4*(1), 1601. DOI: 10.1038/ncomms2615 PMID: 23511466

Jackson, C. J., Preston, N., Burford, M. A., & Thompson, P. J. (2003). Managing the development of sustainable shrimp farming in Australia: The role of sedimentation ponds in treatment of farm discharge water. *Aquaculture (Amsterdam, Netherlands)*, *226*(1-4), 23–34. DOI: 10.1016/S0044-8486(03)00464-2

Jacobs, D. F., Oliet, J. A., Aronson, J., Bolte, A., Bullock, J. M., Donoso, P. J., Landhausser, S. M., Madsen, P., Peng, S., Rey-Benayas, J. M., & Weber, J. C. (2015). Restoring forests: What constitutes success in the twenty-first century? *New Forests*, *46*(5–6), 601–614. DOI: 10.1007/s11056-015-9513-5

Jadaun, J. S., Bansal, S., Sonthalia, A., Rai, K. A., & Singh, S. P. (2022). Biodegradation of plastics for sustainable environment. *Bioresource Technology*, *126697*, 126697. Advance online publication. DOI: 10.1016/j.biortech.2022.126697

Jaga, K., & Dharmani, C. (2003). Sources of exposure to and public health implications of organophosphate pesticides. *Revista Panamericana de Salud Pública = Pan American Journal of Public Health*, *14*(3), 171–185. DOI: 10.1590/S1020-49892003000800004 PMID: 14653904

Jain, C., Rodriguez-R, L. M., Phillippy, A. M., Konstantinidis, K. T., & Aluru, S. (2018). High throughput ANI analysis of 90K prokaryotic genomes reveals clear species boundaries. *Nature Communications*, *9*(1), 5114. DOI: 10.1038/s41467-018-07641-9 PMID: 30504855

Jaipieam, S., Visuthismajarn, P., Sutheravut, P., Siriwong, W., Thoumsang, S., Borjan, M., & Robson, M. (2009). Organophosphate pesticide residues in drinking water from artesian wells and health risk assessment of agricultural communities, Thailand. *Human and Ecological Risk Assessment*, *15*(6), 1304–1316. DOI: 10.1080/10807030903306984 PMID: 20485459

Jaiswal, S., Singh, D. K., & Shukla, P. (2019). Gene Editing and Systems Biology Tools for Pesticide Bioremediation: A review. *Frontiers in Microbiology*, *10*, 87. Advance online publication. DOI: 10.3389/fmicb.2019.00087 PMID: 30853940

Janeeshma, E., Habeeb, H., Sinha, S., Arora, P., Chattaraj, S., Das Mohapatra, P., Panneerselvam, P., & Mitra, D. (2024). Enzymes-mediated solid waste management: A sustainable practice for recycling. *Waste Management Bulletin*, *1*(4), 104–113. DOI: 10.1016/j.wmb.2023.10.007

Janssen, D. B., Dinkla, I. J. T., Poelarends, G. J., & Terpstra, P. (2005). Bacterial degradation of xenobiotic compounds: Evolution and distribution of novel enzyme activities. *Environmental Microbiology*, *7*(12), 1868–1882. DOI: 10.1111/j.1462-2920.2005.00966.x PMID: 16309386

Jeong, S. W., & Choi, Y. J. (2020). Extremophilic microorganisms for the treatment of toxic pollutants in the environment. *Molecules (Basel, Switzerland)*, *25*(21), 4916. DOI: 10.3390/molecules25214916 PMID: 33114255

Jia, J., de Goede, R., Li, Y., Zhang, J., Wang, G., Zhang, J., & Creamer, R. (2025). Unlocking soil health: Are microbial functional genes effective indicators? *Soil Biology & Biochemistry*, *204*, 109768. DOI: 10.1016/j.soilbio.2025.109768

Jiang, Y., Xiong, X., Danska, J., & Parkinson, J. (2016a). Metatranscriptomic analysis of diverse microbial communities reveals core metabolic pathways and microbiome-specific functionality. *Microbiome*, *4*(1), 2. Advance online publication. DOI: 10.1186/s40168-015-0146-x PMID: 26757703

Jin, P. J., Sun, L., Liu, Y. H., Wang, K. K., Narsing Rao, M. P., Mohamad, O. A. A., Fang, B.-Z., Li, L., Gao, L., Li, W.-J., & Wang, S. (2024). Two novel alkaliphilic species isolated from saline-alkali soil in China: Halalkalibacter flavus sp. nov., and Halalkalibacter lacteus sp. nov. *Microorganisms*, *12*(5), 950. DOI: 10.3390/microorganisms12050950 PMID: 38792780

Johnson, D. B. (2008). Biodiversity and interactions of acidophiles: Key to understanding and optimizing microbial processing of ores and concentrates. *Transactions of Nonferrous Metals Society of China*, *18*(6), 1367–1373. DOI: 10.1016/S1003-6326(09)60010-8

Jomova, K., Alomar, S. Y., Nepovimova, E., Kuca, K., & Valko, M. (2025). Heavy metals: Toxicity and human health effects. *Archives of Toxicology*, *99*(1), 153–209. DOI: 10.1007/s00204-024-03903-2 PMID: 39567405

Jones, A. M., & Harrison, R. M. (2004). The effects of meteorological factors on atmospheric bioaerosol concentrations—A review. *The Science of the Total Environment*, *326*(1-3), 151–180. DOI: 10.1016/j.scitotenv.2003.11.021 PMID: 15142773

Joos, L., & De Tender, C. (2022). Soil under stress: The importance of soil life and how it is influenced by (micro) plastic pollution. *Computational and Structural Biotechnology Journal, 20*, 1554–1566. DOI: 10.1016/j.csbj.2022.03.041 PMID: 35422972

Jorth, P., Turner, K. H., Gumus, P., Nizam, N., Buduneli, N., & Whiteley, M. (2014). Metatranscriptomics of the human oral microbiome during health and disease. *mBio, 5*(2), e01012–e01014. DOI: 10.1128/mBio.01012-14 PMID: 24692635

Junaid, M., Siddiqui, J. A., Sadaf, M., Liu, S., & Wang, J. (2022). Enrichment and dissemination of bacterial pathogens by microplastics in the aquatic environment. *The Science of the Total Environment, 830*, 154720. DOI: 10.1016/j.scitotenv.2022.154720 PMID: 35337880

Kafaei, R., Arfaeinia, H., Savari, A., Mahmoodi, M., Rezaei, M., Rayani, M., Sorial, G. A., Fattahi, N., & Ramavandi, B. (2020). Organochlorine pesticides contamination in agricultural soils of southern Iran. *Chemosphere, 240*, 124983. DOI: 10.1016/j.chemosphere.2019.124983 PMID: 31726590

Kaiser, D., Kowalski, N., & Waniek, J. J. (2017). Effects of biofouling on the sinking behavior of microplastics. *Environmental Research Letters, 12*(12), 124003. DOI: 10.1088/1748-9326/aa8e8b

Kalalian, C., Grira, A., Illmann, J. N., Patroescu-Klotz, I., Dib, G. E., & Coddeville, P.. (2022). Experimental and theoretical studies of Trans-2-Pentenal atmospheric ozonolysis. *Atmosphere (Basel), 13*(2), 291. DOI: 10.3390/atmos13020291

Kalia, A., & Gosal, S. K. (2011). Effect of pesticide application on soil microorganisms. *Archives of Agronomy and Soil Science, 57*(6), 569–596. DOI: 10.1080/03650341003787582

Kanekar, P. P., & Kanekar, S. P. (2022). Alkaliphilic, alkalitolerant microorganisms. In *Diversity and Biotechnology of Extremophilic Microorganisms from India* (pp. 71–116). Springer Nature Singapore. DOI: 10.1007/978-981-19-1573-4_3

Kang, J., Xue, Y., Chen, X., & Han, B. Z. (2022). Integrated multi-omics approaches to understand microbiome assembly in Jiuqu, a mixed-culture starter. *Comprehensive Reviews in Food Science and Food Safety, 21*(5), 4076–4107. DOI: 10.1111/1541-4337.13025 PMID: 36038529

Kannaujiya, M. C., Mandal, T., Mandal, D. D., & Mondal, M. K. (2019). Treatment of leather industry wastewater and recovery of valuable substances to solve waste management problem in environment. In *Environmental contaminants: ecological implications and management* (pp. 311–340). Springer Singapore. DOI: 10.1007/978-981-13-7904-8_14

Kansal, A. (2008). Sources and reactivity of NMHCs and VOCs in the atmosphere: A review. *Journal of Hazardous Materials, 166*(1), 17–26. DOI: 10.1016/j.jhazmat.2008.11.048 PMID: 19136203

Karigar, C. S., & Rao, S. S. (2011). Role of microbial enzymes in the bioremediation of pollutants: A review. *Enzyme Research, 2011*, 1–11. DOI: 10.4061/2011/805187 PMID: 21912739

Karimi, B., Habibi, M., & Esvand, M. (2015). Biodegradation of naphthalene using Pseudomonas aeruginosa by up-flow anoxic–aerobic continuous flow combined bioreactor. *Journal of Environmental Health Science & Engineering, 13*(1), 26. Advance online publication. DOI: 10.1186/s40201-015-0175-1 PMID: 25859393

Karr, J. R., Fausch, K. D., Angermeier, P. L., Yant, P. R., & Schlosser, I. J. (2000). Assessing ecological integrity in running waters: A method and its rationale. *Special Publication of the Illinois Natural History Survey, 5*, 1-19. https://www.researchgate.net/publication/245507787

Karunarathna, B. S., Gajasinghe, S. T., Diwankara, U. T., Hunupolagama, M. D., & Chen, X. (2025). Fungal-mediated biodegradation of synthetic plastic: Mechanisms, enzymes, and strategies for enhanced plastic waste management. *New Zealand Journal of Botany*, •••, 1–27. DOI: 10.1080/0028825X.2025.2522207

Kashtan, N., Roggensack, S. E., Rodrigue, S., Thompson, J. W., Biller, S. J., Coe, A., & Chisholm, S. W. (2014). Single-cell genomics reveals hundreds of coexisting subpopulations in wild Prochlorococcus. *Science, 344*(6182), 416–420. DOI: 10.1126/science.1248575 PMID: 24763590

Kato, N. N., Arini, G. S., Silva, R. R., Bichuette, M. E., Bitencourt, J. A. P., & Lopes, N. P. (2024). The world of cave microbiomes: Biodiversity, ecological interactions, chemistry, and the multi-omics integration. Journal of the Brazilian Chemical Society, 35(11), Article e-20230148.

Kato, H., Miura, D., Kato, M., & Shimizu, M. (2024). Metabolic mechanism of lignin-derived aromatics in white-rot fungi. *Applied Microbiology and Biotechnology, 108*(1), 532. DOI: 10.1007/s00253-024-13371-4 PMID: 39661194

Kaushal, J., Khatri, M., & Arya, S. K. (2021). A treatise on Organophosphate pesticide pollution: Current strategies and advancements in their environmental degradation and elimination. *Ecotoxicology and Environmental Safety, 207*, 111483. DOI: 10.1016/j.ecoenv.2020.111483 PMID: 33120277

Kawai, F., Oda, M., Tamashiro, T., Waku, T., Tanaka, N., Yamamoto, M., Mizushima, H., Miyakawa, T., & Tanokura, M. (2014). A novel Ca^{2+} activated, thermostabilized polyesterase capable of hydrolyzing polyethylene terephthalate from *Saccharomonospora viridis* AHK190. *Applied Microbiology and Biotechnology, 98*(24), 10053–10064. DOI: 10.1007/s00253-014-5860-y PMID: 24929560

Kebede, G., Tafese, T., Abda, E. M., Kamaraj, M., & Assefa, F. (2021). Factors influencing the bacterial bioremediation of hydrocarbon contaminants in the soil: Mechanisms and impacts. *Journal of Chemistry, 2021*(1), 9823362. DOI: 10.1155/2021/9823362

Keiblinger, K. M., Fuchs, S., Zechmeister-Boltenstern, S., & Riedel, K. (2016). Soil and leaf litter metaproteomics—A brief guideline from sampling to understanding. *FEMS Microbiology Ecology, 92*(2), fiw008. DOI: 10.1093/femsec/fiw180 PMID: 27549116

Kensa, V. M. (2011). Bioremediation-an overview. *Journal of Industrial Pollution Control, 27*(2), 161–168.

Kerrigan, Z., & D'Hondt, S. (2022). Patterns of relative bacterial richness and community composition in seawater and marine sediment are robust for both operational taxonomic units and amplicon sequence variants. *Frontiers in Microbiology, 13*, 876543. DOI: 10.3389/fmicb.2022.796758 PMID: 35197949

Kettner, M. T., Oberbeckmann, S., Labrenz, M., & Grossart, H. P. (2019). The eukaryotic life on microplastics in brackish ecosystems. *Frontiers in Microbiology, 10*, e00538. DOI: 10.3389/fmicb.2019.00538 PMID: 30949147

Kettner, M. T., Rojas-Jimenez, K., Oberbeckmann, S., Labrenz, M., & Grossart, H. P. (2017). Microplastics alter composition of fungal communities in aquatic ecosystems. *Environmental Microbiology, 19*(11), 4447–4459. DOI: 10.1111/1462-2920.13891 PMID: 28805294

Khajezadeh, M., Abbaszadeh-Goudarzi, K., Pourghadamyari, H., & Kafilzadeh, F. (2020). A newly isolated Streptomyces rimosus strain capable of degrading deltamethrin as a pesticide in agricultural soil. *Journal of Basic Microbiology, 60*(5), 435–443. DOI: 10.1002/jobm.201900263 PMID: 32128846

Khalili, B., Ogunseitan, O. A., Goulden, M. L., & Allison, S. D. (2016). Interactive effects of precipitation manipulation and nitrogen addition on soil properties in California grassland and shrubland. *Applied Soil Ecology, 107*, 144–153. DOI: 10.1016/j.apsoil.2016.05.018

Khan, T. F., & Faruque, M. G. (2023). Degradation of polymer materials in environment. Ecology Journal, 237.

Khan, B., & Saha, P. (2026). Challenges, adaptations, and biotechnological potential of oligophilic bacteria. *Antonie van Leeuwenhoek, 119*(1), 24. DOI: 10.1007/s10482-025-02236-z PMID: 41454939

Khan, H. A., Sher, S., Bukhari, D. A., & Rehman, A. (2025). Bioremediation of heavy metals from electronic waste dumping sites with bacteria. *Current Research in Biotechnology, 10*, 100309. DOI: 10.1016/j.crbiot.2025.100309

Khan, M., Wibowo, A., Karim, Z., Posoknistakul, P., Matsagar, B., Wu, K., & Sakdaronnarong, C. (2024). Wastewater Treatment Using Membrane Bioreactor Technologies: Removal of Phenolic Contaminants from Oil and Coal Refineries and Pharmaceutical Industries. *Polymers, 16*(3), 443. DOI: 10.3390/polym16030443 PMID: 38337332

Khan, S., Nadir, S., Shah, Z. U., Shah, A. A., Karunarathna, S. C., Xu, J., Khan, A., Munir, S., & Hasan, F. (2017). Biodegradation of polyester polyurethane by *Aspergillus tubingensis. Environmental Pollution, 225*, 469–480. DOI: 10.1016/j.envpol.2017.03.012 PMID: 28318785

Khasanova, S., Alieva, E., & Shemilkhanova, A. (2023). Environmental pollution: types, causes and consequences. In *BIO Web of Conferences* (Vol. 63, p. 07014). EDP Sciences. DOI: 10.1051/bioconf/20236307014

Kim, H., Jeon, H., Shin, G., Lee, M., Jegal, J., Hwang, S. Y., Oh, D. X., Koo, J. M., Eom, Y., & Park, J. (2021). Biodegradable nanocomposite of poly(ester-*co*-carbonate) and cellulose nanocrystals for tough tear-resistant disposable bags. *Green Chemistry, 23*(6), 2293–2299. DOI: 10.1039/D0GC04072J

Kim, K., Song, I. G., Yoon, H., & Park, J. W. (2023). Sub-micron microplastics affect nitrogen cycling by altering microbial abundance and activities in a soil-legume system. *Journal of Hazardous Materials, 460*, 132504. DOI: 10.1016/j.jhazmat.2023.132504 PMID: 37703725

Kleindienst, S., Paul, J. H., & Joye, S. B. (2015). Using dispersants after oil spills: Impacts on the composition and activity of microbial communities. *Nature Reviews. Microbiology, 13*(6), 388–396. DOI: 10.1038/nrmicro3452 PMID: 25944491

Kleinheinz, G. T., Bagley, S. T., St John, W. P., Rughani, J. R., & McGinnis, G. D. (1999). Characterization of Alpha-Pinene-Degrading microorganisms and application to a Bench-Scale biofiltration system for VOC degradation. *Archives of Environmental Contamination and Toxicology*, *37*(2), 151–157. DOI: 10.1007/s002449900500 PMID: 10398764

Kloepper, J. W., Ryu, C. M., & Zhang, S. (2004). Induced systemic resistance and promotion of plant growth by *Bacillus* spp. *Phytopathology*, *94*(11), 1259–1266. DOI: 10.1094/PHYTO.2004.94.11.1259 PMID: 18944464

Kloppenburg, J., Calderón, C. I., & Ané, J. M. (2024). The Nagoya Protocol and nitrogen-fixing maize: Close encounters between Indigenous Oaxacans and the men from Mars (Inc.). *Elementa*, *12*(1), 00115. DOI: 10.1525/elementa.2023.00115

Kochhar, N., Shrivastava, S., Ghosh, A., Rawat, V. S., Sodhi, K. K., & Kumar, M. (2022). Perspectives on the microorganism of extreme environments and their applications. *Current Research in Microbial Sciences*, *3*, 100134. DOI: 10.1016/j.crmicr.2022.100134 PMID: 35909612

Köhl, J., Kolnaar, R., & Willem, R. (2019). Mode of action of microbial biological control agents against plant diseases: Relevance beyond efficacy. *Frontiers in Plant Science*, *10*, 845. DOI: 10.3389/fpls.2019.00845 PMID: 31379891

Kolmert, Å., & Johnson, D. B. (2001). Remediation of acidic waste waters using immobilised, acidophilic sulfate-reducing bacteria. *Journal of Chemical Technology & Biotechnology: International Research in Process. Environmental & Clean Technology*, *76*(8), 836–843.

Konstantinidis, K. T., Rosselló-Móra, R., & Amann, R. (2017). Uncultivated microbes in need of their own taxonomy. *The ISME Journal*, *11*(11), 2399–2406. DOI: 10.1038/ismej.2017.113 PMID: 28731467

Koonin, E. V., Dolja, V. V., & Krupovic, M. (2015). Origins and evolution of viruses of eukaryotes: The ultimate modularity. *Virology*, *479-480*, 2–25. DOI: 10.1016/j.virol.2015.02.039 PMID: 25771806

Koskey, G., Mburu, S. W., Awino, R., Njeru, E. M., & Maingi, J. M. (2021). Potential use of beneficial microorganisms for soil amelioration, phytopathogen biocontrol, and sustainable crop production in smallholder agroecosystems. *Frontiers in Sustainable Food Systems*, *5*, 606308. DOI: 10.3389/fsufs.2021.606308

Koul, B., Chaudhary, R., & Taak, P. (2021). Extremophilic microbes and their application in bioremediation of environmental contaminants. In *Microbe mediated remediation of environmental contaminants* (pp. 115–128). Woodhead Publishing. DOI: 10.1016/B978-0-12-821199-1.00012-2

Kour, D., Kaur, T., Devi, R., Yadav, A., Singh, M., Joshi, D., & Saxena, A. K. (2021). Beneficial microbiomes for bioremediation of diverse contaminated environments for environmental sustainability: Present status and future challenges. *Environmental Science and Pollution Research International*, *28*(20), 24917–24939. DOI: 10.1007/s11356-021-13252-7 PMID: 33768457

Kour, D., Khan, S. S., Kour, H., Kaur, T., Devi, R., Judy, C., Rai, P. K., McQuestion, C., Spells, S., Bianchi, A., Mohan, R., Rai, A. K., & Yadav, A. N. (2022). Microbe-mediated bioremediation: Current research and future challenges. *Journal of Applied Biology and Biotechnology*, •••, 6–24. DOI: 10.7324/JABB.2022.10s202

Krause, T., Wassan, J. T., Mc Kevitt, P., Wang, H., Zheng, H., & Hemmje, M. (2021). Analyzing large microbiome datasets using machine learning and big data. *BioMedInformatics*, *1*(2), 138–165. DOI: 10.3390/biomedinformatics1030010

Krokidis, M. G., Koumadorakis, D. E., Lazaros, K., Ivantsik, O., Exarchos, T. P., Vrahatis, A. G., Kotsiantis, S., & Vlamos, P. (2025). AlphaFold3: An overview of applications and performance insights. *International Journal of Molecular Sciences*, *26*(8), 3671. DOI: 10.3390/ijms26083671 PMID: 40332289

Kuiper, I., Lagendijk, E. L., Bloemberg, G. V., & Lugtenberg, B. J. J. (2004). Rhizoremediation: A beneficial plant-microbe interaction. *Molecular Plant-Microbe Interactions*, *17*(1), 6–15. DOI: 10.1094/MPMI.2004.17.1.6 PMID: 14714863

Kulkarni, A., Anderson, A. G., Merullo, D. P., & Konopka, G. (2019). Beyond bulk: A review of single cell transcriptomics methodologies and applications. *Current Opinion in Biotechnology*, *58*, 129–136. DOI: 10.1016/j.copbio.2019.03.001 PMID: 30978643

Kumar, A., Alam, A., Tripathi, D., Rani, M., Khatoon, H., Pandey, S., & Hasnain, S. E. (2018, December). Protein adaptations in extremophiles: An insight into extremophilic connection of mycobacterial proteome. [). Academic Press.]. *Seminars in Cell & Developmental Biology*, *84*, 147–157. DOI: 10.1016/j.semcdb.2018.01.003 PMID: 29331642

Kumar, A., & Chandra, R. (2020). Ligninolytic enzymes and its mechanisms for degradation of lignocellulosic waste in environment. *Heliyon*, *6*(2), e03170. DOI: 10.1016/j.heliyon.2020.e03170 PMID: 32095645

Kumar, A., Mishra, S., Pandey, R., Yu, Z. G., Kumar, M., Khoo, K. S., Thakur, T. K., & Show, P. L. (2023). Microplastics in terrestrial ecosystems: Un-ignorable impacts on soil characterises, nutrient storage and its cycling. *Trends in Analytical Chemistry*, *158*, 116869. DOI: 10.1016/j.trac.2022.116869

Kumar, R., Sharma, P., Verma, A., Jha, P. K., Singh, P., Gupta, P. K., Chandra, R., & Prasad, P. V. (2021). Effect of physical characteristics and hydrodynamic conditions on transport and deposition of microplastics in riverine ecosystem. *Water (Basel)*, *13*(19), 2710. DOI: 10.3390/w13192710

Kuppan, N., Padman, M., Mahadeva, M., Srinivasan, S., & Devarajan, R. (2024). A comprehensive review of sustainable bioremediation techniques: Eco friendly solutions for waste and pollution management. *Waste Management Bulletin*, *2*(3), 154–171. DOI: 10.1016/j.wmb.2024.07.005

Kuźma, Ł., Dąbrowski, E. J., Kurasz, A., Bachórzewska-Gajewska, H., & Dobrzycki, S. (2020). The 10-Year study of the impact of particulate matters on mortality in two transit cities in North-Eastern Poland (PL-PARTICLES). *Journal of Clinical Medicine*, *9*(11), 3445. DOI: 10.3390/jcm9113445 PMID: 33120927

Kvenvolden, K. A., & Cooper, C. K. (2003). Natural Seepage of Oils into the Marine Environment. *Geo-Marine Letters*, *23*(3-4), 140–146. DOI: 10.1007/s00367-003-0135-0

Kwon, J. H., Ji, M. K., Kumar, R., Islam, M. M., Khan, M. A., Park, Y. K., Yadav, K. K., Vaziri, R., Hwang, J.-H., Lee, W. H., Ahn, Y.-T., & Jeon, B. H. (2023). Recent advancement in enhanced soil flushing for remediation of petroleum hydrocarbon-contaminated soil: A state-of-the-art review. *Reviews in Environmental Science and Biotechnology*, *22*(3), 679–714. DOI: 10.1007/s11157-023-09657-0

Kye, H., Kim, J., Ju, S., Lee, J., Lim, C., & Yoon, Y. (2023). Microplastics in water systems: A review of their impacts on the environment and their potential hazards. *Heliyon*, *9*(3), e14359. DOI: 10.1016/j.heliyon.2023.e14359 PMID: 36950574

la Cecilia, D., Malvolti, G., Despini, F., Teggi, S., Bigi, A., & Bach, P. (2020). Urban heat island mitigation scenarios by blue-green infrastructures modelling. *Proceedings of International Conference on Environmental Science and Technology 18*. DOI: 10.30955/gnc2023.00287

Lakshmanan, V., Ray, P., & Craven, K. D. (2017). Rhizosphere sampling protocols for microbiome (16S/18S/ITS rRNA) library preparation and enrichment for the isolation of drought tolerance-promoting microbes. In *Plant stress tolerance: Methods and protocols* (pp. 349–362). Springer. DOI: 10.1007/978-1-4939-7136-7_23

Lakshmanan, V., Selvaraj, G., & Bais, H. P. (2014). Functional soil microbiome: Belowground solutions to an aboveground problem. *Plant Physiology*, *166*(2), 689–700. DOI: 10.1104/pp.114.245811 PMID: 25059708

Landa-Acuña, D., Acosta, R. A. S., Hualpa Cutipa, E., Vargas de la Cruz, C., & Luis Alaya, B. (2020). Bioremediation: A low-cost and clean-green technology for environmental management. In Shah, M. (Ed.), *Microbial bioremediation & biodegradation*. Springer., DOI: 10.1007/978-981-15-1812-6_7

Lane, C., Al Shoffe, Y., Schafran, P., Li, F.-W., Kao-Kniffin, J., & Watkins, C. B. (2024). Amplicon sequencing and shotgun metagenomics reveal functional impacts of aminoethoxyvinylglycine-mediated ripening and cold storage on the microbiome of 'NY1' apples. *Postharvest Biology and Technology*, *213*, 112969. DOI: 10.1016/j.postharvbio.2024.112969

Lang-Yona, N., Flores, J. M., Haviv, R., Alberti, A., Poulain, J., Belser, C., ... (2021). A cross-biomes bacterial diversity shed light on ocean-atmosphere microbial transmission. bioRxiv *(Cold Spring Harbor Laboratory)*. https://doi.org/DOI: 10.1101/2021.06.06.445733

Lapidus, A. L., & Korobeynikov, A. I. (2021). Metagenomic data assembly – The way of decoding unknown microorganisms. *Frontiers in Microbiology*, *12*, 613791. DOI: 10.3389/fmicb.2021.613791 PMID: 33833738

Lau, W. W. Y., Shiran, Y., Bailey, R. M., Cook, E., Stuchtey, M. R., Keskella, J., Velis, C. A., Godfrey, L., Boucher, J., Murphy, M. B., Thompson, R. C., Jankowska, E., Castillo, A. C., Pilditch, T. D., Dixon, B., Koerselman, L., Kosior, E., Favoino, E., Gutberlet, J., & Palardy, J. E. (2020). Evaluating scenarios toward zero plastic pollution. *Science*, *369*(6510), 1455–1461. DOI: 10.1126/science.aba9475 PMID: 32703909

Lawal, A. T. (2017). Polycyclic aromatic hydrocarbons. A review. *Cogent Environmental Science*, *3*(1), 1339841. DOI: 10.1080/23311843.2017.1339841

Ławniczak, Ł., Woźniak-Karczewska, M., Loibner, A. P., Heipieper, H. J., & Chrzanowski, Ł. (2020). Microbial degradation of hydrocarbons—basic principles for bioremediation: A review. *Molecules (Basel, Switzerland)*, *25*(4), 856. DOI: 10.3390/molecules25040856 PMID: 32075198

Law, S. R., Serrano, A. R., Daguerre, Y., Sundh, J., Schneider, A. N., Stangl, Z. R., Castro, D., Grabherr, M., Näsholm, T., Street, N. R., & Hurry, V. (2022). Metatranscriptomics captures dynamic shifts in mycorrhizal coordination in boreal forests. *Proceedings of the National Academy of Sciences of the United States of America*, *119*(2), e2118852119. DOI: 10.1073/pnas.2118852119 PMID: 35727987

Lawson, C. E., & Lücker, S. (2018). Complete ammonia oxidation: An important control on nitrification in engineered ecosystems? *Current Opinion in Biotechnology*, *50*, 158–165. DOI: 10.1016/j.copbio.2018.01.015 PMID: 29414055

Leahy, J. G., & Colwell, R. R. (1990). microbial degradation of hydrocarbons in the environment. *Microbiological Reviews*, *54*(3), 305–315. DOI: 10.1128/mr.54.3.305-315.1990 PMID: 2215423

Lebrero, R., Rodríguez, E., Estrada, J. M., García-Encina, P. A., & Muñoz, R. (2012). Odor abatement in biotrickling filters: Effect of the EBRT on methyl mercaptan and hydrophobic VOCs removal. *Bioresource Technology*, *109*, 38–45. DOI: 10.1016/j.biortech.2012.01.052 PMID: 22310214

Lecca-Caballero, D., Vega-Moreno, E., Cabanillas-Chirinos, L., Diaz Del Aguila, K., Rojas-Villacorta, W., Salvatierra-Espinola, W., Naveda, R. N., Rojas-Flores, S., & De La Cruz-Noriega, M. (2023). An i*n vitro* study of the effects of temperature and pH on lead bioremoval using *Serratia marcescens. Sustainability (Basel)*, *15*(19), 14048. DOI: 10.3390/su151914048

Lee, J., Hong, S., Song, Y. K., Hong, S. H., Jang, Y. C., Jang, M., Heo, N. W., Han, G. M., Lee, M. J., Kang, D., & Shim, W. J. (2013). Relationships among the abundances of plastic debris in different size classes on beaches in South Korea. *Marine Pollution Bulletin*, *77*(1-2), 349–354. DOI: 10.1016/j.marpolbul.2013.08.013 PMID: 24054782

Lema, N. K., Gemeda, M. T., & Woldesemayat, A. A. (2023). Recent advances in metagenomic approaches, applications, and challenges. *Current Microbiology*, *80*(10), 347. DOI: 10.1007/s00284-023-03451-5 PMID: 37733134

Lennon, J. T., Abramoff, R. Z., Allison, S. D., Burckhardt, R. M., DeAngelis, K. M., Dunne, J. P., Frey, S. D., Friedlingstein, P., Hawkes, C. V., Hungate, B. A., Khurana, S., Kivlin, S. N., Levine, N. M., Manzoni, S., Martiny, A. C., Martiny, J. B. H., Nguyen, N. K., Rawat, M., Talmy, D., & Zakem, E. J. (2024). Priorities, opportunities, and challenges for integrating microorganisms into Earth system models for climate change prediction. *mBio*, *15*(5), e00455–e24. DOI: 10.1128/mbio.00455-24 PMID: 38526088

Lennon, J. T., & Jones, S. E. (2011). Microbial seed banks: The ecological and evolutionary implications of dormancy. *Nature Reviews. Microbiology*, *9*(2), 119–130. DOI: 10.1038/nrmicro2504 PMID: 21233850

Lensch, A., Lindfors, H. A., Duwenig, E., Fleischmann, T., Hjort, C., Kärenlampi, S. O., McMurtry, L., Melton, E., Andersen, M. R., Skinner, R., Wyss, M., & van Kranenburg, R. (2024). Safety aspects of microorganisms deliberately released into the environment. *EFB Bioeconomy Journal, 4*, 100061. DOI: 10.1016/j.bioeco.2023.100061

Lenton, T. M., Held, H., Kriegler, E., Hall, J. W., Lucht, W., Rahmstorf, S., & Schellnhuber, H. J. (2008). Tipping elements in the Earth's climate system. *Proceedings of the National Academy of Sciences of the United States of America, 105*(6), 1786–1793. DOI: 10.1073/pnas.0705414105 PMID: 18258748

Li, C., Cui, Q., Li, Y., Zhang, K., Lu, X., & Yong, Z. (2022). Effect of different additions of low-density polyethylene and microplastics polyadipate/butylene terephthalate on soil bacterial community structure. *Journal of Hazardous Materials*. 429:10, 128353. DOI: DOI: 10.1016/j.jhazmat.2022.128353

Liang, C., Tian, J., & Liao, H. (2013). Proteomics dissection of plant responses to mineral nutrient deficiency. *Proteomics, 13*(3–4), 624–636. DOI: 10.1002/pmic.201200263 PMID: 23193087

Liang, M., Luo, G., & Lei, P. (2025). Recent advances in bioelectrochemical approaches for sustainable environmental remediation. *International Journal of Electrochemical Science, 20*(8), 101077. DOI: 10.1016/j.ijoes.2025.101077

Li, B., Mao, Z., Xue, J., Xing, P., & Wu, Q. L. (2024). Metabolic versatility of aerobic methane-oxidizing bacteria under anoxia in aquatic ecosystems. *Environmental Microbiology Reports, 16*(5), e70002. DOI: 10.1111/1758-2229.70002 PMID: 39232853

Li, M.-R., Zhan, F.-D., Chen, J.-J., Zu, Y.-Q., & Li, Y. (2020). Atrazine degradation pathway and genes of Arthrobacter sp. FM326. *Polish Journal of Environmental Studies, 29*(5), 3683–3689. DOI: 10.15244/pjoes/115326

Lin, B., Van Verseveld, H. W., & Röling, W. F. M. (2002). Microbial aspects of anaerobic BTEX degradation. *PubMed, 15*(2), 130–144. https://pubmed.ncbi.nlm.nih.gov/12244754

Lin, Z., Jin, T., Zou, T., Xu, L., Xi, B., Xu, D., He, J., Xiong, L., Tang, C., Peng, J., Zhou, Y., & Fei, J. (2022). Current progress on plastic/microplastic degradation: Fact influences and mechanism. *Environmental Pollution, 304*, 119159. DOI: 10.1016/j.envpol.2022.119159 PMID: 35304177

Li, P., Wang, X., Su, M., Zou, X., Duan, L., & Zhang, H. (2021). Characteristics of plastic pollution in the environment: A review. *Bulletin of Environmental Contamination and Toxicology*, *107*(4), 577–584. DOI: 10.1007/s00128-020-02820-1 PMID: 32166334

Liu, C., Cai, J., Chen, R., Sera, F., Guo, Y., Tong, S., Li, S., Lavigne, E., Correa, P. M., Ortega, N. V., Orru, H., Maasikmets, M., Jaakkola, J. J. K., Ryti, N., Breitner, S., Schneider, A., Katsouyanni, K., Samoli, E., Hashizume, M., & Kan, H. (2022). Coarse particulate air pollution and daily mortality: A global study in 205 cities. *American Journal of Respiratory and Critical Care Medicine*, *206*(8), 999–1007. DOI: 10.1164/rccm.202111-2657OC PMID: 35671471

Liu, F., Wang, T., Yang, W., Zhang, Y., Gong, Y., Fan, X., Wang, G., Lu, Z., & Wang, J. (2023). Current advances in the structural biology and molecular engineering of PETase. *Frontiers in Bioengineering and Biotechnology*, *11*, 1263996. DOI: 10.3389/fbioe.2023.1263996 PMID: 37795175

Liu, L., Li, W., Song, W., & Guo, M. (2018). Remediation techniques for heavy metal-contaminated soils: Principles and applicability. *The Science of the Total Environment*, *633*, 206–219. DOI: 10.1016/j.scitotenv.2018.03.161 PMID: 29573687

Liu, X., & Mattes, T. E. (2016). Epoxyalkane:Coenzyme M Transferase Gene Diversity and Distribution in Groundwater Samples from Chlorinated-Ethene-Contaminated Sites. *Applied and Environmental Microbiology*, *82*(11), 3269–3279. DOI: 10.1128/AEM.00673-16 PMID: 27016563

Liu, X., Sathishkumar, K., Zhang, H., Saxena, K. K., Zhang, F., Naraginti, S., & Guo, X. (2024). Frontiers in environmental cleanup: Recent advances in remediation of emerging pollutants from soil and water. *Journal of Hazardous Materials Advances*, *16*, 100461. DOI: 10.1016/j.hazadv.2024.100461

Liu, X., Zhang, Z., Tan, C., Ai, Y., Liu, H., Li, Y., Yang, J., & Song, Y. (2025). Global trends in machine learning applications for single-cell transcriptomics research. *Hereditas*, *162*(1), 164. DOI: 10.1186/s41065-025-00528-y PMID: 40819059

Liu, Y., Tang, T., Zhu, Z., Zeng, X., Zou, Q., & Li, K. (2025). Quality scores compression of genomic sequencing data: A comprehensive review and performance evaluation. *IEEE Transactions on Computational Biology and Bioinformatics*, *22*(3), 817–831. DOI: 10.1109/TCBBIO.2025.3539629 PMID: 40811383

Li, X., & Lu, H. (2025). Enhanced metagenomic strategies for elucidating the complexities of gut microbiota: A review. *Frontiers in Microbiology*, *16*, 1626002. DOI: 10.3389/fmicb.2025.1626002 PMID: 40933132

Li, Y. Y., Song, W. Y., Han, X. Y., Wang, Y. C., Rao, S. Q., Zhang, Q., Zhou, J. W., Li, J. H., Liu, S., & Du, G. C. (2022). Recent progress in key lignocellulosic enzymes: Enzyme discovery, molecular modifications, production, and enzymatic biomass saccharification. *Bioresource Technology*, *363*, 127986. DOI: 10.1016/j.biortech.2022.127986 PMID: 36126851

Li, Z., Gao, Y., Lu, Z., Xie, J., Liu, Y., Wang, G., & Cheng, X. (2023). Strengthening and microbial regulation mechanism of *Bacillus* on purification device for grass carp culture wastewater. *Frontiers in Environmental Science*, *11*, 1128329. Advance online publication. DOI: 10.3389/fenvs.2023.1128329

Lokesh, P., Shobika, R., Omer, S., Reddy, M., Saravanan, P., Rajeshkannan, R., Saravanan, V., & Venkatkumar, S. (2023). Bioremediation of plastics by the help of microbial tool: A way for control of plastic pollution. *Sustainable Chemistry for the Environment*, *3*, 100027. DOI: 10.1016/j.scenv.2023.100027

Loreau, M., Naeem, S., Inchausti, P., Bengtsson, J., Grime, J. P., Hector, A., Hooper, D. U., Huston, M. A., Raffaelli, D., Schmid, B., Tilman, D., & Wardle, D. A. (2001). Biodiversity and ecosystem functioning: Current knowledge and future challenges. *Science*, *294*(5543), 804–808. DOI: 10.1126/science.1064088 PMID: 11679658

Louca, S., Polz, M. F., Mazel, F., Albright, M. B. N., Huber, J. A., O'Connor, M. I., Ackermann, M., Hahn, A. S., Srivastava, D. S., Crowe, S. A., Doebeli, M., & Parfrey, L. W. (2018). Function and functional redundancy in microbial systems. *Nature Ecology & Evolution*, *2*(6), 936–943. DOI: 10.1038/s41559-018-0519-1 PMID: 29662222

Loveniers, P., Devlieghere, F., & Sampers, I. (2024). Towards tailored guidelines for microbial air quality in the food industry. *International Journal of Food Microbiology*, *421*, 110779. DOI: 10.1016/j.ijfoodmicro.2024.110779 PMID: 38852216

Lukas, M., Schwidetzky, R., Kunert, A. T., Pöschl, U., Fröhlich-Nowoisky, J., Bonn, M., & Meister, K. (2020). Electrostatic interactions control the functionality of bacterial ICE nucleators. *Journal of the American Chemical Society*, *142*(15), 6842–6846. DOI: 10.1021/jacs.9b13069 PMID: 32223131

Lv, Y., Wang, L., Liu, X., Chen, B., & Zhang, M. (2022). Degradation kinetics of aromatic VOCs polluted wastewater by functional bacteria at laboratory scale. *Scientific Reports*, *12*(1), 19053. Advance online publication. DOI: 10.1038/s41598-022-21356-4 PMID: 36351963

Mackelprang, R., Waldrop, M. P., DeAngelis, K. M., David, M. M., Chavarria, K. L., Blazewicz, S. J., Rubin, E. M., & Jansson, J. K. (2011). Metagenomic analysis of a permafrost microbial community reveals a rapid response to thaw. *Nature, 480*(7377), 368–371. DOI: 10.1038/nature10576 PMID: 22056985

Madhav, S., Ahamad, A., Singh, A. K., Kushawaha, J., Chauhan, J. S., Sharma, S., & Singh, P. (2019). Water pollutants: Sources and impact on the environment and human health. In Sensors in water pollutants monitoring: Role of material (pp. 43–62).

Madhuri, R. J., Saraswathi, M., Gowthami, K., Bhargavi, M., Divya, Y., & Deepika, V. (2019). Recent approaches in the production of novel enzymes from environmental samples by enrichment culture and metagenomic approach. In Buddolla, V. (Ed.), *Recent developments in Applied Microbiology and Biochemistry* (pp. 251–262). Academic Press., DOI: 10.1016/B978-0-12-816328-3.00019-2

Madigan, M. T., Bender, K. S., Buckley, D. H., Sattley, W. M., & Stahl, D. A. (2018). *Brock Biology of Microorganisms* (15th ed.). Pearson Education.

Maestri, S., Scalzo, D., Damaggio, G., Zobel, M., Besusso, D., & Cattaneo, E. (2025). Navigating triplet repeats sequencing: Concepts, methodological challenges and perspective for Huntington's disease. *Nucleic Acids Research, 53*(3), gkae1155. DOI: 10.1093/nar/gkae1155 PMID: 39676657

Maglione, G., Zinno, P., Tropea, A., Mussagy, C. U., Dufossé, L., Giuffrida, D., & Mondello, A. (2024). Microbes' role in environmental pollution and remediation: A bioeconomy focus approach. *AIMS Microbiology, 10*(3), 723–755. DOI: 10.3934/microbiol.2024033 PMID: 39219757

Ma, II., Zhang, Y., Shen, R., & Jia, Y. (2025). Droplet-based microfluidics in single-bacterium analysis: Advancements in cultivation, detection, and application. *Biosensors (Basel), 15*(9), 535. DOI: 10.3390/bios15080535 PMID: 40862995

Mahalle, S., Bhende, R. S., Bokade, P., Bajaj, A., & Dafale, N. A. (2025). Emerging microbial remediation methods for rejuvenation of pesticide-contaminated sites. *Total Environment Microbiology, 1*(3), 100026. DOI: 10.1016/j.temicr.2025.100026

Mahjadeen, I. (2023). Airborne Microorganisms in the Environment. *Research Gate.* DOI: DOI: 10.13140/RG.2.2.27961.58729

Maity, S., Banerjee, S., Biswas, C., Guchhait, R., Chatterjee, A., & Pramanick, K. (2021). Functional interplay between plastic polymers and microbes: A comprehensive review. *Biodegradation, 32*(5), 487–510. DOI: 10.1007/s10532-021-09954-x PMID: 34086181

Malard, L. A., & Pearce, D. A. (2022). Bacterial colonisation: From airborne dispersal to integration within the soil community. *Frontiers in Microbiology, 13*, 782789. Advance online publication. DOI: 10.3389/fmicb.2022.782789 PMID: 35615521

Malik, S., Dhasmana, A., Preetam, S., Mishra, Y. K., Chaudhary, V., Bera, S. P., & Rajput, V. D. (2022). Exploring microbial-based green nanobiotechnology for wastewater remediation: A sustainable strategy. *Nanomaterials (Basel, Switzerland), 12*(23), 4187. DOI: 10.3390/nano12234187 PMID: 36500810

Mallawaarachchi, V., Wickramarachchi, A., & Lin, Y. (2020). GraphBin: Refined binning of metagenomic contigs using assembly graphs. *Bioinformatics (Oxford, England), 36*(11), 3307–3313. DOI: 10.1093/bioinformatics/btaa180 PMID: 32167528

Manisalidis, I., Stavropoulou, E., Stavropoulos, A., & Bezirtzoglou, E. (2020). Environmental and Health Impacts of Air Pollution: A review. *Frontiers in Public Health, 8*, 14. Advance online publication. DOI: 10.3389/fpubh.2020.00014 PMID: 32154200

Manni, M., Berkeley, M. R., Seppey, M., & Zdobnov, E. M. (2021). BUSCO: Assessing genomic data quality and beyond. *Current Protocols, 1*(12), e323. DOI: 10.1002/cpz1.323 PMID: 34936221

Maqsood, Q., Hussain, N., Sumrin, A., Ali, S. W., Tariq, M. R., & Mahnoor, M. (2024). Monitoring and abatement of synthetic pollutants using engineered microbial systems. *Discover Life, 54*(1), 9. DOI: 10.1007/s11084-024-09652-7

Margesin, R., & Schinner, F. (2001). Potential of halotolerant and halophilic microorganisms for biotechnology. *Extremophiles, 5*(2), 73–83. DOI: 10.1007/s007920100184 PMID: 11354458

Martias, C., Baroukh, N., Mavel, S., Blasco, H., Lefèvre, A., Roch, L., Montigny, F., Gatien, J., Schibler, L., Dufour-Rainfray, D., Nadal-Desbarats, L., & Emond, P. (2021). Optimization of sample preparation for metabolomics exploration of urine, feces, blood and saliva in humans using combined NMR and UHPLC-HRMS platforms. *Molecules (Basel, Switzerland), 26*(14), 4111. DOI: 10.3390/molecules26144111 PMID: 34299389

Martinez, B., Tomkins, J., Wackett, L. P., Wing, R., & Sadowsky, M. J. (2001). Complete nucleotide sequence and organization of the atrazine catabolic plasmid pADP-1 from *Pseudomonas* sp. ADP. *Journal of Bacteriology, 183*(19), 5684–5697. DOI: 10.1128/JB.183.19.5684-5697.2001 PMID: 11544232

Martínez-García, E., Nikel, P. I., Aparicio, T., & De Lorenzo, V. (2014). Pseudomonas 2.0: Genetic upgrading of *P. putida* KT2440 as an enhanced host for heterologous gene expression. *Microbial Cell Factories*, *13*(1), 159. Advance online publication. DOI: 10.1186/s12934-014-0159-3 PMID: 25384394

Martinez, J. L. (2008). Antibiotics and antibiotic resistance genes in natural environments. *Science*, *321*(5887), 365–367. DOI: 10.1126/science.1159483 PMID: 18635792

Marycz, M., Brillowska-Dąbrowska, A., Muñoz, R. A., & Gębicki, J. (2022). State of the art review on the use of fungi in biofiltration to remove volatile hydrophobic pollutants. *Reviews in Environmental Science and Biotechnology*, *21*(1), 225–246. DOI: 10.1007/s11157-021-09608-7

Mateo-Sagasta, J., Zadeh, S. M., Turral, H., & Burke, J. (2017). Water pollution from agriculture: A global review. Executive summary.

Mathuria, A., Jain, K., Saini, A., Verma, C., & Mani, I. (2024). Metatranscriptomics, metaproteomics, and metabolomics approaches for microbiome characterization. In *Multi-omics analysis of the human microbiome: From technology to clinical applications* (pp. 115–147). Springer. DOI: 10.1007/978-981-97-1844-3_6

Matlock, M., Hopfer, S., & Ogunseitan, O. A. (2019). Communicating risk for a climate-sensitive disease: A case study of Valley Fever in Central California. *International Journal of Environmental Research and Public Health, 16*(17), 3254. DOI: 10.3390/ijerph16183254 PMID: 31491843

Matsumura, Y., Yamamoto, M., Gomi, R., Tsuchido, Y., Shinohara, K., Noguchi, T., & Nagao, M. (2025). Integrating whole genome sequencing into antimicrobial resistance surveillance: Methodologies, challenges, and perspectives. *Clinical Microbiology Reviews*, *38*(4), e00140–e22. DOI: 10.1128/cmr.00140-22 PMID: 40910632

Maurya, A., Sharma, D., Partap, M., Kumar, R., & Bhargava, B. (2023). Microbially-assisted phytoremediation toward air pollutants: Current trends and future directions. *Environmental Technology & Innovation*, *31*, 103140. DOI: 10.1016/j.eti.2023.103140

Mawarda, P. C., Le Roux, X., Acosta, M. U., van Elsas, J. D., & Salles, J. F. (2023). The impact of protozoa addition on the survivability of *Bacillus* inoculants and soil microbiome dynamics. *ISME Communications*, *2*(1), 82. DOI: 10.1038/s43705-022-00166-9 PMID: 37938668

Mayans, B., Antón-Herrero, R., García-Delgado, C., Delgado-Moreno, L., Guirado, M., Pérez-Esteban, J., Escolástico, C., & Eymar, E. (2024). Bioremediation of petroleum hydrocarbons polluted soil by spent mushroom substrates: Microbiological structure and functionality. *Journal of Hazardous Materials*, *473*, 134650. DOI: 10.1016/j.jhazmat.2024.134650 PMID: 38776816

McNulty, R., Sritharan, D., Pahng, S. H., Meisch, J. P., Liu, S., Brennan, M. A., Saxer, G., Hormoz, S., & Rosenthal, A. Z. (2023). Probe-based bacterial single-cell RNA sequencing predicts toxin regulation. *Nature Microbiology*, *8*(5), 934–945. DOI: 10.1038/s41564-023-01348-4 PMID: 37012420

McReynolds, E., Elshahed, M. S., & Youssef, N. H. (2025). An ecological-evolutionary perspective on the genomic diversity and habitat preferences of the Acidobacteriota. *Microbial Genomics*, *11*(1), 001344. DOI: 10.1099/mgen.0.001344 PMID: 39879090

Mdeni, N. L., Adeniji, A. O., Okoh, A. I., & Okoh, O. O. (2022). Analytical evaluation of carbamate and organophosphate pesticides in human and environmental matrices: A review. *Molecules (Basel, Switzerland)*, *27*(3), 618. DOI: 10.3390/molecules27030618 PMID: 35163876

Mearns, A. J. (1997). Cleaning oiled shores: Putting bioremediation to test. *Spill Science & Technology Bulletin*, *4*(4), 209–217. DOI: 10.1016/S1353-2561(98)00026-7

Megharaj, M., & Naidu, R. (2017). Soil and brownfield bioremediation. *Microbial Biotechnology*, *10*(5), 1244–1249. DOI: 10.1111/1751-7915.12840 PMID: 28834380

Megharaj, M., Ramakrishnan, B., Venkateswarlu, K., Sethunathan, N., & Naidu, R. (2011). Bioremediation approaches for organic pollutants: A critical perspective. *Environment International*, *37*(8), 1362–1375. DOI: 10.1016/j.envint.2011.06.003 PMID: 21722961

Meier-Kolthoff, J. P., Carbasse, J. S., Peinado-Olarte, R. L., & Göker, M. (2022). TYGS and LPSN: A database tandem for fast and reliable genome-based classification and nomenclature of prokaryotes. *Nucleic Acids Research*, *50*(D1), D801–D807. DOI: 10.1093/nar/gkab902 PMID: 34634793

Meier-Kolthoff, J. P., & Göker, M. (2019). TYGS is an automated high-throughput platform for state-of-the-art genome-based taxonomy. *Nature Communications*, *10*(1), 2182. DOI: 10.1038/s41467-019-10210-3 PMID: 31097708

Mekonnen, B. A., Aragaw, T. A., & Genet, M. B. (2024). Bioremediation of petroleum hydrocarbon contaminated soil: A review on principles, degradation mechanisms, and advancements. *Frontiers in Environmental Science*, *12*, 1354422. DOI: 10.3389/fenvs.2024.1354422

Mendes, R., Garbeva, P., & Raaijmakers, J. M. (2013). The rhizosphere microbiome: Significance of plant beneficial, plant pathogenic, and human pathogenic microorganisms. *FEMS Microbiology Reviews, 37*(5), 634–663. DOI: 10.1111/1574-6976.12028 PMID: 23790204

Meng, Q., Liu, S., Guo, Y., Hu, Y., Yu, Z., Bello, A., Wang, Z., Xu, W., & Xu, X. (2023). The co-occurrence network patterns and keystone species of microbial communities in cattle manure-corn straw composting. *Environmental Science and Pollution Research International, 30*(8), 20265–20276. DOI: 10.1007/s11356-022-23599-0 PMID: 36251182

Meron, D., Atias, E., Iasur Kruh, L., Elifantz, H., Minz, D., Fine, M., & Banin, E. (2011). The impact of reduced pH on the microbial community of the coral *Acropora eurystoma. The ISME Journal, 5*(1), 51–60. DOI: 10.1038/ismej.2010.102 PMID: 20668489

Mihalakakou, G., Souliotis, M., Papadaki, M., Menounou, P., Dimopoulos, P., Kolokotsa, D., Gobakis, K., Alexandri, E., Koronaki, I., Paravantis, J., Giannakopoulos, E., Kalaitzoglou, A., Karagiozis, D., & Santamouris, M. (2023). Green roofs as a nature-based solution for improving urban sustainability: Progress and perspectives. *Renewable & Sustainable Energy Reviews, 180*, 113306. DOI: 10.1016/j.rser.2023.113306

Miller, M. E., Hamann, M., & Kroon, F. J. (2020). Bioaccumulation and biomagnification of microplastics in marine organisms: A review and meta-analysis of current data. *PLoS One, 15*(10), e0240792. DOI: 10.1371/journal.pone.0240792 PMID: 33064755

Minhas, W. R., Saleemi, H., Raheem, M. F., Khalid, S. S., Aslam, H., & Ahmad, W. (2024). CRISPR-Engineered Microbes for Enhanced biodegradation of recalcitrant pollutants a genomic approach to environmental remediation. *Frontiers in Health Informatics*, •••, 8281–8304. DOI: 10.63682/fhi787

Miralles, I., Ortega, R., & Comeau, A. M. (2021). Functional and taxonomic effects of organic amendments on the restoration of semiarid quarry soils. *mSystems, 6*(6), e00752–e21. DOI: 10.1128/mSystems.00752-21 PMID: 34812648

Mirete, S., Sánchez-Costa, M., Díaz-Rullo, J., González de Figueras, C., Martínez-Rodríguez, P., & González-Pastor, J. E. (2025). Metagenome-assembled genomes (MAGs): Advances, challenges, and ecological insights. *Microorganisms, 13*(5), 985. DOI: 10.3390/microorganisms13050985 PMID: 40431158

Mishra, P. N., Jain, S., Bore, T., Chang, I., Kwon, Y., Wang, Y., Dash, H. R., Kumar, A., Tiwari, S., Jiang, N., Das, S. K., & Scheuermann, A. (2024). Biological perspectives in geotechnics: Application and monitoring. *Journal of Rock Mechanics and Geotechnical Engineering*, *16*(7), 2854–2878. DOI: 10.1016/j.jrmge.2023.10.007

Mishra, T., Tiwari, P. B., Kanchan, S., & Kesheri, M. (2025). Advances in microbial bioremediation for effective wastewater treatment. *Water (Basel)*, *17*(22), 3196. DOI: 10.3390/w17223196

Mitra, S., Chakraborty, A. J., Tareq, A. M., Emran, T. B., Nainu, F., Khusro, A., Idris, A. M., Khandaker, M. U., Osman, H., Alhumaydhi, F. A., & Simal-Gandara, J. (2022). Impact of heavy metals on the environment and human health: Novel therapeutic insights to counter toxicity. *Journal of King Saud University. Science*, *34*(3), 101865. DOI: 10.1016/j.jksus.2022.101865

Mock, J., Schühle, K., Linne, U., Mock, M., & Heider, J. (2024). A Synthetic Pathway for the Production of Benzylsuccinate in *Escherichia coli. Molecules (Basel, Switzerland)*, *29*(2), 415. DOI: 10.3390/molecules29020415 PMID: 38257328

Mohamadi, S., Madadi, R., Rakib, M. R. J., De-la-Torre, G. E., & Idris, A. M. (2023). Abundance and characterization of personal protective equipment (PPE) polluting Kish Island, *Persian Gulf. The Science of the Total Environment*, *854*, 158678. DOI: 10.1016/j.scitotenv.2022.158678 PMID: 36099950

Mohan, A. J., Sekhar, V. C., Bhaskar, T., & Nampoothiri, K. M. (2016). Microbial assisted high impact polystyrene (HIPS) degradation. *Bioresource Technology*, *213*, 204–207. DOI: 10.1016/j.biortech.2016.03.021 PMID: 26993201

Mohan, L., Cherian, E., & Joshi, T. J. (2022). Emerging issues and challenges for plastic bioremediation. In *Microbes and Microbial Biotechnology for Green Remediation* (pp. 589–600). Elsevier., DOI: 10.1016/B978-0-323-90452-0.00010-4

Mongkolthanaruka, W., & Dharmsthiti, S. (2002). Biodegradation of lipid-rich wastewater by a mixed bacterial consortium. *International Biodeterioration & Biodegradation*, *50*(2), 101–103. DOI: 10.1016/S0964-8305(02)00057-4

Monteux, S., Weedon, J. T., Blume-Werry, G., Gavazov, K., Jassey, V. E. J., Johansson, M., Keuper, F., Olid, C., & Dorrepaal, E. (2018). Long-term in situ permafrost thaw effects on bacterial communities and potential aerobic respiration. *The ISME Journal*, *12*(9), 2129–2141. DOI: 10.1038/s41396-018-0176-z PMID: 29875436

Morakinyo, O. M., & Ana, G. R. E. E. (2025). Seasonal levels and size fractions of indoor air bioaerosols as predictors of respiratory morbidities among school pupils in Ibadan, Nigeria. *Frontiers in Public Health, 13*, 1611167. DOI: 10.3389/fpubh.2025.1611167 PMID: 41112650

Morán, A. C., Müller, A., Manzano, M., & González, B. (2010). Simazine treatment history determines a significant herbicide degradation potential in soils that is not improved by bioaugmentation with *Pseudomonas* sp. ADP. *Journal of Applied Microbiology, 101*(1), 26–35. DOI: 10.1111/j.1365-2672.2006.02990.x PMID: 16834588

Mor, R., & Sivan, A. (2008). Biofilm formation and partial biodegradation of polystyrene by the actinomycete *Rhodococcus ruber*: Biodegradation of polystyrene. *Biodegradation, 19*(6), 851–858. DOI: 10.1007/s10532-008-9188-0 PMID: 18401686

Morris, C. E., Sands, D. C., Glaux, C., Samsatly, J., Asaad, S., Moukahel, A. R., Gonçalves, F. L. T., & Bigg, E. K. (2013). Urediospores of rust fungi are ice nucleation active at > −10 °C and harbor ice nucleation active bacteria. *Atmospheric Chemistry and Physics, 13*(8), 4223–4233. DOI: 10.5194/acp-13-4223-2013

Mudge, S. M. (1995). Deleterious effects from accidental spillages of vegetable oils. *Spill Science & Technology Bulletin, 2*(2-3), 187–191. DOI: 10.1016/S1353-2561(96)00019-9

Mu, F., Li, Y., Lu, R., Qi, Y., Xie, W., & Bai, W. (2019). Source identification of airborne bacteria in the mountainous area and the urban areas. *Atmospheric Research, 231*, 104676. DOI: 10.1016/j.atmosres.2019.104676

Müller, R. J., Schrader, H., Profe, J., Dresler, K., & Deckwer, W. D. (2005). Enzymatic degradation of poly (ethylene terephthalate): Rapid hydrolyse using a hydrolase from T. fusca. *Macromolecular Rapid Communications, 26*(17), 1400–1405. DOI: 10.1002/marc.200500410

Mustapha, M. U., & Halimoon, N. (2019). An overview on biodegradation of carbamate pesticides by soil bacteria. *Pertanika Journal of Science & Technology, 27*(2).

Nasrabadi, T., Nabi Bidhendi, G., Karbassi, A., Grathwohl, P., & Mehrdadi, N. (2011). Impact of major organophosphate pesticides used in agriculture to surface water and sediment quality (Southern Caspian Sea basin, Haraz River). *Environmental Earth Sciences, 63*(4), 873–883. DOI: 10.1007/s12665-010-0757-2

National Volatile Organic Compound Emission Standards for Aerosol Coatings: Interim Final rule. (2025, July 2). Federal Register. https://www.federalregister.gov/documents/2025/07/02/2025-12323/national-volatile-organic-compound-emission-standards-for-aerosol-coatings-interim-final-rule

Nebauer, D. J., Pearson, L. A., & Neilan, B. A. (2024). Critical steps in an environmental metaproteomics workflow. *Environmental Microbiology*, *26*(6), e16637. DOI: 10.1111/1462-2920.16637 PMID: 38760994

Nguyen, Q.-H., Nguyen, H., Oh, E. C., & Nguyen, T. (2024). Current approaches and outstanding challenges of functional annotation of metabolites: A comprehensive review. *Briefings in Bioinformatics*, *25*(6), bbae498. DOI: 10.1093/bib/bbae498 PMID: 39397425

Nihart, A. J., Garcia, M. A., Hyek, E. E., Liu, R., Olewine, M., Kingston, J. D., Castillo, E. F., Gullapalli, R. R., Howard, T., Bleske, B., Scott, J., Gonzalez-Estrella, J., Gross, J. M., Spilde, M., Adolphi, N. L., Gallego, D. F., Jarrell, H. S., Dvorscak, G., Zuluaga-Ruiz, M. E., & Campen, M. J. (2025). Bioaccumulation of plastics in decent human brains. *Nature Medicine*, *31*(4), 1114–1119. DOI: 10.1038/s41591-024-03453-1 PMID: 39901044

Nikiema, J., Mateo-Sagasta, J., Asiedu, Z., Saad, D., & Lamizana, B. (2020). Water pollution by plastics and microplastics: A review of technical solutions from source to sea.

Nizamani, M. M., Hughes, A. C., Qureshi, S., Zhang, Q., Tarafder, E., Das, D., Acharya, K., Wang, Y., & Zhang, Z.-G. (2024). Microbial biodiversity and plant functional trait interactions in multifunctional ecosystems. *Applied Soil Ecology*, *201*, 105515. DOI: 10.1016/j.apsoil.2024.105515

Noecker, C., Chiu, H.-C., McNally, C. P., & Borenstein, E. (2019). Defining and evaluating microbial contributions to metabolite variation in microbiome-metabolome association studies. *mSystems*, *4*(1), e00579–e19. DOI: 10.1128/msystems.00579-19 PMID: 31848305

Nogales, B., Moore, E. R. B., Abraham, W. R., & Timmis, K. N. (1999). Identification of the metabolically-active members of a bacterial community in a polychlorinated biphenyl polluted Moorland soil. *Environmental Microbiology*, *1*(3), 199–212. DOI: 10.1046/j.1462-2920.1999.00024.x PMID: 11207739

Nwankwegu, A. S., Zhang, L., Xie, D., Onwosi, C. O., Muhammad, W. I., Odoh, C. K., Sam, K., & Idenyi, J. N. (2021). Bioaugmentation as a green technology for hydrocarbon pollution remediation. Problems and prospects. *Journal of Environmental Management*, *304*, 114313. DOI: 10.1016/j.jenvman.2021.114313 PMID: 34942548

Oberbeckman, S., Kreikemeyer, B., & Labrenz, M. (2018). Environmental factors support the function of specific bacterial assemblages on microplastics. *Frontiers in Microbiology*, *8*, 2709. DOI: 10.3389/fmicb.2017.02709 PMID: 29403454

Ochman, H., Lawrence, J. G., & Groisman, E. A. (2000). Lateral gene transfer and the nature of bacterial innovation. *Nature, 405*(6784), 299–304. DOI: 10.1038/35012500 PMID: 10830951

Oda, K., & Wlodawer, A. (2024). Development of enzyme-based approaches for recycling PET on an industrial scale. *Biochemistry, 63*(4), 369–401. DOI: 10.1021/acs.biochem.3c00554 PMID: 38285602

Ogunseitan, O. A. (2005). *Microbial diversity* (Foreword by L. Margulis). Wiley-Blackwell Publishing. Print ISBN:9780632047086, Online ISBN:9780470750490

Ogunseitan, O. A. (2019). Embracing global warmth and climate resilience through green chemistry legislation. *Hastings Environmental Law Journal, 25*(2), 301–318. https://repository.uclawsf.edu/hastings_environmental_law_journal/vol25/iss2/5

Ogunseitan, O. A. (2026). Bacterial diversity, introduction to. In J. B. Wolf & C. A. de M. Russo (Eds.), *Encyclopedia of evolutionary biology* (2nd ed., Vol. 5, pp. 42–47). Elsevier, Academic Press. https://doi.org/DOI: 10.1016/B978-0-443-15750-9.00073-2

Ogunseitan, O. A. (1993). Direct extraction of proteins from environmental samples. *Journal of Microbiological Methods, 17*(4), 273–281. DOI: 10.1016/0167-7012(93)90056-N

Ogunseitan, O. A. (1996). Protein profile variation in cultivated and native freshwater microorganisms exposed to chemical environmental pollutants. *Microbial Ecology, 31*(3), 291–304. DOI: 10.1007/BF00171573 PMID: 8661530

Ogunseitan, O. A. (1998). Protein method for investigating mercuric reductase gene expression in aquatic environments. *Applied and Environmental Microbiology, 64*(2), 695–702. DOI: 10.1128/AEM.64.2.695-702.1998 PMID: 9464410

Ogunseitan, O. A. (2000). Microbial proteins as biomarkers of ecosystem health. In Scow, K., Fogg, G., Hinton, D., & Johnson, M. L. (Eds.), *Integrated assessment of ecosystem health* (pp. 207–222). CRC Press., DOI: 10.1201/9780429104404

Ogunseitan, O. A. (2008). Genetic transduction in freshwater ecosystems. *Freshwater Biology, 53*(6), 1228–1239. DOI: 10.1111/j.1365-2427.2007.01857.x

Ogunseitan, O. A. (2016). Power failure: The battered legacy of leaded batteries. *Environmental Science & Technology, 50*(17), 8997–8998. DOI: 10.1021/acs.est.6b03174 PMID: 27467531

Ogunseitan, O. A. (2017). Mercury safety reform in the 21st century: Advancing the new framework for toxic substances control. *Environment, 59*(4), 4–13. DOI: 10.1080/00139157.2017.1325299

Ogunseitan, O. A. (2022). One Health and the environment: From conceptual framework to implementation science. *Environment, 64*(2), 4–24. DOI: 10.1080/00139157.2022.2021792

Ogunseitan, O. A. (2023). Chemicals management approach to sustainable development of materials. *MRS Bulletin, 48*(5), 491–497. DOI: 10.1557/s43577-023-00518-3

Ogunseitan, O. A. (2025). Perspective—Planetary boundaries for recalcitrant materials and toxic chemical pollutants: Specifications for sustainable safe operating zones. *Frontiers in Environmental Science, 13*, 1593947. DOI: 10.3389/fenvs.2025.1593947

Ogunseitan, O. A., Gussin, G., Jimah, T., Wang, S., Fenny, A., Bada-Alambedji, R., & Andre, B. (2022). *The ten commandments of antibiotic stewards.* CABI One Health Cases., DOI: 10.1079/onehealthcases.2022.0003

Ogunseitan, O. A., & LeBlanc, J. (2005). Environmental proteomics: Methods and applications for aquatic ecosystems. *Molecular Microbial Ecology Manual, 4*(10), 1027–1046.

Ogunseitan, O. A., LeBlanc, J., & Dalmasso, E. (2001). Microbial community proteomics. In Rochelle, P. A. (Ed.), *Environmental molecular microbiology* (pp. 125–140). Horizon Scientific Press.

Ogunseitan, O. A., & Matlock, M. (2022). *Mouldy planet: Fungi and One Health.* CABI One Health Case Studies., DOI: 10.1079/onehealthcases.2022.0004

Ogunseitan, O. A., Yang, S. L., & Scheinbach, E. (1999). The delta-aminolevulinate dehydratase of marine *Vibrio alginolyticus* is resistant to lead (Pb). *The Biological Bulletin, 197*(2), 283–284. DOI: 10.2307/1542653 PMID: 28281822

Ogunyemi, A. K., Buraimoh, O. M., Ogunyemi, B. C., Samuel, T. A., Ilori, M. O., & Amund, O. O. (2022). Nitrilase gene detection and nitrile metabolism in two bacterial strains associated with waste streams in Lagos, Nigeria. *Bulletin of the National Research Center, 46*(1), 151. DOI: 10.1186/s42269-022-00804-1

Ogunyemi, A. K., Samuel, T. A., Amund, O. O., & Ilori, M. O. (2019a). Cassava wastewater and solid waste leachate as cyanogenic substrates for the growth of nitrile and linamarin-utilizing bacteria. *Journal of Tropical Life Science, 9*(1), 79–87. DOI: 10.11594/jtls.09.01.11

Ogunyemi, A. K., Samuel, T. A., Amund, O. O., & Ilori, M. O. (2019b). Nitrile-metabolizing bacterial strains associated with municipal waste tips in the Lagos metropolis, Nigeria. Polish Journal of Natural Sciences. *Polish Journal of Natural Sciences*, *34*(2), 257–272. https://ir.unilag.edu.ng/handle/123456789/6750

Ogunyemi, A. K., Samuel, T. A., Amund, O. O., & Ilori, M. O. (2019c). Characterization of linamarin-utilizing bacterial strains associated with detoxification of cyanogens in waste effluents. *Malaysian Applied Biology*, *48*(3), 157–165. https://jms.mabjournal.com/index.php/mab/article/view/1851

Ojala, T., Häkkinen, A.-E., Kankuri, E., & Kankainen, M. (2023). Current concepts, advances, and challenges in deciphering the human microbiota with metatranscriptomics. *Trends in Genetics*, *39*(9), 686–702. DOI: 10.1016/j.tig.2023.05.004 PMID: 37365103

Okafor, N., Ijioma, B., & Ijeoma, C. (2018). Microbial diversity and detoxification potential of cassava waste dumpsite soils. *Environmental Research Journal*, *12*(3), 122–130.

Okoh, A. I., Sibanda, T., & Gusha, S. S. (2010). Inadequately treated wastewater as a source of human enteric viruses in the environment. *International Journal of Environmental Research and Public Health*, *7*(6), 2620–2637. DOI: 10.3390/ijerph7062620 PMID: 20644692

Olanrewaju, O. S., Glick, B. R., & Babalola, O. O. (2024). Metabolomics-guided utilization of beneficial microbes for climate-resilient crops. *Current Opinion in Chemical Biology*, *79*, 102427. DOI: 10.1016/j.cbpa.2024.102427 PMID: 38290195

Oliveira, K., Guevara, M., Jorba, O., Petetin, H., Bowdalo, D., Tena, C., Montané Pinto, G., López, F., & Pérez García-Pando, C. (2024). On the uncertainty of anthropogenic aromatic volatile organic compound emissions: Model evaluation and sensitivity analysis. *Atmospheric Chemistry and Physics*, *24*(12), 7137–7177. DOI: 10.5194/acp-24-7137-2024

Omotayo, A. E., & Kukoyi, G. A. (2025). Biocatalytic degradation pathway of atrazine. Canva.com https://www.canva.com

Omotayo, A. E., Jebe-Abdullahi, F. T., & Kukoyi, G. A. (2025). Circulation of materials in urban agroecosystem. Canva.com https://www.canva.com

Omotayo, A. E., Ilori, M. O., Amund, O. O., Ghosh, D., Roy, K., & Radosevich, M. (2011). Establishment and characterization of atrazine degrading cultures from Nigerian agricultural soil using traditional and Bio-Sep bead enrichment techniques. *Applied Soil Ecology*, *48*(1), 63–70. DOI: 10.1016/j.apsoil.2011.01.006

Omotayo, A. E., Ilori, M. O., Obayori, O. S., & Amund, O. O.Omotayo, A. E.Ilori, M. O.Obayori, O. S.Amund, O. O. (2016). Influence of pH, temperature and nutrient addition on the degradation of atrazine by *Nocardioides* spp. isolated from agricultural soil in Nigeria. *Malaysian Journal of Microbiology, 12*(4), 270–278. DOI: 10.21161/mjm.81315

Omotayo, A. E., Ilori, M. O., Radosevich, M., & Amund, O. O. (2013). Metabolism of atrazine in liquid cultures and soil microcosms by *Nocardioides* strains isolated from a contaminated Nigerian agricultural soil. *Soil & Sediment Contamination, 22*(4), 365–375. DOI: 10.1080/15320383.2013.733444

Omotayo, A. E., Shonubi, O. O., Towuru, E. G., Babalola, S. E., Ilori, M. O., & Amund, O. O. (2014). Rhizoremediation of hydrocarbon-contaminated soil by *Paspalum vaginatum* (Sw.) and its associated bacteria. *International Research Journal of Microbiology, 5*(1), 1–7. DOI: 10.14303/irjm.2014.0102

Onyena, A. P., & Sam, K. (2020). A review of the threat of oil exploitation to mangrove ecosystem: Insights from Niger Delta, Nigeria. *Global Ecology and Conservation, 22*, e00961. DOI: 10.1016/j.gecco.2020.e00961

Orellana, R., Macaya, C., Bravo, G., Dorochesi, F., Cumsille, A., Valencia, R., & Seeger, M. (2018). Living at the frontiers of life: Extremophiles in Chile and their potential for bioremediation. *Frontiers in Microbiology, 9*, 2309. DOI: 10.3389/fmicb.2018.02309 PMID: 30425685

Oro, C. E. D., Saorin Puton, B. M., Venquiaruto, L. D., Dallago, R. M., & Tres, M. V. (2024). Effective microbial strategies to remediate contaminated agricultural soils and conserve functions. *Agronomy (Basel), 14*(11), 2637. DOI: 10.3390/agronomy14112637

Orr, I. G., Hadar, Y., & Sivan, A. (2004). Colonization, biofilm formation and biodegradation of polyethylene by a strain of *Rhodococcus ruber. Applied Microbiology and Biotechnology, 65*, 97–104. 101007.s00253-004-1584-8 PMID: 15221232

Orta-Ortiz, M. S., & Geneletti, D. (2023). Prioritizing urban nature-based solutions to support scaling-out strategies: A case study in Las Palmas de Gran Canaria. *Environmental Impact Assessment Review, 102*, 107158. DOI: 10.1016/j.eiar.2023.107158

Osuagwu, E. S., & Olaifa, E. (2018). Effects of oil spills on fish production in the Niger Delta. *PLoS One, 13*(10), e0205114. DOI: 10.1371/journal.pone.0205114 PMID: 30359365

Osunla, C. A., & Okoh, A. I. (2017). Vibrio pathogens: A public health concern in rural water resources in sub-Saharan Africa. *International Journal of Environmental Research and Public Health, 14*(10), 1188. DOI: 10.3390/ijerph14101188 PMID: 28991153

Othman, A. R., Hasan, H. A., Muhamad, M. H., Ismail, N. I., & Abdullah, S. R. S. (2021). Microbial degradation of microplastics by enzymatic processes: A review. *Environmental Chemistry Letters, 19*(4), 3057–3073. DOI: 10.1007/s10311-021-01197-9

Oyebamiji, O. O., Boeing, W. J., Holguin, F. O., Ilori, O. M., & Amund, O. O. (2019). Green microalgae cultured in textile wastewater for biomass generation and biodetoxification of heavy metals and chromogenic substances. *Bioresource Technology Reports, 7*, 100247. DOI: 10.1016/j.biteb.2019.100247

Oyebamiji, O. O., Corcoran, A. A., Pérez, E. N., Ilori, M. O., Amund, O. O., Holguin, F. O., & Boeing, W. J. (2021). Lead tolerance and bioremoval by four strains of green algae from Nigerian fish ponds. *Algal Research, 58*, 102403. DOI: 10.1016/j.algal.2021.102403

Özdemir, S., Akarsu, C., Acer, Ö., Fouillaud, M., Dufossé, L., & Dizge, N. (2022). Isolation of thermophilic bacteria and investigation of their microplastic degradation ability using polyethylene polymers. *Microorganisms, 10*(12), 2441. DOI: 10.3390/microorganisms10122441 PMID: 36557694

Pace, N. R. (1997). A molecular view of microbial diversity and the biosphere. *Science, 276*(5313), 734–740. DOI: 10.1126/science.276.5313.734 PMID: 9115194

Pachaiappan, R., Cornejo-Ponce, L., Rajendran, R., Manavalan, K., Rajan, V. F., & Awad, F. (2022). A review on biofiltration techniques: Recent advancements in the removal of volatile organic compounds and heavy metals in the treatment of polluted water. *Bioengineered, 13*(4), 8432–8477. DOI: 10.1080/21655979.2022.2050538 PMID: 35260028

Padilla-Garfias, F., Araiza-Villanueva, M., Calahorra, M., Sánchez, N. S., & Peña, A. (2024). Advances in the degradation of polycyclic aromatic hydrocarbons by yeasts: A review. *Microorganisms, 12*(12), 2484. DOI: 10.3390/microorganisms12122484 PMID: 39770687

Pande, V., Pandey, S. C., Sati, D., Bhatt, P., & Samant, M. (2022). Microbial interventions in bioremediation of heavy metal contaminants in agroecosystem. *Frontiers in Microbiology, 13*, 824084. DOI: 10.3389/fmicb.2022.824084 PMID: 35602036

Pandey, P., & Yadav, R. (2018). A review on Volatile Organic Compounds (VOCs) as Environmental Pollutants: Fate and Distribution. *International Journal of Plant and Environment*, *4*(02), 14–26. DOI: 10.18811/ijpen.v4i02.2

Pandey, S. B., & Yadav, R. S. (2015). Microbial degradation of cyanide compounds: A review. *International Journal of Advanced Research*, *3*(5), 1223–1231.

Papen, H., & Von Berg, R. (1998). A most probable number method (MPN) for the estimation of cell numbers of heterotrophic nitrifying bacteria in soil. *Plant and Soil*, *199*(1), 123–130. DOI: 10.1023/A:1004243810473

Parashar, N., Mahanty, B., & Hait, S. (2023). Microplastics as carriers of per-and polyfluoroalkyl substances (PFAS) in aquatic environment: Interactions and ecotoxicological effects. *Water Emerging Contaminants & Nanoplastics*, *2*(7), 15. Advance online publication. DOI: 10.20517/wecn.2023.25

Parks, D. H., Chuvochina, M., Chaumeil, P. A., Rinke, C., Mussig, A. J., & Hugenholtz, P. (2020). A complete domain-to-species taxonomy for bacteria and archaea based on a standardized genome phylogeny. *Nature Biotechnology*, *38*(9), 1079–1086. DOI: 10.1038/s41587-020-0501-8 PMID: 32341564

Parks, D. H., Chuvochina, M., Rinke, C., Mussig, A. J., Chaumeil, P. A., & Hugenholtz, P. (2022). GTDB: An ongoing census of bacterial and archaeal diversity through a phylogenetically consistent, rank normalized and complete genome-based taxonomy. *Nucleic Acids Research*, *50*(D1), D785–D794. DOI: 10.1093/nar/gkab776 PMID: 34520557

Parks, D. H., Imelfort, M., Skennerton, C. T., Hugenholtz, P., & Tyson, G. W. (2015). CheckM: Assessing the quality of microbial genomes recovered from isolates, single cells, and metagenomes. *Genome Research*, *25*(7), 1043–1055. DOI: 10.1101/gr.186072.114 PMID: 25977477

Parsaeimehr, A., Miller, C. M., & Ozbay, G. (2023). Microplastics and their interactions with microbiota. *Heliyon*, *9*(4), e15104. DOI: 10.1016/j.heliyon.2023.e15104 PMID: 37089279

Parte, S. G., Mohekar, A. D., & Kharat, A. S. (2017). Microbial degradation of pesticide: A review. *African Journal of Microbiological Research*, *11*(24), 992–1012. DOI: 10.5897/AJMR2016.8402

Pathak, V., & Navneet, . (2017). Review on the current status of polymer degradation: A microbial approach. *Bioresources and Bioprocessing*, *4*(1), 15. DOI: 10.1186/s40643-017-0145-9

Patti, G. J., Yanes, O., & Siuzdak, G. (2012). Innovation: Metabolomics: The apogee of the omics trilogy. *Nature Reviews. Molecular Cell Biology*, *13*(4), 263–269. DOI: 10.1038/nrm3314 PMID: 22436749

Paul, C., Mausz, M. A., & Pohnert, G. (2013). A co-culturing/metabolomics approach to investigate chemically mediated interactions of planktonic organisms reveals influence of bacteria on diatom metabolism. *Metabolomics*, *9*(2), 349–359. DOI: 10.1007/s11306-012-0453-1

Paul, M., White, J. R., Duance, C., Wolf, J., Gregory, T., & Charles, R. M. (2006). Hydrocarbons in a crude oil-contaminated soil. *Water, Air, and Soil Pollution*, *169*(1-4), 207–220. DOI: 10.1007/s11270-006-2194-0

Peeters, J., Thas, O., Shkedy, Z., Kodalci, L., Musisi, C., Owokotomo, O. E., Dyczko, A., Hamad, I., Vangronsveld, J., Kleinewietfeld, M., Thijs, S., & Aerts, J. (2021). Exploring the microbiome analysis and visualization landscape. *Frontiers in Bioinformatics*, *1*, 774631. DOI: 10.3389/fbinf.2021.774631 PMID: 36303773

Pehlivan, N., Gedik, K., Eltem, R., & Terzi, E. (2021). Dynamic interactions of Trichoderma harzianum TS 143 from an old mining site in Turkey for potent metal (oid) s phytoextraction and bioenergy crop farming. *Journal of Hazardous Materials*, *403*, 123609. DOI: 10.1016/j.jhazmat.2020.123609 PMID: 32798794

Pereira, R., Oliveira, J., & Sousa, M. (2020). Bioinformatics and computational tools for next-generation sequencing analysis in clinical genetics. *Journal of Clinical Medicine*, *9*(1), 132. DOI: 10.3390/jcm9010132 PMID: 31947757

Pérez-García, P., Sass, K., Wongwattanarat, S., Amann, J., Feuerriegel, G., Neumann, T., Basc, N., Schmitz, L. S., Dierkes, R. F., Gurschke, M. F., Wypych, A., Bounabi, H., de Divitiis, M., Vollstedt, C., & Streit, W. R. (2025). Microbial plastic degradation: Enzymes, pathways, challenges, and perspectives. *Microbiology and Molecular Biology Reviews : MMBR*, *89*(4), e00087–e24. DOI: 10.1128/mmbr.00087-24 PMID: 40970732

Petrillo, J. E., & Ogunseitan, O. A. (2018). Emerging issues in the environmental context of antibiotic-resistance. *Environment International*, *116*, 39–42. DOI: 10.1016/j.envint.2018.03.049 PMID: 29649775

Petrova, S., Velcheva, I., Nikolov, B., Angelov, N., Hristozova, G., Zaprjanova, P., Mitova, T., & Petrov, P. (2022). Nature-based solutions for the sustainable management of urban soils and quality of life improvements. *Land (Basel)*, *11*(4), 569. DOI: 10.3390/land11040569

Pham, D. N., Clark, L., & Li, M. (2021). Microplastics as hubs enriching antibiotic-resistant bacteria and pathogens in municipal activated sludge. *Journal of Hazardous Materials Letters*, *2*, 100014. DOI: 10.1016/j.hazl.2021.100014

Philippot, L., Raaijmakers, J. M., Lemanceau, P., & Van Der Putten, W. H. (2013). Going back to the roots: The microbial ecology of the rhizosphere. *Nature Reviews. Microbiology*, *11*(11), 789–799. DOI: 10.1038/nrmicro3109 PMID: 24056930

Phulpoto, A. H., Maitlo, M. A., & Kanhar, N. A. (2021). Culture-dependent to culture-independent approaches for the bioremediation of paints: A review. *International Journal of Environmental Science and Technology*, *18*(1), 241–262. DOI: 10.1007/s13762-020-02801-1

Pieper, D. A. (2005). Aerobic degradation of polychlorinated biphenyl. *Applied Microbiology and Biotechnology*, *67*(2), 170–191. DOI: 10.1007/s00253-004-1810-4 PMID: 15614564

Pilapitiya, P. G. C. N. T., & Ratnayake, A. S. (2024). The world of plastic waste: A review. *Cleaner Materials*, *11*, 100220. Advance online publication. DOI: 10.1016/j.clema.2024.100220

Pinto, Y., & Bhatt, A. S. (2024). Sequencing-based analysis of microbiomes. *Nature Reviews. Genetics*, *25*(12), 829–845. DOI: 10.1038/s41576-024-00746-6 PMID: 38918544

Pinu, F. R., Goldansaz, S. A., & Jaine, J. (2019). Translational metabolomics: Current challenges and future opportunities. *Metabolites*, *9*(6), 108. DOI: 10.3390/metabo9060108 PMID: 31174372

Pletz, M. (2022). Ingested microplastics: Do humans eat one credit card per week? *Journal of Hazardous Materials Letters*, *3*, 100071. DOI: 10.1016/j.hazl.2022.100071

Pollack, K., Balasz, K., & Ogunseitan, O. A. (2009). Proteomic assessment of caffeine effects on coral symbionts. *Environmental Science & Technology*, *43*(6), 2085–2091. DOI: 10.1021/es802617f PMID: 19368218

Pouresmaieli, M., Ataei, M., Forouzandeh, P., Azizollahi, P., & Mahmoudifard, M. (2022). Recent progress on sustainable phytoremediation of heavy metals from soil. *Journal of Environmental Chemical Engineering*, *10*(5), 108482. DOI: 10.1016/j.jece.2022.108482

Pozdniakova, S., Uchida, A., Fontal, A., Cañas, L., Santamaria, S., Hui, L. Y., Luhung, I., Schuster, S. C., Rodó, X., & Borràs, S. (2025). Integrating air microbiome for comprehensive air quality analysis. *iScience*, *28*(7), 113015. DOI: 10.1016/j.isci.2025.113015 PMID: 40808715

Prasad, M. P., & Manjunath, K. (2011). Comparative study on biodegradation of lipid-rich wastewater using lipase-producing bacterial species. *Indian Journal of Biotechnology*, *10*(1), 121–124.

Precha, N., Chaisiri, K., Worakhunpiset, S., Limpanont, Y., Yamamoto, N., Suksong, W., Kliengchuay, W., & Tantrakarnapa, K. (2025). Comparison of airborne bacterial communities in PM2.5 between a dry-season haze period and a wet-season non-haze period in thailand. *Scientific Reports*, *15*(1), 12918. Advance online publication. DOI: 10.1038/s41598-025-97966-5 PMID: 40234570

Prenafeta-Boldú, F. X., Guivernau, M., Gallastegui, G., Viñas, M., Hoog, G. S., & Elías, A. (2012). Fungal/bacterial interactions during the biodegradation of TEX hydrocarbons (toluene, ethylbenzene and p-xylene) in gas biofilters operated under xerophilic conditions. *FEMS Microbiology Ecology*, *80*(3), 722–734. DOI: 10.1111/j.1574-6941.2012.01344.x PMID: 22380456

Prince, R. C. (1993). Petroleum spill bioremediation in marine environments. *Critical Reviews in Microbiology*, *19*(4), 217–242. DOI: 10.3109/10408419309113530 PMID: 8305136

Priya, A. K., Gnanasekaran, L., Dutta, K., Rajendran, S., Balakrishnan, D., & Soto-Moscoso, M. (2022). Biosorption of heavy metals by microorganisms: Evaluation of different underlying mechanisms. *Chemosphere*, *307*, 135957. DOI: 10.1016/j.chemosphere.2022.135957 PMID: 35985378

Prosser, J. I. (2015). Dispersing misconceptions and identifying opportunities for the use of 'omics' in soil microbial ecology. *Nature Reviews. Microbiology*, *13*(7), 439–446. DOI: 10.1038/nrmicro3468 PMID: 26052662

Prussin, A. J.II, & Marr, L. C. (2015). Sources of airborne microorganisms in the built environment. *Microbiome*, *3*(1), 78. Advance online publication. DOI: 10.1186/s40168-015-0144-z PMID: 26694197

Puniya, B. L., Verma, M., Damiani, C., Bakr, S., & Dräger, A. (2024). Perspectives on computational modeling of biological systems and the significance of the SysMod community. *Bioinformatics Advances*, *4*(1), vbae090. DOI: 10.1093/bioadv/vbae090 PMID: 38948011

Püttker, S., Kohrs, F., Benndorf, D., Heyer, R., Rapp, E., & Reichl, U. (2015). Metaproteomics of activated sludge from a wastewater treatment plant—A pilot study. *Proteomics*, *15*(20), 3596–3601. DOI: 10.1002/pmic.201400559 PMID: 26201837

Qamar, S. A., Bhatt, P., Ghotekar, S., & Bilal, M. (2022). Nanomaterials for bioremediation of air pollution. In *Elsevier eBooks* (pp. 243–261). https://doi.org/DOI: 10.1016/B978-0-12-823962-9.00008-8

Qattan, S. Y. A. (2025). Harnessing bacterial consortia for effective bioremediation: Targeted removal of heavy metals, hydrocarbons and persistent pollutants. *Environmental Sciences Europe*, *37*(1), 85. DOI: 10.1186/s12302-025-01103-y

Qayyum, H., Talib, M. S., Ali, A., & Kayani, M. U. R. (2025). Evaluating the potential of assembler-binner combinations in recovering low-abundance and strain-resolved genomes from human metagenomes. *Heliyon*, *11*(5), e41938. DOI: 10.1016/j.heliyon.2025.e41938 PMID: 39897886

Qi, Y., Li, Y., Xie, W., Lu, R., Mu, F., Bai, W., & Du, S. (2019). Temporal-spatial variations of fungal composition in PM2.5 and source tracking of airborne fungi in mountainous and urban regions. *The Science of the Total Environment*, *708*, 135027. DOI: 10.1016/j.scitotenv.2019.135027 PMID: 31787277

Quince, C., Walker, A. W., Simpson, J. T., Loman, N. J., & Segata, N. (2017). Shotgun metagenomics, from sampling to analysis. *Nature Biotechnology*, *35*(9), 833–844. DOI: 10.1038/nbt.3935 PMID: 28898207

Quinn, L. P., De Vos, B. J., Fernandes-Whaley, M., Roos, C., Bouwman, H., Kylin, H., & van den Berg, J. (2011). Pesticide use in South Africa: One of the largest importers of pesticides in Africa. *Pesticides in the Modern World-Pesticides Use and Management*, *4*, 50–95. DOI: 10.5772/16995

Rafeeq, H., Afsheen, N., Rafique, S., Arshad, A., Intisar, M., Hussain, A., Bilal, M., & Iqbal, H. M. N. (2023). Genetically engineered microorganisms for environmental remediation. *Chemosphere*, *310*, 136751. DOI: 10.1016/j.chemosphere.2022.136751 PMID: 36209847

Raffa, C. M., & Chiampo, F. (2021). Bioremediation of agricultural soils polluted with pesticides: A review. *Bioengineering (Basel, Switzerland)*, *8*(7), 92. DOI: 10.3390/bioengineering8070092 PMID: 34356199

Rahman, M. A., Dey, B., Halim, M. A., & Ahmed, R. (2025). Mobilizing Microbes for Bioremediation Strategies in the Context of Climate Change. In *Sustainable Remediation for Pollution and Climate Resilience* (pp. 315–346). https://doi.org/DOI: 10.1007/978-981-96-5674-5_13

Rahmati, F., Lajayer, B. A., Shadfar, N., Van Bodegom, P. M., & Van Hullebusch, E. D. (2022). A review on biotechnological approaches applied for marine hydrocarbon spills remediation. *Microorganisms, 10*(7), 1289. DOI: 10.3390/microorganisms10071289 PMID: 35889007

Rajendran, S., Al-Samydai, A., Palani, G., Trilaksana, H., Sathish, T., Giri, J., Saravanan, R., Lalvani, J. I. J., & Nasri, F. (2025). Replacement of petroleum based products with plant-based materials, green and sustainable energy—A review. *Engineering Reports, 7*(4), e70108. DOI: 10.1002/eng2.70108

Rakib, M. R. J., Sarker, A., Ram, K., Uddin, M. G., Walker, T. R., Chowdhury, T., Uddin, J., Khandaker, M. U., Rahman, M. M., & Idris, A. M. (2023). Microplastic toxicity in aquatic organisms and aquatic ecosystems: A review. *Water, Air, and Soil Pollution, 234*(1), 52. DOI: 10.1007/s11270-023-06062-9

Ramanathan, S. S., Gannon, T. W., Everman, W. J., & Locke, A. M. (2022). Atrazine, mesosulfuron-methyl, and topramezone persistence in North Carolina soils. *Agronomy Journal, 114*(2), 1068–1079. DOI: 10.1002/agj2.21041

Ramezani, M., Enayati, M., Ramezani, M., & Ghorbani, A. (2021). A study of different strategical views into heavy metal (oid) removal in the environment. *Arabian Journal of Geosciences, 14*(21), 2225. DOI: 10.1007/s12517-021-08572-4

Ramond, P., Galand, P. E., & Logares, R. (2025). Microbial functional diversity and redundancy: Moving forward. *FEMS Microbiology Reviews, 49*, fuae031. DOI: 10.1093/femsre/fuae031 PMID: 39689915

Rampelotto, P. H. (2013). Extremophiles and extreme environments. *Life (Chicago, Ill.), 3*(3), 482–485. PMID: 25369817

Rashid, A., Schutte, B. J., Ulery, A., Deyholos, M. K., Sanogo, S., Lehnhoff, E. A., & Beck, L. (2023). Heavy metal contamination in agricultural soil: Environmental pollutants affecting crop health. *Agronomy (Basel), 13*(6), 1521. DOI: 10.3390/agronomy13061521

Rashid, N., Ur Rehman, M. S., Han, J. I., & Park, J. Y. (2016). Current status, issues, and developments in cyanide biodegradation. *Biotechnology Advances, 34*(5), 874–881. PMID: 27188926

Rastogi, R. P., & Madamwar, D. (2021). Biotechnological potential of microbial consortia for sustainable environmental management. *Bioresource Technology, 337*, 125411.

Raupp, P., Carrillo, Y., & Nielsen, U. N. (2024). Soil health to enhance ecological restoration and conservation. *Journal of Sustainable Agriculture and Environment, 3*(4), e70022. DOI: 10.1002/sae2.70022

Ray, S., & Cooney, R. P. (2018). Thermal degradation of polymer and polymer composites. In *Handbook of Environmental Degradation of Materials* (pp. 185–206). William Andrew publishing., DOI: 10.1016/B978-0-323-52472-8.00009-5

Rekadwad, B. N. (2024). *Microbe hunting: Unveiling the secrets of microorganisms through assessment, sequencing, and bioinformatics analysis.* CRC Press. DOI: 10.1201/9781003492221

Renella, G., Ogunseitan, O. A., Giagnoni, L., & Arenella, M. (2014). Environmental proteomics: A long march in the pedosphere. *Soil Biology & Biochemistry, 69,* 34–37. DOI: 10.1016/j.soilbio.2013.10.035

Ren, L., Zhang, J., Geng, B., Zhao, J., Jia, W., & Cheng, L. (2025). Ecological Shifts and Functional Adaptations of Soil Microbial Communities Under Petroleum Hydrocarbon Contamination. *Water (Basel), 17*(8), 1216. DOI: 10.3390/w17081216

Richter, M., & Rosselló-Móra, R. (2009). Shifting the genomic gold standard for the prokaryotic species definition. *Proceedings of the National Academy of Sciences of the United States of America, 106*(45), 19126–19131. DOI: 10.1073/pnas.0906412106 PMID: 19855009

Rinke, C., Lee, J., Nath, N., Goudeau, D., Thompson, B., Poulton, N., Dmitrieff, E., Malmstrom, R., Stepanauskas, R., & Woyke, T. (2014). Obtaining genomes from uncultivated environmental microorganisms using FACS-based single-cell genomics. *Nature Protocols, 9*(5), 1038–1048. DOI: 10.1038/nprot.2014.067 PMID: 24722403

Rodríguez, E., García-Encina, P. A., Stams, A. J. M., Maphosa, F., & Sousa, D. Z. (2015). Meta-omics approaches to understand and improve wastewater treatment systems. *Reviews in Environmental Science and Biotechnology, 14*(3), 385–406. DOI: 10.1007/s11157-015-9370-x

Rodríguez-Rojas, F., Tapia, P., Castro-Nallar, E., Undabarrena, A., Muñoz-Díaz, P., Arenas-Salinas, M., & Vásquez, C. (2016). Draft genome sequence of a multi-metal resistant bacterium *Pseudomonas putida* ATH-43 isolated from Greenwich Island, Antarctica. *Frontiers in Microbiology, 7,* 1777. DOI: 10.3389/fmicb.2016.01777 PMID: 27877169

Rojo, F. (2009). Degradation of alkanes by bacteria. *Environmental Microbiology, 11*(10), 2477–2490. DOI: 10.1111/j.1462-2920.2009.01948.x PMID: 19807712

Rong, L., Zhao, L., Zhao, L., Cheng, Z., Yao, Y., Yuan, C., Wang, L., & Sun, H. (2021). LDPE microplastics affect soil microbial communities and nitrogen cycling. *The Science of the Total Environment*, *773*, 145640. DOI: 10.1016/j.scitotenv.2021.145640 PMID: 33582358

Rosato, A., Barone, M., Negroni, A., Brigidi, P., Fava, F., Biagi, E., Candela, M., & Zanaroli, G. (2022). Bacterial colonization dynamics of different microplastic types in an anoxic salt marsh sediment and impact of adsorbed polychlorinated biphenyls on the plastisphere. *Environmental Pollution*, *315*, 120411. DOI: 10.1016/j.envpol.2022.120411 PMID: 36240963

Rosato, A., Barone, M., Negroni, A., Brigidi, P., Fava, F., Xu, P., Candela, M., & Zanaroli, G. (2020). Microbial colonization of different microplastic types and biotransformation of sorbed PCBs by a marine anaerobic bacterial community. *The Science of the Total Environment*, *705*, 135790. DOI: 10.1016/j.scitotenv.2019.135790 PMID: 31972939

Roth, E., Tape, C., Michonneau, P., & Chakir, A. (2025). Time-dependent fluxes and emission factors of volatile organic compounds from pig manure and solid/liquid digestates. *The Science of the Total Environment*, *996*, 180147. DOI: 10.1016/j.scitotenv.2025.180147 PMID: 40743754

Rouf, A., Kanojia, V., Naik, H., Naseer, B., & Qadri, T. (2017). An overview of microbial cell culture. *Journal of Pharmacognosy and Phytochemistry*, *6*(1), 1923–1928.

Ruan, W., Lyu, Y., Zhang, J., Cai, J., Shu, P., Ge, Y., Lu, Y., Gao, S., Wang, Y., Wang, P., Zhao, L., Wang, T., Liu, Y., Fang, L., Liu, Z., Liu, Z., Li, Y., Wu, Z., Chen, J., & Liu, T. (2026). Large language models for bioinformatics. *Quantitative Biology*, *14*(1), e70014. DOI: 10.1002/qub2.70014 PMID: 41676319

Russo, A., Escobedo, F. J., Cirella, G. T., & Zerbe, S. (2017). Edible green infrastructure: An approach and review of provisioning ecosystem services and disservices in urban environments. *Agriculture, Ecosystems & Environment*, *242*, 53–66. DOI: 10.1016/j.agee.2017.03.026

Sabir, M., Sohail, A., Saeed, A., Abbas, S., Khan, K. A., & Khalil, M. (2025). Bioinformatics applied to microbiology. In *Microbiology in the era of artificial intelligence* (pp. 209–243). CRC Press. DOI: 10.1201/9781003410164-12

Saeed, M. U., Hussain, N., Sumrin, A., Shahbaz, A., Noor, S., Bilal, M., Aleya, L., & Iqbal, H. M. (2022). Microbial bioremediation strategies with wastewater treatment potentialities–A review. *The Science of the Total Environment*, *818*, 151754. DOI: 10.1016/j.scitotenv.2021.151754 PMID: 34800451

Sagova-Mareckova, M., Boenigk, J., Bouchez, A., Cermakova, K., Chonova, T., Cordier, T., Eisendle, U., Elersek, T., Fazi, S., Fleituch, T., Frühe, L., Gajdosova, M., Graupner, N., Haegerbaeumer, A., Kelly, A.-M., Kopecky, J., Leese, F., Nõges, P., Orlic, S., & Stoeck, T. (2021). Expanding ecological assessment by integrating microorganisms into routine freshwater biomonitoring. *Water Research, 191*, 116767. DOI: 10.1016/j.watres.2020.116767 PMID: 33418487

Saheb, K. S., Almeida, A., Segre, J. A., & Finn, R. D. (2021). Recovering prokaryotic genomes from host-associated, short-read shotgun metagenomic sequencing data. *Nature Protocols*, *16*(5), 2520–2541. DOI: 10.1038/s41596-021-00508-2 PMID: 33864056

Saini, S., Tewari, S., Dwivedi, J., & Sharma, V. (2023). Biofilm-mediated wastewater treatment: A comprehensive review. *Materials Advances*, *4*(6), 1415–1443. DOI: 10.1039/D2MA00945E

Sakai, Y. (2023). Methane monooxygenases: Physiology, biochemistry and structural insights. *Chemical Reviews*, *123*(8), 5432–5468. DOI: 10.1021/acs.chemrev.2c00712

Sakshi, S., Singh, S. K., & Haritash, A. K. (2019). Polycyclic aromatic hydrocarbons: Soil pollution and remediation. *International Journal of Environmental Science and Technology*, *16*(10), 6489–6512. DOI: 10.1007/s13762-019-02414-3

Sala, O. E., Chapin, F. S., Armesto, J. J., Berlow, E., Bloomfield, J., Dirzo, R., Huber-Sanwald, E., Huenneke, L. F., Jackson, R. B., Kinzig, A., Leemans, R., Lodge, D. M., Mooney, H. A., Oesterheld, M., Poff, N. L., Sykes, M. T., Walker, B. H., Walker, M., & Wall, D. H. (2000). Global biodiversity scenarios for the year 2100. *Science*, *287*(5459), 1770–1774. DOI: 10.1126/science.287.5459.1770 PMID: 10710299

Samuel, M. S., Jose, S., Selvarajan, E., Mathimani, T., & Pugazhendhi, A. (2020). Biosynthesized silver nanoparticles using Bacillus amyloliquefaciens; Application for cytotoxicity effect on A549 cell line and photocatalytic degradation of p-nitrophenol. *Journal of Photochemistry and Photobiology. B, Biology*, *202*, 111642. DOI: 10.1016/j.jphotobiol.2019.111642 PMID: 31734434

Šantl-Temkiv, T., Amato, P., Casamayor, E. O., Lee, P. K. H., & Pointing, S. B. (2022). Microbial ecology of the atmosphere. *FEMS Microbiology Reviews*, *46*(4), fuac009. Advance online publication. DOI: 10.1093/femsre/fuac009 PMID: 35137064

Sarojini, S., Balasubramanian, B., Jayaram, S., Nagaraj, V., Restine, A., Priyadarshini, P., Kaur, S., Rajeev, T. K., Pappuswamy, M., Kamyab, H., & Chelliapan, S. (2025). Invisible threats: Microplastics in milk and their implications for human and animal health. *Water, Air, and Soil Pollution*, *236*(10), 638. DOI: 10.1007/s11270-025-08267-6

Sattler, B., Puxbaum, H., & Psenner, R. (2001). Bacterial growth in supercooled cloud droplets. *Geophysical Research Letters*, *28*(2), 239–242. DOI: 10.1029/2000GL011684

Satya, S., Sharma, S., Choudhary, G., & Kaushik, G. (2024). Advances in environmental microbiology: A multi-omic perspective. In Kesheri, M., Kanchan, S., Salisbury, T. B., & Sinha, R. P. (Eds.), *Microbial omics in environment and health* (pp. 175–204). Springer Nature Singapore. DOI: 10.1007/978-981-97-1769-9_7

Sazzad, I. (2024). Advancing sustainable lubricating oil management: Re-refining techniques, market insights, innovative enhancements, and conversion to fuel. *Heliyon*, *10*(20), e39248. DOI: 10.1016/j.heliyon.2024.e39248 PMID: 39640706

Scheffer, M., & Carpenter, S. R. (2003). Catastrophic regime shifts in ecosystems: Linking theory to observation. *Trends in Ecology & Evolution*, *18*(12), 648–656. DOI: 10.1016/j.tree.2003.09.002

Scheffer, M., Carpenter, S., Foley, J. A., Folke, C., & Walker, B. (2001). Catastrophic shifts in ecosystems. *Nature*, *413*(6856), 591–596. DOI: 10.1038/35098000 PMID: 11595939

Schimel, J. P. (2018). Life in dry soils: Effects of drought on soil microbial communities and processes. *Annual Review of Ecology, Evolution, and Systematics*, *49*(1), 409–432. DOI: 10.1146/annurev-ecolsys-110617-062614

Schindler, D. W. (2006). Recent advances in the understanding and management of eutrophication. *Limnology and Oceanography*, *51*(1, Part 2), 356–363. DOI: 10.4319/lo.2006.51.1_part_2.0356

Schrimpe-Rutledge, A. C., Codreanu, S. G., Sherrod, S. D., & McLean, J. A. (2016). Untargeted metabolomics strategies—Challenges and emerging directions. *Journal of the American Society for Mass Spectrometry*, *27*(12), 1897–1905. DOI: 10.1007/s13361-016-1469-y PMID: 27624161

Schulz, S., Brankatschk, R., Dumig, A., Kogel-Knabner, I., Schloter, M., & Zeyer, J. (2013). The role of microorganisms at different stages of ecosystem development for soil formation. *Biogeosciences*, *10*(6), 3983–3996. DOI: 10.5194/bg-10-3983-2013

Schumann, Y., Gocke, A., & Neumann, J. E. (2025). Computational methods for data integration and imputation of missing values in omics datasets. *Proteomics*, *25*(3), e202400100. DOI: 10.1002/pmic.202400100 PMID: 39740174

Schuur, E. A. G., McGuire, A. D., Schädel, C., Grosse, G., Harden, J. W., Hayes, D. J., Hugelius, G., Koven, C. D., Kuhry, P., Lawrence, D. M., Natali, S. M., Olefeldt, D., Romanovsky, V. E., Schaefer, K., Turetsky, M. R., Treat, C. C., & Vonk, J. E. (2015). Climate change and the permafrost carbon feedback. *Nature*, *520*(7546), 171–179. DOI: 10.1038/nature14338 PMID: 25855454

Schwacke, L. H., Smith, C. R., Townsend, F. I., Wells, R. S., Hart, L. B., Balmer, B. C., Collier, T. K., De Guise, S., Fry, M. M., Guillette, L. J.Jr, Lamb, S. V., Lane, S. M., McFee, W. E., Place, N. J., Tumlin, M. C., Ylitalo, G. M., Zolman, E. S., & Rowles, T. K. (2014). Health of common bottlenose dolphins (Tursiops truncatus) in Barataria Bay, Louisiana, following the Deepwater Horizon oil spill. *Environmental Science & Technology*, *48*(1), 93–103. DOI: 10.1021/es403610f PMID: 24350796

Schwarzenbach, R. P., Escher, B. I., Fenner, K., Hofstetter, T. B., Johnson, C. A., von Gunten, U., & Wehrli, B. (2006). The challenge of micropollutants in aquatic systems. *Science*, *313*(5790), 1072–1077. DOI: 10.1126/science.1127291 PMID: 16931750

Sebald, M., & Hauser, D. (1995). Pasteur, oxygen and the anaerobes revisited. *Anaerobe*, *1*(1), 11–16. DOI: 10.1016/S1075-9964(95)80353-X PMID: 16887501

Seeley, M. E., Song, B., Passie, R., & Hale, R. C. (2020). Microplastics affect sedimentary microbial communities and nitrogen cycling. *Nature Communications*, *11*(1), 2372. DOI: 10.1038/s41467-020-16235-3 PMID: 32398678

Seemann, T. (2014). Prokka: Rapid prokaryotic genome annotation. *Bioinformatics (Oxford, England)*, *30*(14), 2068–2069. DOI: 10.1093/bioinformatics/btu153 PMID: 24642063

Senko, O., Maslova, O., Stepanov, N., Aslanli, A., Lyagin, I., & Efremenko, E. (2024). Role of humic substances in the (bio)degradation of synthetic polymers under environmental conditions. *Microorganisms*, *12*(10), 2024. DOI: 10.3390/microorganisms12102024 PMID: 39458333

Sesay, F., Sesay, R. E. V., Kamara, M., Li, X., & Niu, C. (2025). Biodegradation of pharmaceutical contaminants in wastewater using microbial consortia: Mechanisms, applications, and challenges. *Journal of Environmental Management*, *384*, 125564. DOI: 10.1016/j.jenvman.2025.125564 PMID: 40306218

Sewe, S. O., Silva, G., Sicat, P., Seal, S. E., & Visendi, P. (2022). Trimming and validation of Illumina short reads using trimmomatic, trinity assembly, and assessment of RNA-seq data. In *Plant bioinformatics: Methods and protocols* (pp. 211–232). Springer. DOI: 10.1007/978-1-0716-2067-0_11

Shade, A., Dunn, R. R., Blowes, S. A., Keil, P., Bohannan, B. J. M., Herrmann, M., Küsel, K., Lennon, J. T., Sanders, N. J., Storch, D., & Chase, J. (2018). Macroecology to unite all life, large and small. *Trends in Ecology & Evolution, 33*(10), 731–744. DOI: 10.1016/j.tree.2018.08.005 PMID: 30209011

Shade, A., Jones, S. E., Caporaso, J. G., Handelsman, J., Knight, R., Fierer, N., & Gilbert, J. A. (2014). Conditionally rare taxa disproportionately contribute to temporal changes in microbial diversity. *mBio*, *5*(4), e01371–e14. DOI: 10.1128/mBio.01371-14 PMID: 25028427

Shade, A., Peter, H., Allison, S. D., Baho, D. L., Berga, M., Bürgmann, H., Huber, D. H., Langenheder, S., Lennon, J. T., Martiny, J. B. H., Matulich, K. L., Schmidt, T. M., & Handelsman, J. (2012). Fundamentals of microbial community resistance and resilience. *Frontiers in Microbiology*, *3*, 417. DOI: 10.3389/fmicb.2012.00417 PMID: 23267351

Shah, A. A., Hasan, F., Hameed, A., & Ahmed, S. (2008). Biological degradation of plastics: A comprehensive review. *Biotechnology Advances*, *26*(3), 246–265. DOI: 10.1016/j.biotechadv.2007.12.005 PMID: 18337047

Shah, B. A., Malhotra, H., Papade, S. E., Dhamale, T., Ingale, O. P., Kasarlawar, S. T., & Phale, P. S. (2024). Microbial degradation of contaminants of emerging concern: Metabolic, genetic and omics insights for enhanced bioremediation. *Frontiers in Bioengineering and Biotechnology*, *12*, 1470522. Advance online publication. DOI: 10.3389/fbioe.2024.1470522 PMID: 39364263

Shahnawaz, M., Sangale, M. K., & Ade, A. B. (2019). Ex situ bioremediation technology for plastic degradation. In *Bioremediation Technology for Plastic Waste* (pp. 77–83). Springer Singapore., DOI: 10.1007/978-981-13-7492-0_7

Shahpiri, A., & Mohammadzadeh, A. (2018). Mercury removal by engineered *Escherichia coli* cells expressing different rice metallothionein isoforms. *Annals of Microbiology*, *68*(3), 145–152. DOI: 10.1007/s13213-018-1326-2

Shakya, M., Lo, C.-C., & Chain, P. S. G. (2019). Advances and challenges in metatranscriptomic analysis. *Frontiers in Genetics*, *10*, 904. Advance online publication. DOI: 10.3389/fgene.2019.00904 PMID: 31608125

Sharafi, K., Moradi, M., Azari, A., Sharafi, H., & Pirsaheb, M. (2016). Comparative evaluation of parasitic removal in municipal wastewater using constructed wetland and extended aeration–activated sludge system in Kermanshah province, Iran. *International Journal of Health and Life Sciences*, *2*(1), 16–21.

Sharma, M., & Kashyap, P. (2022). Pollution and its types. Journal Global Values, 13.

Sharma, P. R., Chopra, A., Cameotra, S. S., & Suri, C. R. (2010). Efficient biotransformation of herbicide diuron by bacterial strain *Micrococcus* sp. PS-1. *Biodegradation, 21*(6), 979–987. https://api.semanticscholar.org/CorpusID:29519247. DOI: 10.1007/s10532-010-9357-9 PMID: 20401520

Sharma, P., Bano, A., Singh, S. P., Dubey, N. K., Chandra, R., & Iqbal, H. M. N. (2022). Recent advancements in microbial-assisted remediation strategies for toxic contaminants. *Cleaner Chemical Engineering, 2*, 100020. DOI: 10.1016/j.clce.2022.100020

Sheela, B., Khasim, S., & Rao, O. (2014). Bioremediation of ammonia using ammonia oxidizing bacteria isolated from sewage. *International Journal of Environmental Bioremediation & Biodegradation, 2*, 146–150.

Shen, M., Song, B., Zhou, C., Almatrafi, E., Hu, T., Zeng, G., & Zhang, Y. (2022). Recent advances in impacts of microplastics on nitrogen cycling in the environment: A review. *The Science of the Total Environment, 815*, 152740. DOI: 10.1016/j.scitotenv.2021.152740 PMID: 34974017

Shen, Y., Qian, Q., Ding, L., Qu, W., Zhang, T., Song, M., Huang, Y., Wang, M., Xu, Z., Chen, J., Dong, L., Chen, H., Shen, E., Zheng, S., Chen, Y., Liu, J., Fan, L., & Wang, Y. (2025). High-throughput single-microbe RNA sequencing reveals adaptive state heterogeneity and host-phage activity associations in human gut microbiome. *Protein & Cell, 16*(3), 211–226. DOI: 10.1093/procel/pwae027 PMID: 38779805

Sheoran, K., Siwal, S. S., Kapoor, D., Singh, N., Saini, A. K., Alsanie, W. F., & Thakur, V. K. (2022). Air pollutants removal using biofiltration technique: A challenge at the frontiers of sustainable environment. *ACS Engineering Au, 2*(5), 378–396. DOI: 10.1021/acsengineeringau.2c00020 PMID: 36281334

Shin, G., Park, S. A., Koo, J. M., Kim, M., Lee, M., Jegal, J., Hwang, S. Y., Park, J., & Oh, D. X. (2021). A micro-spray-based high-throughput screening system for bioplastic-degrading microorganisms. *Green Chemistry, 23*(15), 5429–5436. rsc.li/greenchem. DOI: 10.1039/D1GC01916C

Shittu, O. E., Enagbonma, B. J., & Babalola, O. O. (2025). Deciphering the influence of fertilization systems on the Allium ampeloprasum rhizosphere microbial diversity and community structure through a shotgun metagenomics profiling approach. *Environmental Microbiome, 20*(1), 126. Advance online publication. DOI: 10.1186/s40793-025-00771-w PMID: 41053880

Shrestha, H. K., Appidi, M. R., Villalobos Solis, M. I., Wang, J., Carper, D. L., Burdick, L., Pelletier, D. A., Doktycz, M. J., Hettich, R. L., & Abraham, P. E. (2021). Metaproteomics reveals insights into microbial structure, interactions, and dynamic regulation in defined communities as they respond to environmental disturbance. *BMC Microbiology, 21*(1), 308. DOI: 10.1186/s12866-021-02370-4 PMID: 34749649

Shuai, J., Kim, S., Ryu, H., Park, J., Lee, C. K., Kim, G., Ultra, V. U.Jr, & Yang, W. (2018). Health risk assessment of volatile organic compounds exposure near Daegu dyeing industrial complex in South Korea. *BMC Public Health, 18*(1), 528. Advance online publication. DOI: 10.1186/s12889-018-5454-1 PMID: 29678134

Siaperas, R., Taxeidis, G., Nikolaivits, E., & Topakas, E. (2024). Multi-omics insights into the response of Aspergillus parasiticus to long-chain alkanes in relation to polyethylene degradation [Preprint]. *bioRxiv*. https://doi.org/DOI: 10.1101/2024.12

Siddiqui, M. F. (2024). The role of bioinformatics in molecular biology: Tools and techniques for data analysis. *Multidisciplinary Journal of Biochemistry, 1*(1), 64–73.

Siliakus, M. F., van der Oost, J., & Kengen, S. W. (2017). Adaptations of archaeal and bacterial membranes to variations in temperature, pH and pressure. *Extremophiles, 21*(4), 651–670. DOI: 10.1007/s00792-017-0939-x PMID: 28508135

Silva, J. R., Quinta-Ferreira, R. M., & Castro, L. M. (2025). Biological Treatments for VOC-Contaminated Off-Gas: Advances, Challenges, and Energetic Valorization Opportunities. *Sustainability (Basel), 17*(11), 4802. DOI: 10.3390/su17114802

Silvestrini, F., Fratini, F., & Magini, R. (2021). Planning with nature: Sustainable urban prototypes for Portuense district in Rome, Italy. *WIT Transactions on Ecology and the Environment, 253*, 15–26. Advance online publication. DOI: 10.2495/SC210021

Singh, D., Geat, N., Mehriya, M., Rajawat, M. V. S., Prasanna, R., Kumar, A., Kumari, G., & Jha, M. N. (2021). Omics (genomics, proteomics, metabolomics, etc.) tools to study the environmental microbiome and bioremediation. In Waste to energy: Prospects and applications (pp. 235–260). Springer.

Singha, L. P., & Pandey, P. (2021). Rhizosphere assisted bioengineering approaches for the mitigation of petroleum hydrocarbons contamination in soil. *Critical Reviews in Biotechnology, 41*(5), 749–766. DOI: 10.1080/07388551.2021.1888066 PMID: 33626996

Singh, B. K., & Walker, A. (2006). Microbial degradation of organophosphorus compounds. *FEMS Microbiology Reviews, 30*(3), 428–471. DOI: 10.1111/j.1574-6976.2006.00018.x PMID: 16594965

Singh, J. S., & Gupta, V. K. (2018). Soil microbial biomass: A key soil driver in management of ecosystem functioning. *The Science of the Total Environment, 634*, 497–500. DOI: 10.1016/j.scitotenv.2018.03.373 PMID: 29635193

Singh, M., Singh, D., Gupta, A., Pandey, K. D., Singh, P. K., & Kumar, A. (2019). Plant growth promoting rhizobacteria: application in biofertilizers and biocontrol of phytopathogens. In *PGPR Amelioration in Sustainable Agriculture* (pp. 41–66). Woodhead Publishing. DOI: 10.1016/B978-0-12-815879-1.00003-3

Singh, P., Jain, K., Desai, C., Tiwari, O., & Madamwar, D. (2019). Microbial community dynamics of extremophiles/extreme environments. In *Microbial diversity in the genomic era* (pp. 323–332). Academic Press. DOI: 10.1016/B978-0-12-814849-5.00018-6

Singh, P., Suri, C. R., & Cameotra, S. S. (2004). Isolation of a member of *Acinetobacter* species involved in atrazine degradation. *Biochemical and Biophysical Research Communications, 317*(3), 697–702. DOI: 10.1016/j.bbrc.2004.03.112 PMID: 15081396

Singh, S., Kumar, V., Chauhan, A., Datta, S., Wani, A. B., Singh, N., & Singh, J. (2018). Toxicity, degradation and analysis of the herbicide atrazine. *Environmental Chemistry Letters, 16*(1), 211–237. DOI: 10.1007/s10311-017-0665-8

Siripattanakul, S., Wirojanagud, W., McEvoy, J., Limpiyakorn, T., & Khan, E. (2010). Atrazine degradation by stable mixed cultures enriched from agricultural soil and their characterization. *Journal of Applied Microbiology, 106*(3), 986–992. DOI: 10.1111/j.1365-2672.2008.04075.x PMID: 19191954

Smets, W., Moretti, S., Denys, S., & Lebeer, S. (2016). Airborne bacteria in the atmosphere: Presence, purpose, and potential. *Atmospheric Environment, 139*, 214–221. DOI: 10.1016/j.atmosenv.2016.05.038

Smidt, H., & de Vos, W. M. (2004). Anaerobic microbial dehalogenation. *Annual Review of Microbiology, 58*(1), 43–73. DOI: 10.1146/annurev.micro.58.030603.123600 PMID: 15487929

Smith, S. E., & Read, D. J. (2008). *Mycorrhizal Symbiosis* (3rd ed.). Academic Press.

Sobhani, Z., Fang, C., Naidu, R., & Megharaj, M. (2021). Microplastics as a vector of toxic chemicals in soil: Enhanced uptake of perfluorooctane sulfonate and perfluorooctanoic acid by earthworms through sorption and reproductive toxicity. *Environmental Technology & Innovation, 22*, 101476. DOI: 10.1016/j.eti.2021.101476

Sofo, A., & Ricciuti, P. (2019). A standardized method for estimating the functional diversity of soil bacterial community by Biolog® EcoPlatesTM assay—The case study of a sustainable olive orchard. *Applied Sciences (Basel, Switzerland)*, *9*(19), 4035. DOI: 10.3390/app9194035

Solaja, O. M., Osifo, O. K., & Adegoke, D. (2025). Integrating recycled PET plastics into Nigeria's textile industry: Environmental, ethical, and circular economy perspectives. *Circular Economy and Sustainability*, *5*(6), 1–29. DOI: 10.1007/s43615-025-00625-0

Sooriyakumar, P., Bolan, N., Kumar, M., Singh, L., Yu, Y., Li, Y., Weralupitiya, C., Vithanage, M., Ramamayaka, S., Sarkar, B., Wang, F., Gleeson, D. B., Zhang, D., Kirkham, M. B., Rinklebe, J., & Siddique, K. H. M. (2022). Biofilm formation and its implications on the properties and fate of microplastics in aquatic environments: A review. *Journal of Hazardous Materials Advances*, *100077*, 100077. Advance online publication. DOI: 10.1016/j.hazadv.2022.100077

Sousa, A. M., Machado, I., Nicolau, A., & Pereira, M. O. (2013). Improvements on colony morphology identification towards bacterial profiling. *Journal of Microbiological Methods*, *95*(3), 327–335. DOI: 10.1016/j.mimet.2013.09.020 PMID: 24121049

Srivastava, A., Akhter, Y., & Verma, D. (2024). A step-by-step procedure for analysing the 16S rRNA-based microbiome diversity using QIIME 2 and comprehensive PICRUSt2 illustration for functional prediction. *Archives of Microbiology*, *206*(11), 467. DOI: 10.1007/s00203-024-04177-z PMID: 39540937

Srivastava, N., Singh, R., Sing, P., Ahmad, I., Singh, R. P., Rai, A. K., Asiri, M., & Gupta, V. K. (2023). Recent advances on lignocellulosic bioresources and their valorization in biofuels production: Challenges and viability assessment. *Environmental Technology & Innovation*, *29*, 103037. DOI: 10.1016/j.eti.2023.103037

Starr, A. E., Deeke, S. A., Li, L., Zhang, X., Daoud, R., Ryan, J., Ning, Z., Cheng, K., Nguyen, L. V. H., Abou-Samra, E., Lavallée-Adam, M., & Figeys, D. (2018). Proteomic and metaproteomic approaches to understand host–microbe interactions. *Analytical Chemistry*, *90*(1), 86–109. DOI: 10.1021/acs.analchem.7b04340 PMID: 29061041

Steffen, W., Richardson, K., Rockström, J., Cornell, S. E., Fetzer, I., Bennett, E. M., Biggs, R., Carpenter, S. R., de Vries, W., de Wit, C. A., Folke, C., Gerten, D., Heinke, J., Mace, G. M., Persson, L. M., Ramanathan, V., Reyers, B., & Sörlin, S. (2015). Planetary boundaries: Guiding human development on a changing planet. *Science*, *347*(6223), 1259855. DOI: 10.1126/science.1259855 PMID: 25592418

Storz, G., & Hengge-Aronis, R. (Eds.). (2000). *Bacterial stress responses*. ASM Press. Print ISBN:9781119737841, Online ISBN:9781683671213

Strong, L. C., Rosendahl, C., Johnson, G., Sadowsky, M. J., & Wackett, L. P. (2002). *Arthrobacter aurescens* TC1 metabolizes diverse s-triazine ring compounds. *Applied and Environmental Microbiology*, *68*(12), 5973–5980. DOI: 10.1128/AEM.68.12.5973-5980.2002 PMID: 12450818

Strozzi, F., Janssen, R., Wurmus, R., Crusoe, M. R., Githinji, G., Di Tommaso, P., Belhachemi, D., Möller, S., Smant, G., De Ligt, J., & Prins, P. (2019). Scalable workflows and reproducible data analysis for genomics. In Dutheil, N. (Ed.), *Evolutionary genomics: Statistical and computational methods* (pp. 723–745). Springer. DOI: 10.1007/978-1-4939-9074-0_24

Su, C., Jiang, S., & Dai, J. (2025). Development of a versatile synthetic microbe for the concurrent degradation of organic pollutants. *Advanced Biotechnology*, *3*(2), 16. Advance online publication. DOI: 10.1007/s44307-025-00072-w PMID: 40408009

Sudarikov, K., Tyakht, A., & Alexeev, D. (2017). Methods for the metagenomic data visualization and analysis. *Current Issues in Molecular Biology*, *24*, 37–58. DOI: 10.21775/cimb.024.037 PMID: 28686567

Suding, K., Higgs, E., Palmer, M., Callicott, J. B., Anderson, C. B., Baker, M., Gutrich, J. J., Hondula, K. L., LaFevor, M. C., Larson, B. M. H., Randall, A., Ruhl, J. B., & Schwartz, K. Z. S. (2015). Committing to ecological restoration. *Science*, *348*(6235), 638–640. DOI: 10.1126/science.aaa4216 PMID: 25953995

Sui, B., Wang, T., Fang, J., Hou, Z., Shu, T., Lu, Z., Liu, F., & Zhu, Y. (2023). Recent advances in the biodegradation of polyethylene terephthalate with cutinase-like enzymes. *Frontiers in Microbiology*, *14*, 1265139. DOI: 10.3389/fmicb.2023.1265139 PMID: 37849919

Sumampouw, O. J., & Risjani, Y. (2014). Bacteria as indicators of environmental pollution. *Environment*, *51*(52), 10–5923.

Sun, A., & Wang, W. X. (2023). Human exposure to microplastics and its associated health risks. *Environment & Health*, *1*(3), 139–149. DOI: 10.1021/envhealth.3c00053 PMID: 39473618

Sun, C., Yuan, J., Xu, H., Huang, S., Wen, X., Tong, N., & Zhang, Y. (2019). Simultaneous removal of nitric oxide and sulfur dioxide in a biofilter under micro-oxygen thermophilic conditions: Removal performance, competitive relationship and bacterial community structure. *Bioresource Technology*, *290*, 121768. DOI: 10.1016/j.biortech.2019.121768 PMID: 31323510

Sun, S., Sidhu, V., Rong, Y., & Zheng, Y. (2018). Pesticide pollution in agricultural soils and sustainable remediation methods: A review. *Current Pollution Reports*, *4*(3), 240–250. DOI: 10.1007/s40726-018-0092-x

Sun, X., & Sanchez, A. (2023). Synthesizing microbial biodiversity. *Current Opinion in Microbiology*, *75*, 102348. DOI: 10.1016/j.mib.2023.102348 PMID: 37352679

Sun, Y., Chen, Q., Fan, G., Sun, Q., Zhou, Q., Zhang, J., Nie, J., Ma, J., & Wu, L. (2025). gcMeta 2025: A global repository of metagenome-assembled genomes enabling cross-ecosystem microbial discovery and function research. *Nucleic Acids Research*, •••, gkaf1115. PMID: 41171134

Sun, Y., Wang, X., Xia, S., & Zhao, J. (2022). Cu (II) adsorption on poly (lactic acid) microplastics: Significance of microbial colonization and degradation. *Chemical Engineering Journal*, *429*, 132306. DOI: 10.1016/j.cej.2021.132306

Surendran, U., Jayakumar, M., Raja, P., Gopinath, G., & Chellam, P. V. (2023). Microplastics in terrestrial ecosystem: Sources and migration in soil environment. *Chemosphere*, *318*, 137946. DOI: 10.1016/j.chemosphere.2023.137946 PMID: 36708782

Sutar, R. S., Motghare, V. M., Kollur, S. C., Parikh, Y., & Asolekar, S. R. (2018). Significance of addressing persistence of pathogens and micropollutants to enhance reuse of treated sewages using constructed wetlands. In *Advances in waste management: Select proceedings of Recycle 2016* (pp. 355–367). Springer Singapore.

Suyal, D. C., Soni, R., Sai, S., & Goel, R. (2016). Microbial inoculants as biofertilizer. In *Microbial inoculants in sustainable agricultural productivity* (pp. 311–318). Springer. DOI: 10.1007/978-81-322-2647 5_18

Syranidou, E., Karkanorachaki, K., Amorotti, F., Avgeropoulos, A., Kolvenbach, B., Zhou, N., Fava, F., Corvini, P., & Kalogerakis, N. (2019). Biodegradation of mixture of plastic films by tailored marine consortia. *Journal of Hazardous Materials*, *375*, 33–42. DOI: 10.1016/j.jhazmat.2019.04.078 PMID: 31039462

Syuhadah, B. H., & Raikhan, N. H. N. (2022). *Pseudomonas aeruginosa* NR.22 - local hydrocarbon degrader species originated from crude oil non-exploration site: Isolation, survivability and hydrocarbon degradation study. *Desalination and Water Treatment*, *257*, 260–269. DOI: 10.5004/dwt.2022.28473

Szabó, G., Eloe-Fadrosh, E. A., Pett-Ridge, J., & Woyke, T. (2025). A genomic view of Earth's biomes. *Nature Reviews. Genetics*, *27*(1), 13–31. DOI: 10.1038/s41576-025-00888-1 PMID: 40954201

Tahir, L., Ali, M. I., Zia, M., Atiq, N., Hasan, F., & Ahmed, S. (2013). Production and characterization of esterase in *Lantinus tigrinus* for degradation of polystyrene. *Polish Journal of Microbiology*, *62*(1), 101–108. DOI: 10.33073/pjm-2013-015 PMID: 23829086

Takeshita, R., Bursian, S. J., Colegrove, K. M., Collier, T. K., Deak, K., Dean, K. M., De Guise, S., DiPinto, L. M., Elferink, C. J., Esbaugh, A. J., Griffitt, R. J., Grosell, M., Harr, K. E., Incardona, J. P., Kwok, R. K., Lipton, J., Mitchelmore, C. L., Morris, J. M., Peters, E. S., & Hall, A. J. (2021). A review of the toxicology of oil in vertebrates: What we have learned following the Deepwater Horizon oil spill. *Journal of Toxicology and Environmental Health. Part B, Critical Reviews*, *24*(8), 355–394. DOI: 10.1080/10937404.2021.1975182 PMID: 34542016

Tan, K. H., & Gill, C. O. (1985). Batch Growth of *Saccaromycopsis lipolytica* on animal fats. *Applied Microbiology and Biotechnology*, *21*(5), 292–298. DOI: 10.1007/BF00252707

Tartaglia, M., Bastida, F., Sciarrillo, R., & Guarino, C. (2020). Soil metaproteomics for the study of the relationships between microorganisms and plants: A review of extraction protocols and ecological insights. *International Journal of Molecular Sciences*, *21*(22), 8455. DOI: 10.3390/ijms21228455 PMID: 33187080

Terra, L. A., Klepa, M. S., Nogueira, M. A., & Hungria, M. (2025). Pangenome analysis indicates evolutionary origins and genetic diversity: Emphasis on the role of nodulation in symbiotic Bradyrhizobium. *Frontiers in Plant Science*, *16*, 1539151. DOI: 10.3389/fpls.2025.1539151 PMID: 40241821

Teseo, S., Otani, S., Brinch, C., Leroy, S., Ruiz, P., Desvaux, M., Forano, E., Aarestrup, F. M., & Sapountzis, P. (2022). A global phylogenomic and metabolic reconstruction of the large intestine bacterial community of domesticated cattle. *Microbiome*, *10*(1), 155. DOI: 10.1186/s40168-022-01357-1 PMID: 36155629

Tett, P., Benjamins, S., Black, K., Coulson, M., Davidson, K., Fernandes, T. F., Fox, C., Hart, M., Hicks, N., Hughes, A., & Hunter, D. C. (2018). Review of the environmental impacts of salmon farming in Scotland.

Tettelin, H., & Medini, D. (Eds.). (2020). *The pangenome: A new frontier in microbiology*. Springer International Publishing., DOI: 10.1007/978-3-030-38281-0

Thakur, A., & Koul, B. (2022). Impact of oil exploration and spillage on marine environments. In *Advances in oil-water separation* (pp. 115–135). Elsevier. DOI: 10.1016/B978-0-323-89978-9.00018-5

Thauer, R. K., Kaster, A. K., Seedorf, H., Buckel, W., & Hedderich, R. (2008). Methanogenic archaea: Ecologically relevant differences in energy conservation. *Nature Reviews. Microbiology*, *6*(8), 579–591. DOI: 10.1038/nrmicro1931 PMID: 18587410

Thomas, C. M., & Nielsen, K. M. (2005). Mechanisms of, and barriers to, horizontal gene transfer between bacteria. *Nature Reviews. Microbiology*, *3*(9), 711–721. DOI: 10.1038/nrmicro1234 PMID: 16138099

Thompson, R. R., Moore, C. J., vom Saal, F. S., & Swan, S. H. (2009). Plastics, the environment and human health: Current consensus and future trends. *Philosophical Transactions of the Royal Society of London. Series B, Biological Sciences*, *364*(1526), 2153–2166. DOI: 10.1098/rstb.2009.0053 PMID: 19528062

Toledo, M., & Muñoz, R. (2025). Odour prevention strategies in wastewater treatment and composting plants: A review. *Journal of Environmental Management*, *375*, 124402. DOI: 10.1016/j.jenvman.2025.124402 PMID: 39908607

Topp, E., Zhu, H., Nour, S. M., Houot, S., Lewis, M., & Cuppels, D. (2000). Characterization of an atrazine-degrading *Pseudaminobacter* sp. isolated from Canadian and French agricultural soils. *Applied and Environmental Microbiology*, *66*(7), 2773–2782. DOI: 10.1128/AEM.66.7.2773-2782.2000 PMID: 10877767

Tortoli, E., Fedrizzi, T., Meehan, C. J., Trovato, A., Grottola, A., Giacobazzi, E., Serpini, G. F., Tagliazucchi, S., Fabio, A., Bettua, C., Bertorelli, R., Frascaro, F., De Sanctis, V., Pecorari, M., Jousson, O., Segata, N., & Cirillo, D. M. (2017). The new phylogeny of the genus Mycobacterium: The old and the new. *Infection, Genetics and Evolution*, *56*, 19–25. DOI: 10.1016/j.meegid.2017.10.013 PMID: 29030295

Trego, A., Keating, C., Nzeteu, C., Graham, A., O'Flaherty, V., & Ijaz, U. Z. (2022). Beyond basic diversity estimates—Analytical tools for mechanistic interpretations of amplicon sequencing data. *Microorganisms*, *10*(10), 1961. DOI: 10.3390/microorganisms10101961 PMID: 36296237

Ulfat, M., Abad, Z., Ali, N., Sarwar, S., Jabeen, K., & Abrar, A. (2024). Screening, biochemical characterization and antibiotics resistance/susceptibility of bacteria isolated from native soil and water samples. *Brazilian Journal of Biology*, *84*, e254016. DOI: 10.1590/1519-6984.254016 PMID: 35352774

Urbieta, M. S., Donati, E. R., Chan, K. G., Shahar, S., Sin, L. L., & Goh, K. M. (2015). Thermophiles in the genomic era: Biodiversity, science, and applications. *Biotechnology Advances*, *33*(6), 633–647. DOI: 10.1016/j.biotechadv.2015.04.007 PMID: 25911946

Valavanidis, A. (2024). Plastic pollution, unsustainable environmental problem? Industry and academic research must co-operate to sort out the pervasive global problem of plastic waste. *Scientific Review (Singapore)*, *1*, 1–42.

Vali, G., & Schnell, R. (1976). Biogenic Ice Nuclei: Part I. Terrestrial and Marine Sources. *Journal of the Atmospheric Sciences*, *33*(8), 1554–1564. DOI: 10.1175/1520-0469(1976)033<1565:BINPIB>2.0.CO;2

Valois, L., Brachet, A., Schiopu, N., & Barot, S. (2023). Performance assessment of the ecosystem services provided by urban nature-based solutions: Focus on rainwater management. *IOP Conference Series. Earth and Environmental Science*, *1196*(1), 012028. DOI: 10.1088/1755-1315/1196/1/012028

Van Beilen, J. B., & Funhoff, E. G. (2007). Alkane hydroxylases involved in microbial alkane degradation. *Applied Microbiology and Biotechnology*, *74*(1), 13–21. DOI: 10.1007/s00253-006-0748-0 PMID: 17216462

Van Den Bossche, T., Armengaud, J., Benndorf, D., Blakeley-Ruiz, J. A., Brauer, M., Cheng, K., Creskey, M., Figeys, D., Grenga, L., Griffin, T. J., Henry, C., Hettich, R. L., Holstein, T., Jagtap, P. D., Jehmlich, N., Kim, J., Kleiner, M., Kunath, B. J., Malliet, X., & Initiative, T. M. (2025a). The microbiologist's guide to metaproteomics. *iMeta*, *4*(1), e70031. DOI: 10.1002/imt2.70031 PMID: 40469504

van den Hoogen, J., Geisen, S., Routh, D., Ferris, H., Traunspurger, W., Wardle, D. A., de Goede, R. G. M., Adams, B. J., Ahmad, W., Andriuzzi, W. S., Bardgett, R. D., Bonkowski, M., Campos-Herrera, R., Cares, J. E., Caruso, T., de Brito Caixeta, L., Chen, X., Costa, S. R., Creamer, R., & Crowther, T. W. (2019). Soil nematode abundance and functional group composition at a global scale. *Nature*, *572*(7768), 194–198. DOI: 10.1038/s41586-019-1418-6 PMID: 31341281

Van Rossum, T., Ferretti, P., Maistrenko, O. M., & Bork, P. (2020). Diversity within species: Interpreting strains in microbiomes. *Nature Reviews. Microbiology*, *18*(9), 491–506. DOI: 10.1038/s41579-020-0368-1 PMID: 32499497

Vannucchi, F., Bretzel, F., Pini, R., & Rumble, H. (2021). Less is more: Soil and substrate quality as an opportunity for urban greening and biodiversity conservation. In Catalano, C., Andreucci, M. B., Guarino, R., Bretzel, F., Leone, M., & Pasta, S. (Eds.), *Urban services to ecosystems. Future city* (Vol. 17, pp. 207–224). Springer., DOI: 10.1007/978-3-030-75929-2_11

Varjani, S. J. (2017). Microbial degradation of petroleum hydrocarbons. *Bioresource Technology*, *223*, 277–286. DOI: 10.1016/j.biortech.2016.10.037 PMID: 27789112

Vass, M., Ramasamy, K. P., & Andersson, A. (2024). Microbial hitchhikers on microplastics: The exchange of aquatic microbes across distinct aquatic habitats. *Environmental Microbiology*, *26*(4), e16618. Advance online publication. DOI: 10.1111/1462-2920.16618 PMID: 38561820

Veloso, M., Waldisperg, A., Arros, P., Berríos-Pastén, C., Acosta, J., Colque, H., Varas, M. A., Allende, M. L., Orellana, L. H., & Marcoleta, A. E. (2023). Diversity, taxonomic novelty, and encoded functions of Salar de Ascotán microbiota, as revealed by metagenome-assembled genomes. *Microorganisms*, *11*(12), 2819. DOI: 10.3390/microorganisms11112819 PMID: 38004830

Ventura, E., Marín, A., Gámez-Pérez, J., & Cabedo, L. (2024). Recent advances in the relationships between biofilms and microplastics in natural environments. *World Journal of Microbiology & Biotechnology*, *40*(7), 220. DOI: 10.1007/s11274-024-04021-y PMID: 38809290

Vergara-Fernández, A., Yánez, D., Morales, P., Scott, F., Aroca, G., Diaz-Robles, L., & Moreno-Casas, P. (2017). Biofiltration of benzo[α]pyrene, toluene and formaldehyde in air by a consortium of Rhodococcus erythropolis and Fusarium solani: Effect of inlet loads, gas flow and temperature. *Chemical Engineering Journal*, *332*, 702–710. DOI: 10.1016/j.cej.2017.09.095

Victorio, L., Gilbride, K. A., Allen, D. G., & Liss, S. N. (1996). Phenotypic fingerprinting of microbial communities in wastewater treatment systems. *Water Research*, *30*(5), 1077–1086. DOI: 10.1016/0043-1354(95)00269-3

Viwatpanich, K., Siritanon, T., & Tongsook, S. (2013). Detoxification of cassava wastewater using microbial fermentation. *Journal of Environmental Science and Health. Part B, Pesticides, Food Contaminants, and Agricultural Wastes*, *48*(6), 460–466.

Vogt, C., Kleinsteuber, S., & Richnow, H. H. (2011). Anaerobic benzene degradation by bacteria. *Microbial Biotechnology*, *4*(6), 710–724. DOI: 10.1111/j.1751-7915.2011.00260.x PMID: 21450012

Wahsha, M., Nadimi-Goki, M., Fornasier, F., Al-Jawasreh, R., Hussein, E. I., & Bini, C. (2017). Microbial enzymes as an early warning management tool for monitoring mining site soils. *Catena, 148*, 40-45.Wang, Z., Luo, P., Zha, X., Xu, C., Kang, S., Zhou, M., ... & Wang, Y. (2022). Overview assessment of risk evaluation and treatment technologies for heavy metal pollution of water and soil. *Journal of Cleaner Production*, *379*, 134043.

Wall, D. H., Bardgett, R. D., Behan-Pelletier, V., Herrick, J. E., Jones, T. H., Ritz, K., Six, J., Strong, D. R., & van der Putten, W. H. (Eds.). (2012). *Soil ecology and ecosystem services*. Oxford University Press., DOI: 10.1093/acprof:oso/9780199575923.001.0001

Walters, S. A., & Midden, K. S. (2018). Sustainability of urban agriculture: Vegetable production on green roofs. *Agriculture*, *8*(11), 168. DOI: 10.3390/agriculture8110168

Wang, C., Wu, W., Pang, Z., Liu, J., Qiu, J., Luan, T., Deng, J, & Fang, Z. (2023). Polystyrene microplastics significantly facilitate influenza A virus infection of host cells. *Journal of Hazardous Materials*, 15(446):130617. DOI: . 130617DOI: 10.1016/j.jharmat.2022

Wang, C., Zhao, L., Lim, M. K., Chen, W.-Q., & Sutherland, J. W. (2020). Structure of the global plastic waste trade network and the impact of China's import ban. Resource Conservatory. *Resources, Conservation and Recycling*, *153*, 104591. DOI: 10.1016/j.resconrec.2019.104591

Wang, F., Xiang, L., Leung, K. S. Y., Elsner, M., Zhang, Y., & Guo, Y.. (2024). Emerging contaminants: A one health perspective. *Innovation (Cambridge (Mass.))*, *5*(4), 100612. Advance online publication. DOI: 10.1016/j.xinn.2024.100612 PMID: 38756954

Wang, H., Liu, X., Wu, C., & Lin, G. (2024). Regional to global distributions, trends, and drivers of biogenic volatile organic compound emission from 2001 to 2020. *Atmospheric Chemistry and Physics*, *24*(5), 3309–3328. DOI: 10.5194/acp-24-3309-2024

Wang, L., Cheng, G., Ren, Y., Dai, Z., Zhao, Z. S., Liu, F., & Tang, B. (2015). Degradation of intact chicken feathers by Thermoactinomyces sp. CDF and characterization of its keratinolytic protease. *Applied Microbiology and Biotechnology*, *99*(9), 3949–3959. DOI: 10.1007/s00253-014-6207-4 PMID: 25412577

Wang, M., Ma, L., Zhang, D., Yang, F., Yang, X., Wang, B., Li, X., Ge, C., Luo, J., & Zhang, Y. (2025). Dissolved organic matter derived from long-term photodegradation of plastics alters microbial methane conversion in mangrove sediments. *Bioresource Technology*, *132677*, 132677. Advance online publication. DOI: 10.1016/j.biortech.2025.132677 PMID: 40378974

Wang, S., Ang, H., & Tade, M. O. (2007). Volatile organic compounds in indoor environment and photocatalytic oxidation: State of the art. *Environment International*, *33*(5), 694–705. DOI: 10.1016/j.envint.2007.02.011 PMID: 17376530

Wang, Y., Bi, R., Ge, J., & Xu, P. (2025). VOCs absorption using ionic liquids. *Separation and Purification Technology, 132265*, 132265. Advance online publication. DOI: 10.1016/j.seppur.2025.132265

Wang, Y., Hayatsu, M., & Fujii, T. (2012). Extraction of bacterial RNA from soil: Challenges and solutions. *Microbes and Environments, 27*(2), 111–121. DOI: 10.1264/jsme2.ME11304 PMID: 22791042

Wang, Z., Liang, J.-L., Huang, L.-N., Mengoni, A., & Shu, W.-S. (2021). Metagenomic assembly: Reconstructing genomes from metagenomes. In *Bacterial pangenomics: Methods and protocols* (pp. 139–152). Springer. DOI: 10.1007/978-1-0716-1099-2_9

Wang, Z., Zhao, Z., & Liu, J. (2025). Homoploid hybrid speciation: Advancements and challenges. *Plant Ecology & Diversity, 18*(3–4), 141–158. DOI: 10.1080/17550874.2025.2550015

Webster, N. S., Negri, A. P., Botté, E. S., Laffy, P. W., Flores, F., Noonan, S., Schmidt, C., & Uthicke, S. (2016). Host-associated coral reef microbes respond to the cumulative pressures of ocean warming and ocean acidification. *Scientific Reports, 6*(1), 19324. DOI: 10.1038/srep19324 PMID: 26758800

Weinstein, J. E., Crocker, B. K., & Gray, A. D. (2016). From macroplastic to microplastic: Degradation of high-density polyethylene, polypropylene, and polystyrene in a salt marsh habitat. *Environmental Toxicology and Chemistry, 35*(7), 1632–1640. DOI: 10.1002/etc.3432 PMID: 26992845

Weinstein, J. E., Viado, H. F.Jr, Leads, R. R., Deloughry, E., Schandera, L., Liddy, K., & Scott, G. I. (2019). Microplastics: A global water pollution problem. *Encyclopedia of Water. Science, Technology & Society*, •••, 1 15. DOI: 10.1002/9781119300762.wsts0203

Wenk, S., Claassens, N. J., & Lindner, S. N. (2022). Synthetic metabolism approaches: A valuable resource for systems biology. *Current Opinion in Systems Biology, 30*, 100417. DOI: 10.1016/j.coisb.2022.100417

Williams, J., & Koppmann, R. (2007). Volatile Organic Compounds in the atmosphere: An overview. In *Volatile Organic Compounds in the Atmosphere* (pp. 1–32). https://doi.org/DOI: 10.1002/9780470988657.ch1

Wittebolle, L., Marzorati, M., Clement, L., Balloi, A., Daffonchio, D., Heylen, K., De Vos, P., Verstraete, W., & Boon, N. (2009). Initial community evenness favours functionality under selective stress. *Nature, 458*(7238), 623–626. DOI: 10.1038/nature07840 PMID: 19270679

Wołejko, E., Jabłońska-Trypuć, A., Wydro, U., Butarewicz, A., & Łozowicka, B. (2020). Soil biological activity as an indicator of soil pollution with pesticides–a review. *Applied Soil Ecology, 147*, 103356. DOI: 10.1016/j.apsoil.2019.09.006

Womack, A. M., Bohannan, B. J. M., & Green, J. L. (2010). Biodiversity and biogeography of the atmosphere. *Philosophical Transactions of the Royal Society of London. Series B, Biological Sciences, 365*(1558), 3645–3653. DOI: 10.1098/rstb.2010.0283 PMID: 20980313

Woyke, T., Doud, D. F. R., & Schulz, F. (2017). The trajectory of microbial single-cell sequencing. *Nature Methods, 14*(11), 1045–1054. DOI: 10.1038/nmeth.4469 PMID: 29088131

Wright, M. A., Taylor, F., Randles, S. J., Brown, D. E., & Higgins, I. J. (1993). Biodegradation of a Synthetic Lubricant by *Micrococcus roseus. Applied and Environmental Microbiology, 59*(4), 1072–1076. DOI: 10.1128/aem.59.4.1072-1076.1993 PMID: 8476283

Wu, L., Wang, X., Li, A., Cui, H., & Li, X. (2025). Discovery of a polyethylene biodegradable laccase from *Acinetobacter dijkshoorniae. AIChE Journal.* 71(10):e70005. DOI:. 70005WU ET AL . 14 of 14DOI: 10.1002/aic

Wu, C. C., Bao, L. J., Liu, L. Y., Shi, L., Tao, S., & Zeng, E. Y. (2017). Impact of polymer colonization on the fate of organic contaminants in sediment. *Environmental Science & Technology, 51*(18), 10555–10561. DOI: 10.1021/acs.est.7b03310 PMID: 28825800

Wu, D., Wei, Z. M., Mohamed, T. A., Zheng, G. G., Qu, F. T., Wang, F., Zhao, Y., & Song, C. H. (2022). Lignocellulose biomass bioconversion during composting: Mechanism of action of lignocellulase, pretreatment methods and future perspectives. *Chemosphere, 286*, 131635. DOI: 10.1016/j.chemosphere.2021.131635 PMID: 34346339

Wu, H., Guo, C., Yin, Z., Quan, Y., & Yin, C. (2018a). Performance and bacterial diversity of biotrickling filters filled with conductive packing material for the treatment of toluene. *Bioresource Technology, 257*, 201–209. DOI: 10.1016/j.biortech.2018.02.108 PMID: 29501953

Wu, H., Yan, H., Quan, Y., Zhao, H., Jiang, N., & Yin, C. (2018b). Recent progress and perspectives in biotrickling filters for VOCs and odorous gases treatment. *Journal of Environmental Management, 222*, 409–419. DOI: 10.1016/j.jenvman.2018.06.001 PMID: 29883876

Wu, Q., Huang, Y., Irga, P., Kumar, P., Li, W., Wei, W., Feng, H., Jiang, Y., Schnelle-Kreis, J., Zimmermann, R., Tang, M., & Xie, X. (2024). Synergistic control of urban heat island and urban pollution island effects using green infrastructure. *Journal of Environmental Management*, *370*, 122985. DOI: 10.1016/j.jenvman.2024.122985 PMID: 39461153

Wu, Y., Zhuang, J., Song, Y., Gao, X., Chu, J., & Han, S. (2024). Advances in single-cell sequencing technology in microbiome research. *Genes & Diseases*, *11*(4), 101129. DOI: 10.1016/j.gendis.2023.101129 PMID: 38545125

Wypych, T. P., Wickramasinghe, L. C., & Marsland, B. J. (2019). The influence of the microbiome on respiratory health. *Nature Immunology*, *20*(10), 1279–1290. DOI: 10.1038/s41590-019-0451-9 PMID: 31501577

Xian, F., Brenek, M., Krisp, C., Urbauer, E., Ravi Kumar, R. K., Aguanno, D., Srikumar, T., Liu, Q., Barry, A. M., Ma, B., Krieger, J., Haller, D., Schmidt, M., & Gómez-Varela, D. (2025). Ultra-sensitive metaproteomics redefines the dark metaproteome, uncovering host-microbiome interactions and drug targets in intestinal diseases. *Nature Communications*, *16*(1), 6644. DOI: 10.1038/s41467-025-61977-7 PMID: 40681571

Xiang, L., Harindintwali, J. D., Wang, F., Redmile-Gordon, M., Chang, S. X., Fu, Y., Feng, Y., Zeng, X. C., Su, S. M., Tian, C., Lammel, D. R., Basiliko, N., Zhang, W., Ok, Y. S., Borken, W., Ludley, K., Xie, S., Sun, H., Wei, Z., & Jiang, X. (2022). Integrating biochar, bacteria, and plants for sustainable remediation of soils contaminated with organic pollutants. *Environmental Science & Technology*, *56*(23), 16546–16566. DOI: 10.1021/acs.est.2c02976 PMID: 36301703

Xie, Y., Yu, X., Ng, N. C., Li, K., & Fang, L. (2018). Exploring the dynamic correlation of landscape composition and habitat fragmentation with surface water quality in the Shenzhen River and deep bay cross-border watershed, China. *Ecological Indicators*, *90*, 231–246. DOI: 10.1016/j.ecolind.2017.11.051

Xu, C., Couvillion, S. P., Sontag, R. L., Isern, N. G., Maezato, Y., Lindemann, S. R., Roy Chowdhury, T., Zhao, R., Morton, B. R., Chu, R. K., Moore, R. J., Jansson, J. K., Bailey, V. L., Mouser, P. J., Romine, M. F., Fredrickson, J. F., & Metz, T. O. (2021). MetFish: A metabolomics pipeline for studying microbial communities in chemically extreme environments. *mSystems*, *6*(2), e01058–e20. DOI: 10.1128/mSystems.01058-20 PMID: 34061574

Xu, M., Yang, M., Sun, H., Gao, M., Wang, Q., & Wu, C. (2022). Bioconversion of biowaste into renewable energy and resources: A sustainable strategy. *Environmental Research*, *214*, 113929. DOI: 10.1016/j.envres.2022.113929 PMID: 35868577

Xu, X., Yang, H., & Li, C. (2022). Theoretical model and actual characteristics of air pollution affecting health cost: A review. *International Journal of Environmental Research and Public Health, 19*(6), 3532. DOI: 10.3390/ijerph19063532 PMID: 35329214

Yaashikaa, P. R., Senthil Kumar, P., Jeevanantham, S., & Saravanan, R. (2022). A review on bioremediation approach for heavy metal detoxification and accumulation in plants. *Environmental Pollution, 301*, 119035. DOI: 10.1016/j.envpol.2022.119035 PMID: 35196562

Yadav, S., Kumar, S., & Haritash, A. K. (2023). A comprehensive review of chlorophenols: Fate, toxicology and its treatment. *Journal of Environmental Management, 342*, 118254. DOI: 10.1016/j.jenvman.2023.118254 PMID: 37295147

Yakovenko, N., Pérez-Serrano, L., Segur, T., Hagelskjaer, O., Margenat, H., Le Roux, G., & Sonke, J. E. (2025). Human exposure to PM10 microplastics in indoor air. *PLoS One, 20*(7), e0328011. DOI: 10.1371/journal.pone.0328011 PMID: 40737229

Yale, R. L., Sapp, M., Sinclair, C. J., & Moir, J. W. (2017). Microbial changes linked to the accelerated degradation of the herbicide atrazine in a range of temperate soils. *Environmental Science and Pollution Research International, 24*(8), 7359–7374. DOI: 10.1007/s11356-017-8377-y PMID: 28108915

Yang, H., He, Y., Liao, J., Li, X., Zhang, J., Liebl, W., & Chen, F. (2022). RNA-Seq transcriptomic analysis reveals gene expression profiles of acetic acid bacteria under high-acidity submerged industrial fermentation process. *Frontiers in Microbiology, 13*, 956729. Advance online publication. DOI: 10.3389/fmicb.2022.956729 PMID: 36246236

Yang, T., Xin, Y., Zhang, L., Gu, Z., Li, Y., Ding, Z., & Shi, G. (2020). Characterization on the aerobic denitrification process of Bacillus strains. *Biomass and Bioenergy, 140*, 105677. DOI: 10.1016/j.biombioe.2020.105677

Yang, W., Li, Y., & Boraschi, D. (2023). Association between microorganisms and microplastics: How does it change the host-pathogen interactions and subsequent immune system? *International Journal of Molecular Sciences, 24*(4), 4065. DOI: 10.3390/ijms24044065 PMID: 36835476

Yang, Y., Zhang, Z. W., Liu, R. X., Ju, H. Y., Bian, X. K., Zhang, W. Z., Zhang, C.-B., Yang, T., Guo, B., Xiao, C.-L., Bai, H., & Lu, W. Y. (2021). Research progress in bioremediation of petroleum pollution. *Environmental Science and Pollution Research International, 28*(34), 46877–46893. DOI: 10.1007/s11356-021-15310-6 PMID: 34254241

Yeoman, A. M., & Lewis, A. C. (2021). Global emissions of VOCs from compressed aerosol products. *Elementa*, *9*(1), 00177. Advance online publication. DOI: 10.1525/elementa.2020.20.00177

Yoon, M. G., Jeon, H. J., & Kim, N. M. (2012). Biodegradation of polyethylene by a soil bacterium and *AlkB* cloned recombinant cell. *Journal of Bioremediation & Biodegradation*, *3*(4), 145. DOI: 10.4172/2155-6199.1000145

Yoshida, S., Hiraga, K., Takehana, T., Taniguchi, I., Yamaji, H., Maeda, Y., Toyohara, K., Miyamoto, K., Kimura, Y., & Oda, K. (2016). A bacterium that degrades and assimilates poly(ethylene terephthalate). *Science*, *351*(6278), 1196–1199. DOI: 10.1126/science.aad6359 PMID: 26965627

Yoshikawa, M., Zhang, M., & Toyota, K. (2016). Integrated Anaerobic-Aerobic biodegradation of multiple contaminants including chlorinated ethylenes, benzene, toluene, and dichloromethane. *Water, Air, and Soil Pollution*, *228*(1), 25. Advance online publication. DOI: 10.1007/s11270-016-3216-1 PMID: 28042183

Yoshikawa, M., Zhang, M., & Toyota, K. (2017). Biodegradation of Volatile Organic Compounds and Their Effects on Biodegradability under Co-Existing Conditions. *Microbes and Environments*, *32*(3), 188–200. DOI: 10.1264/jsme2.ME16188 PMID: 28904262

Yu, H., Qi, W., Cao, X., Hu, J., Li, Y., Peng, J., Hu, C., & Qu, J. (2021). Microplastic residues in wetland ecosystems: Do they truly threaten the plant-microbe-soil system? *Environment International*, *156*, 106708. DOI: 10.1016/j.envint.2021.106708 PMID: 34153891

Zareba, A., Krzeminska, A., & Kozik, R. (2021). Urban vertical farming as an example of nature-based solutions supporting a healthy society living in the urban environment. *Resources*, *10*(11), 109. DOI: 10.3390/resources10110109

Zdarta, J., Jankowska, K., Bachosz, K., Degórska, O., Kaźmierczak, K., Nguyen, L. N., Nghiem, L. D., & Jesionowski, T. (2021). Enhanced Wastewater Treatment by Immobilized Enzymes. *Current Pollution Reports*, *7*(2), 167–179. DOI: 10.1007/s40726-021-00183-7

Zettler, E. R., Mincer, T. J., & Amaral-Zettler, L. A. (2013). Life in the 'plastisphere': Microbial communities on plastic marine debris. *Environmental Science & Technology*, *47*(13), 7137–7146. DOI: 10.1021/es401288x PMID: 23745679

Zhai, X., Zhang, X. H., & Yu, M. (2023). Microbial colonization and degradation of marine microplastics in the plastisphere: A review. *Frontiers in Microbiology*, *14*, 1127308. DOI: 10.3389/fmicb.2023.1127308 PMID: 36876073

Zhang, B., Hu, X., Zhao, D., Wang, Y., Qu, J., Tao, Y., Kang, Z., Yu, H., Zhang, J., & Zhang, Y. (2024). Harnessing microbial biofilms in soil ecosystems: Enhancing nutrient cycling, stress resilience, and sustainable agriculture. *Journal of Environmental Management, 370*, 122973. DOI: 10.1016/j.jenvman.2024.122973 PMID: 39437688

Zhang, B., Xu, J., Sun, M., Yu, P., Ma, Y., Xie, L., & Chen, L. (2023). Comparative secretomic and proteomic analysis reveal multiple defensive strategies developed by *Vibrio cholerae* against the heavy metal (Cd^{2+}, Ni^{2+}, Pb^{2+}, and Zn^{2+}) stresses. *Frontiers in Microbiology, 14*, 1294177. DOI: 10.3389/fmicb.2023.1294177 PMID: 37954246

Zhang, K., Hamidian, A. H., Tubić, A., Zhang, Y., Fang, J. K., Wu, C., & Lam, P. K. (2021a). Understanding plastic degradation and microplastic formation in the environment: A review. *Environmental Pollution, 274*, 116554. DOI: 10.1016/j.envpol.2021.116554 PMID: 33529891

Zhang, K., Shi, H., Peng, J., Xiong, X., Wu, C., & Lam, P. K. S. (2018). Microplastic pollution in China's inland water systems: A review of findings, methods, characteristics, effects and management. *The Science of the Total Environment, 630*, 1641–1653. DOI: 10.1016/j.scitotenv.2018.02.300 PMID: 29554780

Zhang, L., Chen, F., Zeng, Z., Xu, M., Sun, F., Yang, L., Bi, X., Lin, Y., Gao, Y., Hao, H., Yi, W., Li, M., & Xie, Y. (2021a). Advances in metagenomics and its application in environmental microorganisms. *Frontiers in Microbiology, 12*, 766486. DOI: 10.3389/fmicb.2021.766364 PMID: 34975791

Zhang, X., Li, Y., Ouyang, D., Lei, J., Tan, Q., Xie, L., Li, Z., Liu, T., Xiao, Y., Farooq, T. H., Wu, X., Chen, L., & Yan, W. (2021b). Systematic review of interactions between microplastics and microorganisms in the soil environment. *Journal of Hazardous Materials, 418*, 126288. DOI: 10.1016/j.jhazmat.2021.126288 PMID: 34102358

Zhang, X., Ning, Z., Mayne, J., & Figeys, D. (2025). Clinical microbiome analysis by mass spectrometry–based metaproteomics. *Annual Review of Analytical Chemistry (Palo Alto, Calif.), 18*(1), 18. DOI: 10.1146/annurev-anchem-071124-113819 PMID: 39813733

Zhang, X., & Shao, M. (2023). Computational frameworks for functional annotation and data integration in microbial genomics. *Trends in Biotechnology, 41*(6), 711–726. DOI: 10.1016/j.tibtech.2022.12.004

Zhang, Y., Thompson, K. N., Branck, T., Yan, Y., Nguyen, L. H., Franzosa, E. A., & Huttenhower, C. (2021b). Metatranscriptomics for the human microbiome and microbial community functional profiling. *Annual Review of Biomedical Data Science*, *4*(1), 279–311. DOI: 10.1146/annurev-biodatasci-031121-103035 PMID: 34465175

Zhao, J., Rodriguez, J., & Martens-Habbena, W. (2023). Fine-scale evaluation of two standard 16S rRNA gene amplicon primer pairs for analysis of total prokaryotes and archaeal nitrifiers in differently managed soils. Frontiers in Microbiology, 14, Article 1234567. (Note: Volume 14 - 2023.)

Zhao, J., Jin, L., Wu, D., Xie, J., Li, J., Fu, X., Cong, Z., Fu, P., Zhang, Y., Luo, X., Feng, X., Zhang, G., Tiedje, J. M., & Li, X. (2022). Global airborne bacterial community—Interactions with Earth's microbiomes and anthropogenic activities. *Proceedings of the National Academy of Sciences of the United States of America*, *119*(42), e2204465119. Advance online publication. DOI: 10.1073/pnas.2204465119 PMID: 36215495

Zheng, X., Lin, H., Du, D., Li, G., Alam, O., Cheng, Z., Liu, X., Jiang, S., & Li, J. (2024). Remediation of heavy metals polluted soil environment: A critical review on biological approaches. *Ecotoxicology and Environmental Safety*, *284*, 116883. DOI: 10.1016/j.ecoenv.2024.116883 PMID: 39173222

Zhichkina, L., Nosov, V., Zhichkin, K., Zhenzhebir, V., Abramov, Y., & Alborova, M. (2020). Pesticide monitoring of agricultural soil pollution. In *E3S web of conferences* (Vol. 193, p. 01068). EDP Sciences.

Zhong, R., Li, M., Zhu, Z., Fang, X., Wang, X., Gong, H., & Yan, M. (2024). Bacterial degradation of polyethylene and polypropylene microplastics in a mangrove ecosystem. *Chemosphere*, *369*, 143908. DOI: 10.1016/j.chemosphere.2024.143908 PMID: 39643014

Zhou, Y., Kathiresan, N., Yu, Z., Rivera, L. F., Yang, Y., Thimma, M., Manickam, K., Chebotarov, D., Mauleon, R., Chougule, K., Wei, S., Gao, T., Green, C. D., Zuccolo, A., Xie, W., Ware, D., Zhang, J., McNally, K. L., & Wing, R. A. (2024). A high-performance computational workflow to accelerate GATK SNP detection across a 25-genome dataset. *BMC Biology*, *22*(1), 13. DOI: 10.1186/s12915-024-01820-5 PMID: 38273258

Zhou, Z., Liu, Y., Anantharaman, K., & Li, M. (2022). The expanding Asgard archaea invoke novel insights into Tree of Life and eukaryogenesis. *mLife*, *1*(4), 374–381. DOI: 10.1002/mlf2.12048 PMID: 38818484

Zhou, Z., Tran, P. Q., Cowley, E. S., Trembath-Reichert, E., & Anantharaman, K. (2025). Diversity and ecology of microbial sulfur metabolism. *Nature Reviews. Microbiology*, *23*(2), 122–140. DOI: 10.1038/s41579-024-01104-3 PMID: 39420098

Zhuang, X. (2021). Spatially resolved single-cell genomics and transcriptomics by imaging. *Nature Methods*, *18*(1), 18–22. DOI: 10.1038/s41592-020-01037-8 PMID: 33408406

Zinicola, M., Higgins, H., Lima, S., Machado, V., Guard, C., & Bicalho, R. (2015). Shotgun metagenomic sequencing reveals functional genes and microbiome associated with bovine digital dermatitis. *PLoS One*, *10*(7), e0133674. DOI: 10.1371/journal.pone.0133674 PMID: 26193110

Zinsstag, J., Schelling, E., Waltner-Toews, D., & Tanner, M. (2011). From "one medicine" to "one health" and systemic approaches to health and well-being. *Preventive Veterinary Medicine*, *101*(3-4), 148–156. DOI: 10.1016/j.prevetmed.2010.07.003 PMID: 20832879

Коротецкий, И., Кузнецова, Т., Шилов, С., Зубенко, Н., Иванова, Л., & Рева, О. (2025). Methods of metagenomic analysis. Микробиология ж не вирусология, 3, 83–117.

About the Contributors

Olubukola Oluranti Babalola (Pr.Sci.Nat, MASSAF, FTWAS, FAS, FAAS, MAAAS, FISC, FRSB, FAAM) is the Vice President of The World Academy of Sciences (TWAS, Africa region), the Foreign Secretary of the Nigerian Academy of Sciences, quondam Vice President of the Organization for Women in Science for the Developing World (OWSD), former Head of Department and the Director of Research at the North-West University. She is a member of the Think Tank for The Forum for Agricultural Research in Africa (FARA) and an Honorary Visiting Professor at Imperial College London, contributing to global research agendas, policy dialogue, and interdisciplinary collaboration. She is an AAAS-TWAS Science diplomacy alumnus, impacting the world after the eye-opening awareness of policymaking, diplomacy, and diplomatic engagement. She is taking action using the gender lens and forging collaborations with diplomats. She holds advanced academic qualifications culminating in a professorship in Plant Health Management and an MBA in General Management, reflecting a rare integration of scientific depth and strategic leadership. Her leadership vision centers on excellence, inclusivity, and translating science into societal benefit. With an h-index of 90, Prof Babalola ranks among the most influential scientists in agricultural and environmental microbiology. Her research has fundamentally advanced the understanding of plant-microbe interactions, microbial ecology, biofertilizers, biocontrol, and climate-resilient agriculture, shaping sustainable food systems, particularly in Africa and the Global South. She is a distinguished Principal Investigator and mentor, and has supervised 39 PhD and 29 MSc graduates, built a vibrant, high-impact research ecosystem and trained the next generation of scientists. Her scholarly output also includes books and book chapters, underscoring her commitment to knowledge dissemination and capacity building. Prof Babalola is a prolific author with over 450 publications. She serves on the editorial boards of Elsevier, Springer, SAGE, and the American Society for Microbiology (ASM), and is an International Advisor to several others. Her efforts have received many awards, including being a finalist in

GenderInSite 2020, a place in Clarivate's Highly Cited Research Top 1% for the field in 2022 and 2023, and, notably, in 2024, the South African Women in Science Awards (SAWiSA), celebrating her as a trailblazing woman in STEM. In 2025, she rose to an NRF B-rating, and became an internationally acclaimed scientist. Prof Babalola is #1 in Africa for Soil Science and Plant Nutrition. Through her enduring professional trajectory, she continues to redefine scientific leadership, leaving a lasting, transformative imprint on the global STEM landscape.

Olukayode Oladipo Amund is a reputable Microbiologist specializing in Petroleum and Environmental Microbiology. He earned his first degree in Biology graduating in First Class (Hons.) from the University of Lagos in 1977 and a Doctorate degree in Biotechnology from Cranfield University, UK, in 1982. He began his academic career as Graduate Assistant at the University of Lagos in 1978 and rose through the ranks to become a Professor in 1998. He held several leadership positions as Head, Department of Botany and Microbiology, Dean, Faculty of Science, Dean of Student Affairs, Director of the Distance Learning Institute and Member of the Governing Council at the University of Lagos. He was Foundation Vice-Chancellor of Wesley University of Science and Technology, Ondo and later served as the third substantive Vice-Chancellor of Elizade University, Ilara-Mokin, Ondo State, Nigeria. He is a Fellow of the Nigerian Academy of Science (FAS), African Academy of Sciences (FAAS), Institute of Policy Managemnt Development (FIPMD), Royal Society of Biology (FRSB) and African Scientific Institute (FASI). He has published over 150 scientific articles in learned journals, contributed book chapters and presented numerous articles at conferences, while 38 publications were contributed to the database of National Center for Biotechnology Information (NCBI). Emeritus Prof Amund is affiliated with the Department of Microbiology, University of Lagos from where he officially retired in 2023. He secured several research grants from International and local agencies and provided consultancy services to industry. He has received numerous awards for his contributions to science, industry and the society. He is a co-Editor to the book Ecofriendly Frontiers:Harnessing Microbial Applications for Food Security published by Springer in 2025. He is an Alumnus of the Galilee International Management Institute, Nahalal, Galilee, Israel.

Olayinka O. Osuolale embarked on his academic journey with a PhD in Microbiology from the University of Fort Hare in South Africa, further enriching his expertise through postgraduate studies at Loughborough University in the United Kingdom. His research delves deeply into the complex intersections of infectious diseases and environmental health. Dr. Osuolale's expertise spans critical areas such as Leishmaniasis and Dengue virus, antimicrobial resistance, healthcare-associated waterborne infections, microbial water quality monitoring, and the broader field of environmental microbiology and public health. Since 2016, Dr. Osuolale has been a dedicated member of Elizade University's faculty, progressing from Lecturer I to Reader by 2025. Throughout his tenure, he has actively fostered collaborations with both local and international research institutions, championing interdisciplinary approaches to tackle microbial and environmental challenges. Dr. Osuolale's scholarly work has garnered significant recognition, with over 8,661 citations, an h-index of 20, and an i10-index of 22, according to Google Scholar.

***

Moses Abiala is an Environmental Microbiolo-gist with specialization in molecular plant-microbe interactions and microbial genomics geared towards sustainable growth of food crops under stressed conditions. He has conducted international re-search across Nigeria, the USA, Germany, and India, with over 50 peer-reviewed publications and active collaborations in microbiome research. Apart from the fact that Prof. Abiala is a reviewer to some international reputable journals, he also serves on the expert board as one of the reviewers and Ad-hoc assessors for some international grants. His utmost priority is to provide healthy solution to micro-bial-based challenges threatening food security, as well as enhancing capacity building among the upcoming generations of scientists for global impact.

Abiodun Akanmu is a Senior Lecturer in the Department of Animal Science at the University of Pretoria, South Africa. His research focuses on ruminant nutrition, rumen microbiology, methane mitigation, feed efficiency, and the application of microbial and genomic tools to improve livestock productivity and sustainability. His work integrates animal nutrition, metagenomics, and microbial ecology to understand rumen function and develop innovative feeding strategies for dairy and meat-producing animals. Dr Akanmu has published in several peer-reviewed journals and collaborates on international research initiatives related to rumen microbial biotechnology and sustainable livestock systems.

Austin Gray hails from Charleston, SC, where he attended The Citadel, Military College of South Carolina, for his bachelor's and master's degrees. He then attended the University of North Carolina at Greensboro for his Ph.D., becoming the first African American to receive a doctorate in environmental health sciences at UNCG. Currently, Gray is an assistant professor of biological sciences at Virginia Tech, being the first Black tenure-track faculty member in the department. Gray's research priorities focus on addressing questions in environmental toxicology, primarily using physiological and ecological approaches to examine the impacts of legacy and emerging contaminants (heavy metals, personal care products, microplastics, nanoplastics, and pharmaceuticals) from anthropogenic sources and to assess their risks to a variety of freshwater and marine organisms. Currently, thousands of chemicals/pollutants are introduced into waterways. As a result, organisms are exposed to doses that may not be lethal but can cause sublethal effects. Additionally, certain aquatic habitats can be hotspots for biogeochemical processes. As the microbes that govern these processes are exposed to a cocktail of chemicals, it is important to understand how this affects their ability to transform or fixate nutrients. The goal of the work in the Gray lab is to better understand how aquatic ecosystems are changing as a result of human alteration to the surrounding environment and increased pollution. Dr. Gray's work has been supported by the National Science Foundation, the United States Geological Survey, the Sea Grant, and the National Oceanic and Atmospheric Administration

Omobola Okoh is a Professor of Analytical and Environmental Chemistry in the Department of Pure and Applied Chemistry, Faculty of Science and Agriculture, University of Fort Hare, Alice Campus where she teaches Analytical, Organic, Inorganic, Physical, Environmental Chemistry and Chemical Technology and supervises postgraduate students. In 2001 She obtained her BSc Hons Degree from Obafemi Awolowo University, Nigeria and MSc and PhD Degree Chemistry from the University of Fort Hare in 2008 and 2011 respectively. Her academic career took a start in 2003 as a research fellow at the Centre for Energy Research and Development Centre (CERD) at the Obafemi Awolowo University in Nigeria. Due to her passion for teaching and research, she was appointed as a Lecturer in the Department of Chemistry in 2012 and rose to the ranks of Full Professor in 2018. In January 2021, she was appointed as Head of the Department of the Department of Chemistry, a position she holds to date. Professor Omobola Okoh has received and attracted several grant awards such as NRF Thuthuka Funding Instrument (2015-2018), ESKOM /Tertiary Education Support Program (2015-2024) and GMRDC research capacity seed grants (2013-2018). She has also been a recipient of NRF African Scholarship Award for postgraduate studies (2006-2008), GMRDC Doctoral Bursary Award (2008-2010) and NRF Prestigious postdoctoral awards (2011). She is the Leader of the Synthetic, Environmental and Analytical Chemistry Research Group, a member of the Water for Sustainable Development Research Niche Area and the SAMRC Microbial Water Quality Monitoring Centre at the University of Fort Hare. Professor Omobola Okoh has served as a panel member for numerous NRF Grants such as NRF Postdoctoral Fellowship (Geo and Marine Sciences), Joint Institute for Nuclear Research (JINR) and Innovation grants (2019-2021), NRF ERTTG – Applications for funding (2020), DST-NRF Fellowships for Early Career Researchers from United Kingdom (2018), NRF Free standing Innovation and scarce skills, TWAS Postdoctoral Fellowships, NRF funding for DAAD, Chemistry and nanotechnology panel (2016-2017), NRF Postdoctoral fellowship free standing innovation and scarce skill (2015-2018), NRF Doctoral Scholarship Review panel member for bursary application for material Science and Nanotechnology (2013-Date), NRF-SA-and China Joint Research Program (2021), NRF Equipment Related Travel, training and conference application (2021) and NRF BAAP Post-PhD Track Applications (2019). She is an Adjunct Professor in the Industrial Chemistry Department at Bowen University (2021-Date) and collaborates both locally and internationally with eminent academics and researchers. Her research expertise falls within the fields of Analytical and Environmental Chemistry with strong focus on water/wastewater quality, chemical pollutants in water and their health risk assessment, natural products chemistry, renewable energy and nanotechnology. She has successfully supervised 15 Doctoral students, 14 Master's students and hosted 5 postdoctoral fellows, and currently supervising 7 PhD students and 5 Master's students at the University of Fort Hare in perfect sync with the scarce skill capacity development aspiration of the South African Government. Professor Omobola Okoh has published over 106 articles in reputable peer-reviewed journal as well as 5 book chapters and her works have been cited more than 6220 times with an h-index of 42 and i(10) index of 79. She has presented more than 54 papers at national and international conferences, and reviews manuscripts for reputable scientific journals in her areas of research.

Index

A

B

C

E

H

www.ingramcontent.com/pod-product-compliance
Lightning Source LLC
LaVergne TN
LVHW062321300926
849161LV00007B/1521